CAMOUFLAGE AUSTRALIA

ART, NATURE, SCIENCE AND WAR

ANN ELIAS
PREFACE BY ROY R. BEHRENS

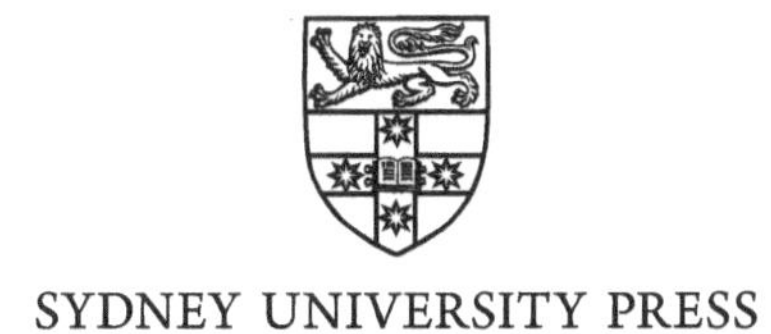

SYDNEY UNIVERSITY PRESS

Published 2011 by Sydney University Press

SYDNEY UNIVERSITY PRESS
University of Sydney Library
sydney.edu.au/sup

Sydney University Press
Fisher Library F03
University of Sydney NSW 2006 AUSTRALIA
Email: sup.info@sydney.edu.au

National Library of Australia Cataloguing-in-Publication entry
Author: Elias, Ann Dirouhi.
Title: Camouflage Australia : art, nature, science and war / Ann Dirouhi Elias
ISBN: 9781920899738 (pbk.)
Notes: Includes bibliographical references and index.
Subjects: Art and camouflage--Australia--History.
Artists--Australia--War work.
Camouflage (Military science)--Australia--History.
World War, 1939-1945--Camouflage.
Dewey Number:
355.410994

Cover image: A cardboard model plane by Frank Hinder, Australian War Memorial Negative Number RC03645

Cover design by Miguel Yamin, the University Publishing Service

In memory of Minas Elias, a makeshift camouflage officer in Burma, World War II

Contents

Preface

To put it simply, this book is the little-known story of how Australian artists made pivotal contributions to military and civilian defence camouflage during World War II.

At the same time, as the author explains, its scope is considerably larger than that. While Australian 'fine artists' (painters, sculptors, printmakers) contributed immeasurably to wartime camouflage, in doing so they worked beside other art- and design-related professionals, such as photographers, graphic designers and architects. They also collaborated with scientists (zoologists, in particular), and with professional soldiers, who often considered concealment to be a sign of cowardice, a subversion of military discipline, and a promotion of battlefield attitudes that were both peculiar and unmanly.

An Australian-based art historian, Ann Elias has written about camouflage for more than a decade. I myself was first drawn to her research when I realised that, unlike so many camouflage scholars, her concern is not merely with military aspects of the subject, but with a vast range of interesting facets, both social and historical. Especially memorable is her incisive analysis of the life and professional work of prominent British-born zoologist W.J. Dakin, the father of Australian camouflage, to whom she devotes a good part of this book.

Ann Elias is a university scholar and art historian who, despite the tendency in our time toward recondite, specialised research, has maintained an astonishing balance between her targeted interests (the trees on academic turf) and their role in a much larger cultural view (the forest of the human race). In part, she is able to do this because she is a skilled and inspiring writer, a person who genuinely writes to be read. Witness these terse yet provocative lines from her introduction: 'Invisibility is simply visibility in disguise. It is the outcome of a process of visual transformations in which the deadly appear innocent and the innocent deadly.'

So, yes—while this book is a close-up look at how Australian artists and designers contributed to World War II camouflage, it is just as much about human vision and the brain; the amazing degree to which form and function deftly fit in the evolved appearance of animals; our persistent susceptibility to racial and gender stereotypes; the military mindset as contrasted with that of civilians; and the cultural headbutts that tend to result from the proximity of artists, scientists and military engineers. But there is much more.

To my mind, Ann Elias' most accomplished achievement is the very act of crossing disciplines. With that in mind, I find it instructive to look closely at the exquisite drawings by British zoologist Hugh B. Cott of the hind limbs of the common frog (reproduced in Elias' introduction as fig. I.4). For Cott, these were iconic examples of what he called 'coincident disruption', occurrences of camouflage in which blending (unit-forming) and disruption (unit-breaking) work together in the same space. As the frog merely folds up its legs, the dark shapes on its surface join to hide the oneness of its limb.

Ann Elias describes this book as 'an Australian arm of a much bigger story about the crossing of discipline boundaries'. Metaphorically, I think of it this way: she has folded up a lengthy limb of scholarly tradition (made up of sacrosanct disciplines like aesthetics, zoology, anthropology and sociology), in order to reveal new zones that are cross-disciplinary.

Roy R. Behrens
Professor of Art and Distinguished Scholar
University of Northern Iowa

Acknowledgments

Over the past decade I have accumulated ideas, inspiration and information from many generous people and institutional sources. I would especially like to thank the University of Sydney for early funding, Craig Judd for his insightful reading of the manuscript, Roy Behrens for stimulating my interest in camouflage, and the team at Sydney University Press, especially Susan Murray-Smith and Agata Mrva-Montoya. Family descendents, associates and friends of camoufleurs, as well as returned servicemen, were enthusiastic about the project, and especially helpful were Dr Isobel Bennett, Renée Free, Enid Hawkins, R. (Bob) D. Leonard, Leigh Purcell, and Margaret Rigg. Of scholars and specialists who shared their knowledge, a special thankyou to Michael Bogle, Julian Holland, and Michael W. Young. Jan Guy and Dominica Lowe were excellent research assistants, Cameron Fargo was always helpful with advice on digital imagery, and the staff at the Sydney College of the Arts Library an invaluable resource.

The Australian War Memorial, the Art Gallery of New South Wales, the National Archives of Australia, the National Gallery of Australia, the State Library of Victoria, the National Gallery of Victoria, the University of Sydney Art Collections, the University of Liverpool Library, Maria Fernanda Cardoso, Debra Dawes, Paul D. Brock and Jack W. Hasenpusch were more than generous with images. Every attempt has been made to trace copyright owners to obtain permissions for the publication of images and to pay usage fees. Earlier versions of parts of chapters 4, 5, 10 and 11 appeared in *Journal of Australian Studies*, *Leonardo*, *History of Photography*, and Prue Ahrens and Chris Dixon (eds), *Coast to coast: case histories of modern Pacific crossings* (2010).

Finally, Greg Poynter and Rose Poynter have loyally accompanied me on research trips from Darwin in the Northern Territory of Australia to Liverpool in the United Kingdom, and always stayed interested in the subject of camouflage.

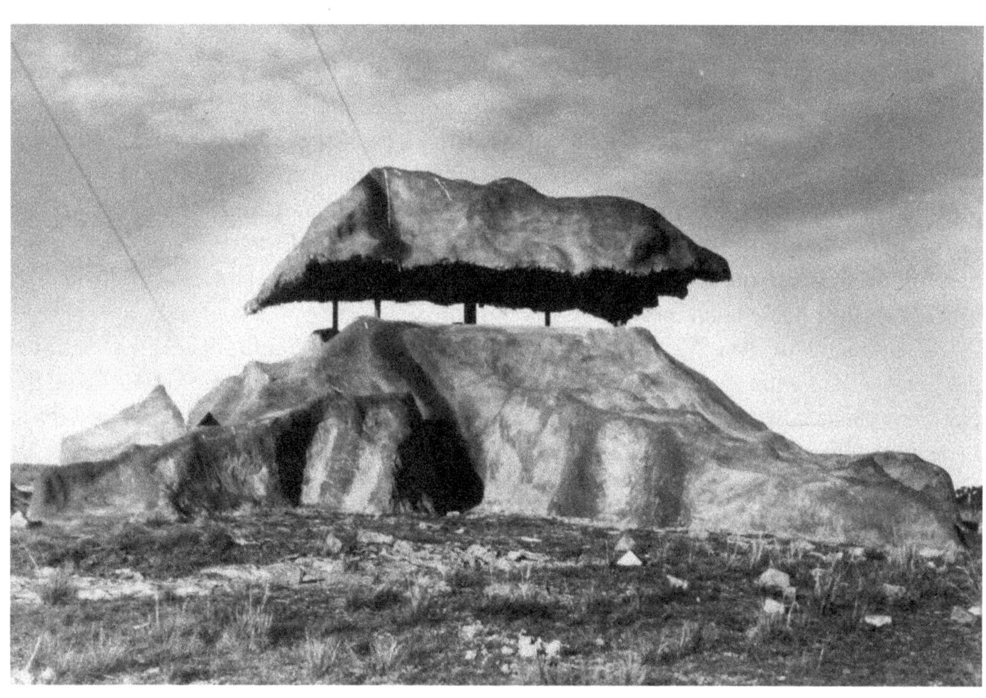

Fig. I.1. Artificial rock. On verso: 'Department Home Security, ACT: *S.L. Post & Sentry etc.*', c. 1943. Collection Art Gallery of New South Wales and Archive.

Abbreviations

ABC	Australian Broadcasting Commission
ACT	Australian Capital Territory
AGNSW	Art Gallery of New South Wales
AIF	Australian Imperial Force
ANGAU	Australian New Guinea Administrative Unit
ANZAC	Australian and New Zealand Army Corps
AWM	Australian War Memorial
CMF	Citizen Military Forces
DCCC	Defence Central Camouflage Committee
DHS	Department of Home Security
NAA	National Archives of Australia
NSW	New South Wales
RAAF	Royal Australian Airforce
RAE	Royal Australian Engineers
RAN	Royal Australian Navy
SMH	*Sydney Morning Herald*
SW Pacific	South West Pacific
US	United States
VDC	Volunteer Defence Corps
WWI	World War I
WWII	World War II

Introduction

> The whole subject of Camouflage is based essentially on what can be seen by the eye direct or through photography, and in practice it is concerned with how to make things invisible altogether or to look like something which is not of a suspicious or obvious nature. (William Dakin)[1]

This book sheds light on the near invisible history of Australian artists and designers who campaigned to make invisibility itself the modern defence for Australia in World War II. They worked in camouflage, for the Australian Department of Home Security (DHS), in secret. To the military they were amateurs working recklessly outside their discipline boundaries, masquerading as real soldiers, and faking a body of knowledge that only war professionals felt they could rightly own. Their leader was a zoologist and a specialist in marine animals, a single-minded, brilliant scientist who called attention to the stripes on fish in motion to teach troops about invisibility, and the play of light and shadow on birds to demonstrate spatial illusion.

But far from being simply a quirky, hidden story relevant only to the remote past, this is one with emotional substance and connection with our contemporary reality. It addresses an historical moment when artists and scientists in WWII were compelled to prove the social value of their fields of expertise, even for political violence, but in the name of national and cultural freedoms from totalitarianism and fascism.[2] Never far from the surface of the story, then, is realisation that in order to fulfil wartime duties the individuals who worked in camouflage had to suppress ideological, psychological and moral positions on art and war, and transform themselves into servants of the war enterprise as camouflage labourers, camouflage designers, and camouflage field officers in the north of Australia and the theatre of war in Papua and New Guinea. And while the story of camouflage in WWII is located in the historical past, it is anything but disconnected from the experiences of Australians and war today. Ethical conflicts and struggles dominate debates on war participation, and camouflage itself, even in an age of nuclear warfare, retains many

1 W.J. Dakin, 'The phenomena of light in relation to camouflage: some notes by the Technical Director of Camouflage' in Camouflage report 1939–1945 (appendix R), Canberra, AWM, Series 81 [77 Part 5], 1947, p. 1.

2 For an example of anti-fascist views expressed in relation to Australian art and art criticism see O'Connor 2006 [1944].

of its historical methods but also controversies: it continues to involve the relatively simple forms of concealment and deception offered by skin paint and clothing patterns developed in WWII; contemporary research by defence analysts still use animal camouflage as a point of reference; and camouflage continues to create dissent among troops. Australians fighting in Afghanistan in 2010 complained, as troops in New Guinea did in 1942, that greasy camouflage paint on the skin causes irritation, and that the wrong colours and patterns on uniforms put soldiers' lives at risk.[3]

What is camouflage? William Dakin, who is central to this story, tried to define it many times, and each time wrote something different. The epigraph for this introduction is one of his many attempts, and a good guide to what camouflage meant to Australians in WWII. Camouflage was predominantly the means by which to deceive vision, and exclusively vision of the enemy. Sound also played a part, but war records on this specialised means of deception are scant and consequently sonic camouflage receives little discussion in these pages. With visual camouflage in WWII it was more important to conceal objects from the camera's vision than from the body's senses since the mechanical eye was sharper and more capable of seeing through disguise, especially the aerial machine-aided eye using infra-red film. Also intimated in Dakin's definition is that, in its objective to create visual confusion, camouflage is cryptic and paradoxical. Invisibility is simply visibility in disguise. It is the outcome of a process of visual transformations in which the deadly appear innocent and the innocent deadly. A photograph of an artificial rock designed by camouflage artists working for William Dakin in the DHS, a rock designed as an observation or sniper post, demonstrates the point: with its top properly in place, it transformed into an unprepossessing object intended for plain sight but to obliterate from view what was being hidden (fig. I.1).

Many examples of camouflage's paradoxical nature are illustrated in the following chapters. As civil defence became more urgent in Australia following the bombing of Darwin by the Japanese in February 1942, so the city landscapes were modified with the erection of simulations of innocent looking domestic houses and public buildings that were really munitions and bomber hideouts placed near airfields disguised as sports arenas, race tracks and market gardens. Intended to deceive enemy cameras in the event of aerial reconnaissance, or to fool the naked eye in case of attack, they demonstrate how camouflage harnessed opposing strategies in a game of measures and countermeasures organised around the dynamics of dissimulation and simulation defined by Jean Baudrillard in the 1980s (when the principles of camouflage became increasingly popular as a conceptual tool for cultural analysis) as the pretence of not having what one has on the one hand, and the pretence of having what one does not on the other.[4]

The last ten years have seen an exponential growth in camouflage studies in all fields of the arts and social sciences including architecture, visual arts, anthropology and sociology. But so rich is camouflage as a metaphor that authors have found its meaning impossible to contain, leaving Michael Taussig to conclude that camouflage is a word 'trying desperately to live up to its name'.[5] Hsuan L. Hsu, on the other hand, invokes camouflage to explain a vast range of physical and social objects and events that involve simulation and dissimulation:

---

3 Oakes 2010a and 2010b.

4 Baudrillard 1997, p. 3.

5 Taussig 2008, p. s107.

> Immigrants, minorities, and refugees are pressured to assimilate to new cultural contexts; both stealth bombers and suicide bombers blend in with their surroundings; antennae and radio towers are designed to simulate trees; along the borders between Gilo and neighboring Palestinian settlements, barrier walls are painted to resemble the landscapes they block out; fashion and advertising continually manipulate the fine line between standing out and blending in; and biologists have begun studying the phenomenon of 'urban speciation,' in which insects and birds mimic and adapt to various aspects of urban environments.[6]

An entire book could be written on definitions of camouflage but in the context of this study, which focuses on WWII, 'camouflage'—the French word that originated on the battlefields of WWI when the first camouflage units in western military history were formalised—refers to the dynamics of concealment and deception for military gain and individual protection.[7]

Primarily this book is concerned with relaying a hidden story about central figures in Australian art history and with filling a gap in the military history of Australia. The deployment of artists for warfare in Australia had precedents in World War I, but the scale in WWII was unparalleled and the phenomenon never repeated. At the same time this study aims to deliver two arguments: that the nation's memory of WWII is incomplete without the inclusion of the camouflage artists who worked for the DHS, a contribution to history that has been overlooked due to their non-military status; and that art history, while demonstrating a keen interest in the anti-war expressions of the moderns, has failed to integrate the war years into its narratives, believing the sociology of war outside its core business and values. Camouflage, it has been noted by Hsu, is a phenomenon that sits ambiguously between violence and aesthetics.[8] The identity of modern art, however, is rooted in the protest of war and violence. This in itself explains why it has been important for me to attempt to get under the skin of the main figures in this story, to understand the circumstances of military-related duties, and to comprehend the emotional demands of war for those who worked in camouflage. And here the task has been difficult, because with the passing of the war generation, the historical context has been less easy to understand and to adequately represent, a situation exacerbated by the nature of the official war record which is immense but largely bureaucratic and offers little on the human cost of war. Luckily William Dakin expressed himself passionately, and Frank Hinder kept the most detailed of diaries. And many chance moments of social and psychological insight slip through otherwise dry reportage. They indicate the emotional toll that working in camouflage took on members of the DHS who were often ridiculed, not least because their title was 'camoufleur' and because camouflage seemed to others a rather effeminate field. A

6 Hsu 2006.

7 A French ministerial order established the first official camouflage unit in 1915 to aid in the concealment of soldiers and military equipment, and the first soldiers assigned this work were artists. Among French artists enlisted in WWI, the idea of working in the camouflage division was attractive, yet it was also competitive and Fernand Léger was disappointed to be refused that opportunity. See Kahn 1984, pp. 1–16.

8 Hsu 2006.

deep-seated suspicion prevailed that camouflage signified passivity and weakness, a mere decoration to the structural necessities of war. As civilians working for the armed forces but not part of them, camoufleurs spent the first years of the war looking highly conspicuous in civilian clothes, and the third section of the book explains how they were desperate to wear uniforms and receive accreditation with the airforce just to blend in with military life. 'We are nobody's business' wrote Charles O'Harte forlornly from the tropical north of Australia.[9]

I claim the symbolic beginning of the national history of the camouflage artists of WWII as the day that Australia's leading modern artist, Margaret Preston, introduced fellow artist Frank Hinder to William Dakin.[10] It was March 1938 and, not only was Dakin a celebrated professor of zoology, he was also a minor patron of the arts. He commissioned Hinder to paint a mural of marine subjects, offering the artist a rare opportunity during the Depression to derive income from creative practice. By the end of 1938, however, and not coincidentally given the imminent outbreak of war and his new acquaintance with Dakin—a scientist with a publication history in biological concealment and deception dating back to WWI—Hinder's diaries turned to the subject of military camouflage. On 23 September 1938, Hinder, who in postwar years became renowned as an artist investigating abstraction using the medium of light, pondered the outbreak of war, declared his interest in camouflage design rather than fighting, speculated about his anticipated role in camouflage and asked whether 'light should play an important part'.[11] This marks the start of an intense wartime collaboration between Dakin and Hinder, one that is explicated and discussed in the majority of chapters ahead, and that involved many of Hinder's friends and acquaintances.

It was in Sydney, New South Wales—even before Britain declared war on Germany on 3 September 1939—that a group of 30 men in the arts and sciences invited members of the army, airforce and navy, and banded together to form the Sydney Camouflage Group. Included were Hinder's close friends, the designers Douglas Annand and Robert E. Curtis, and photographer Russell Roberts as well as other professional acquaintances including Australia's leading modernist photographer, Max Dupain (fig. I.2), and that 'great man' of Australian art patronage and administration—as art historian Bernard Smith referred to him—Sydney Ure Smith.[12] They held meetings chaired by Dakin, conducted experiments, and published a book titled *The art of camouflage* (1941). Many were later seconded to work in the Department of Home Security. A roll call of artists and the nature of their experimentations in Sydney are discussed in the first section of this study. Indeed their mobilisation into the field of camouflage in WWII was part of a much wider international movement of artists intent on defeating the Axis powers through altered vision, dis-

9 Charles O'Harte to D.H. Wilson, 25 February 1943, in Dispersal hangars, Sydney, National Archives of Australia (NAA), Series C1707, Item 4, p. 2.

10 Frank Hinder, 25 March 1938, the Grosvenor Galleries Diary, Frank Hinder Papers, Sydney, Archive of the AGNSW, MS 1995.1.

11 Frank Hinder, 23 September 1938, the Grosvenor Galleries Diary, Frank Hinder Papers, Sydney, Archive of the AGNSW , MS 1995.1.

12 Bernard Smith quoted in Underhill 1991, p. 1.

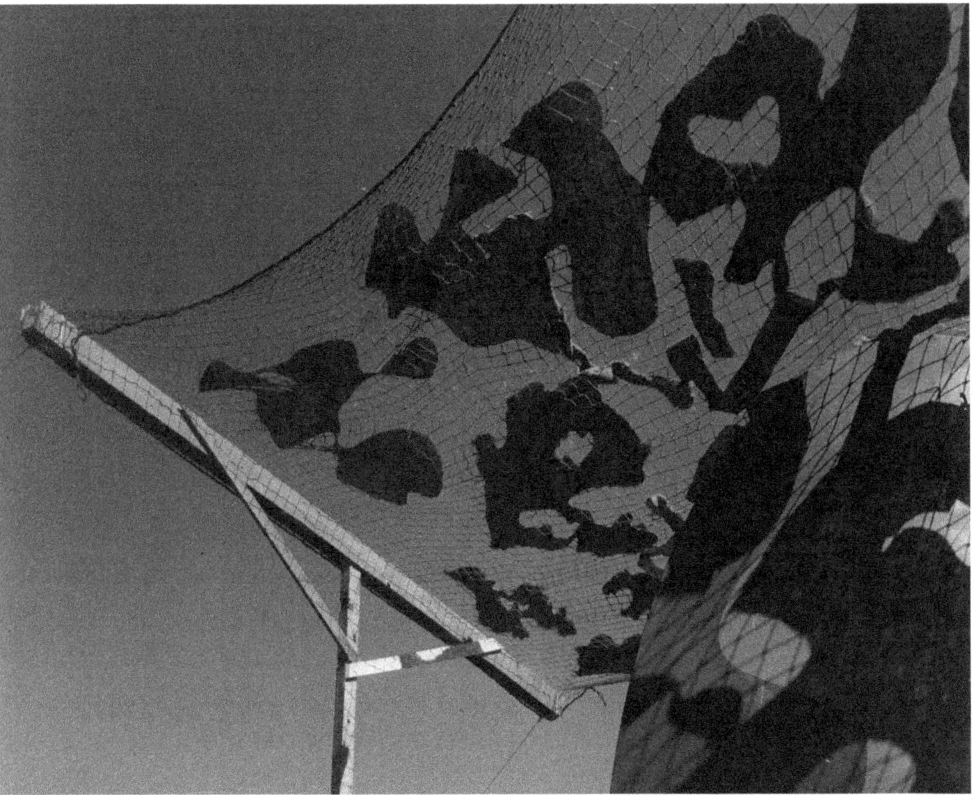

Fig. I.2, Bankstown aerodrome camouflage experiment, c. 1943. Collection of the National Archives of Australia: C1905, 3. Photograph, Max Dupain.

torted perception, illusions and abstractions, and among the Australians' distinguished international counterparts were Arshile Gorky, László Moholy-Nagy, and Roland Penrose. However, as David McCarthy argues in relation to the deployment of American sculptor David Smith for war-related work in 1942, their participation was part of the production of war, and not the act of making modern art.[13]

Frank Hinder investigated the perceptual impact of colours and designs on moving objects. It was something he would return to in the studio in postwar years. Ironically, the war gave Hinder a certain freedom to explore the principles of abstraction at a time when Australian society was hostile to abstraction in modern art. The disruptive patterns he painted on model trucks (fig. I.3) and the dazzle designs he applied to model planes using vibrantly contrasting stripes of orange and yellow, as well as blue and red, do resemble op art. Hinder was not seeking aesthetic knowledge in itself, however, but rather how to induce in the enemy a state of perceptual confusion through visual misinformation.

This is also a story about the crossing of discipline boundaries between art, biological science and military science, and it must be stressed that the mobilisation of artists into camouflage warfare did not proceed without political controversy. The Sydney Camouflage

13 David McCarthy 2010, p. 25.

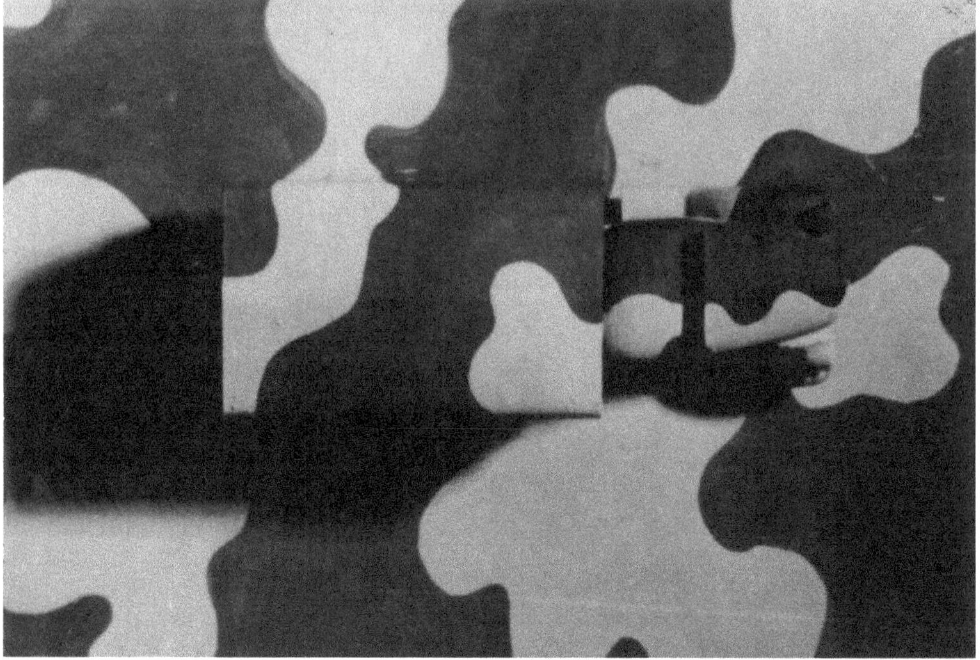

Fig. I.3, Camouflage experiment with model truck by Frank Hinder and team of camoufleurs, c. 1942. AWM, Canberra, Frank Hinder Personal Records AWM 88/133, File 895/4/182.

Group was convinced that the Australian military had neglected the importance of camouflage for national defence and soon many leading artists and administrators throughout Australia, including Daryl Lindsay and Louis McCubbin, figures now at the centre of the nation's history, joined Dakin in persuading then Prime Minister Robert Menzies to initiate a camouflage section in the DHS. Incredibly, given the military's natural dominance in matters of defence in WWII, Menzies appointed Dakin, a civilian scientist, to the top position as Technical Director of Camouflage for the Australian mainland and the SW Pacific, expecting that army, navy and airforce would defer to Dakin's authority. It was a decision many lived to regret. Soon there was a war inside the war, one that Dakin perceived as between the military and civilians.[14] The third section of the book looks closely at the arguments that developed around military perceptions of the value of camouflage specialists, and of camouflage to military operations.

This is therefore not a book about art in itself or about the considered, individualistic responses of artists to political conflict through the expressive means of painted image or sculpted object. Instead it is about the interconnections of art, zoology and military ideas and their application to warfare. More particularly I am interested in the transmutations of theories of camouflage from the field of zoology and the natural sciences into war through the endeavours of artists who in turn modified scientific and military camouflage and

14 W.J. Dakin to Sir Frederick Shedden, 4 July 1944, Establishment of camouflage organisation and procedures, Canberra, NAA, Series A5954, Item 396/2.

introduced their own conceptions. Interchanges of aesthetics and science define the history of camouflage for all warring nations in WWII, and had precedents in WWI. Consequently, with a growing body of international literature on painters, sculptors, photographers, commercial designers and architects (professionals whom I refer to from this point on as 'artists and designers') who were innovators in camouflage—innovators in the art of concealment and deception—this book represents an Australian arm of a much bigger story about the crossing of discipline boundaries.

What readers will notice in the structure of this book is the way it takes a number of turns in order to address the various angles of art, science, social and military histories that feed into the subject. In addition to three separate case studies of William Dakin, Frank Hinder and Max Dupain, there are four sections that demarcate, in the following order: the art community that initiated camouflage activism; the science community that made animal camouflage its model for military methods; the military context where antagonisms between civilians and military communities and conflicting attitudes to war, as well as the value of camouflage, are discussed; and finally, the experiences of camouflage officers within the DHS who were sent on reconnaissance work and operational camouflage duties to Papua and New Guinea. The first chapters of the book therefore focus on the relatively safe, orderly world of civil defence and camouflage on the mainland when the main goal was to protect the coastal cities of Australia from attack. The second half, however, investigates the psychologically fraught context of war, and the dangerous, chaotic life of camouflage officers in war zones when the DHS established a base on Goodenough Island in Papua in late 1943, a location already legendary for camouflage due to a daring deception scheme staged earlier by the Australian army's 'ghost force'. Max Dupain and Bob Curtis were among those forced to confront the westerner's terror of jungle, wild animals, sorcery and savages, fears that brought new intensity to the DHS's mission to convince troops of the imperative to blend with the jungle.

For the reader wondering which artists were members of the DHS and the Sydney Camouflage Group, lists are provided in the appendices. They include many well-known names, especially in the history of Australian design, a fact that was brought to attention by design historian Michael Bogle who was one of the first writers to address the subject of camouflage in his book *Design in Australia: 1880–1970* (1998). Bogle claimed that the participation of designers in concealment and deception represented not only an obscure part of Australian history but also an unexpected one.[15] His text, together with the work of Official War Historian D.P. Mellor who wrote a useful overview of camouflage's organisation in WWII in a volume of *Australia in the war of 1939–1945: the role of science and industry* (1958), laid the Australian groundwork for this research.[16]

For the most part, though, the scope of the historical territory embraced by this study was untouched in Australia, whereas Britain and the United States (US) have comparatively long histories of research and writing in this field. American designer and historian Roy R. Behrens was a leader in camouflage scholarship in the 1980s and has continued to maintain that position with two books published since 2000—*False colors: art, design and modern*

15 Bogle 1998b, p. 94.

16 Mellor 1958, pp. 531–49.

camouflage (2002) and *Camoupedia: a compendium of research on art, architecture and camouflage* (2009). In both he focuses on the interdisciplinarity of camouflage and shows how artists of the stature of Pablo Picasso, Georges Braque and Franz Marc in WWI, and Salvador Dalí, László Moholy-Nagy, Roland Penrose and Arshile Gorky in WWII all played a part in camouflage's history, theory and practice. Among other influential British and American books published since 1970 that place artists working in camouflage in social and military context, or address the crossing-over of art, science and war, are *Camouflage: a history of concealment and deception in war* (1979) by British historian Guy Hartcup; Elizabeth Kahn's *The neglected majority: 'les camoufleurs', art history, and World War I* (1984); Hardy Blechman (ed.) *Disruptive pattern material: an encyclopedia of camouflage* (2004); Henrietta Goodden's account of British camouflage, *Camouflage and art: design for deception in World War 2* (2007); and *Camouflage* (2007) by Tim Newark who, like Blechman, includes the co-option of camouflage aesthetics by popular culture.

As the astonishing visual examples in this array of international publications show, camouflage is intrinsically surreal for the way it creates disorienting confusions between inanimate and animate states, and for the psychology that lies behind visual deceptions. Without doubt perceptual psychology was critical to camouflage design and the point is raised frequently throughout these chapters. However, while Frank Hinder experimented with painted designs on model planes and ships, William Dakin looked at questions of perception through the markings of animals. Dakin's ideas on animal camouflage and their application to war did not emerge from nowhere and he was indebted not only to Charles Darwin but also to American artist and naturalist Abbott H. Thayer and British zoologist Hugh B. Cott whose images and ideas he liberally borrowed for *The art of camouflage* and for the preparation of camouflage teaching materials on principles such as disruptive patterning (fig. I.4). Dakin was part of a network of naturalists and zoologists working in the field of animal camouflage and its application to war, yet, as discussed in the second section of this study, he fails to mention Thayer and Cott. Nevertheless, equipped with knowledge of the behaviours of animals, and armed with passages of Rudyard Kipling's story 'How the leopard got his spots', Dakin thrived on theorising camouflage for jungle warfare. It stimulated a brutish nature, for while he believed that civilisation in peacetime was evidence of evolutionary progression away from savagery, in war he was infatuated with the superiority of primitive life, of instinct and devolution.

My own interest in camouflage developed from research into the uncanny doubles in trompe l'oeil paintings, those cunning illusions that make the observer feel like a fool by tricking the eye into seeing a three-dimensional object where none exists. And fake photographs of WWII air battles constructed in 1942 in Australia by popular artist Ainslee Roberts also aroused my curiosity in camouflage. Roberts contrived a 'documentary' photograph of an air battle between German and British planes by photographing two model planes hung by string off a clothes-horse creating such a successful illusion that it won him a photographic prize.[17] In both cases the images in question demonstrate what happens when artists wage war in the figurative sense with other people's powers of vision.

I begin the book with a chapter about camouflage and the Japanese bombing of Darwin even though, chronologically speaking, the involvement of Australian artists in camouflage

17 Elias 1999, pp. 379–86.

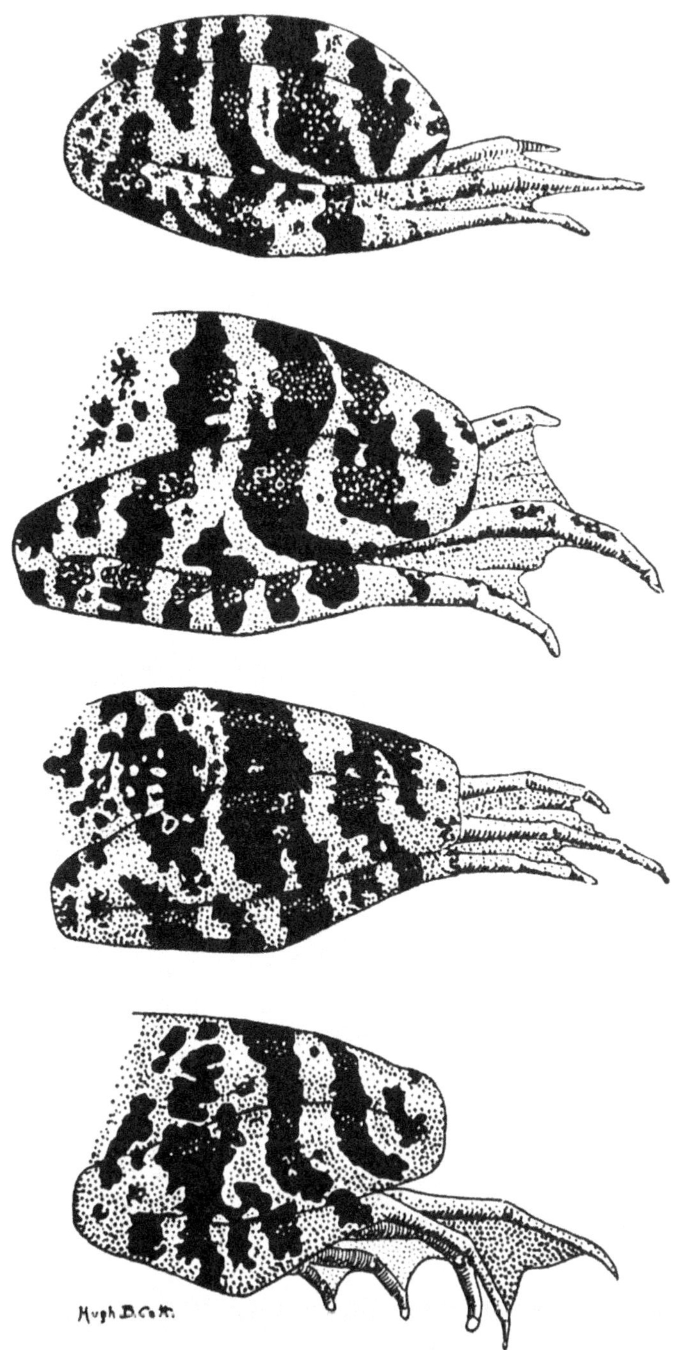

Fig. I.4. Hugh B. Cott, *Hind limbs of the common frog (Rana temporaria)*. From Cott 1957 [1940], p. 71, copied by the DHS onto glass plate negatives (NAA, C1907).

for the DHS properly begins in Sydney in 1939 with the earlier formation of the Sydney Camouflage Group. However, it took the bombing of Darwin to make camouflage design and camouflage preparations accelerate on mainland Australia, and by deferring the Sydney part of the story until chapters 2 and 3 I can better establish the scope of this varied and exciting history. Beginning with Darwin allows me to introduce certain of the main protagonists including William Dakin, Frank Hinder, Max Dupain and Eric Thompson, and establish the essential themes outlined above: interdisciplinary exchanges between art, zoology and military science; animal camouflage as model; civilians in conflict with military authorities; the gendering of camouflage as feminine; links between jungle and tropical camouflage and primitivism; and camouflage as an art and science of visual perception. Beginning with Darwin allows me to start the book with a brief history of the ship *Zealandia,* a vessel that was no stranger to advancements in the role of camouflage patterns for optical confusion in warfare.

PART 1

THE ART COMMUNITY

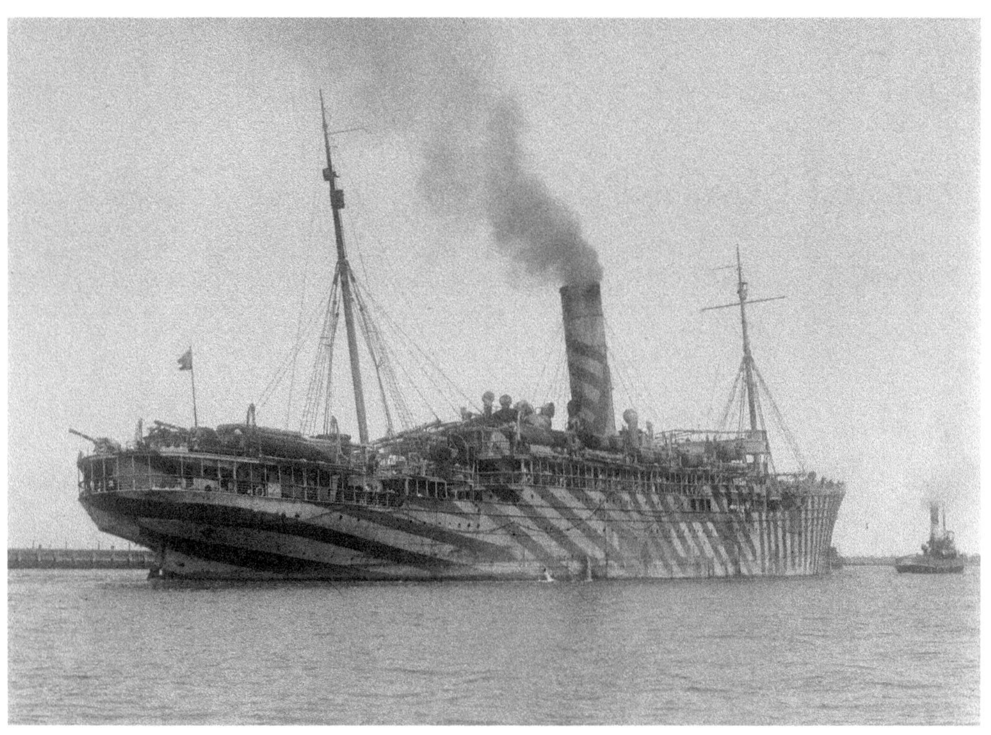

Fig. 1.1. SS *Zealandia* painted in dazzle camouflage in WWI, c. 1917. Picture Collection, State Library of Victoria: H91.325/2150. Photograph, Allan C. Green.

Chapter 1

Darwin

On 19 February 1942, for the first time in its European history, Australia came under attack when a series of Japanese air raids struck the northern port of Darwin in the early stages of the war in the SW Pacific. That morning the SS *Zealandia,* a merchant ship and coastal liner recently returned from transporting troops to Rabaul in the mandated territory of New Guinea, was anchored at Darwin Harbour in the Northern Territory of Australia, waiting to unload a cargo of ammunition and a consignment of camouflage materials.[1] With war escalating to the north, materials for camouflaging military installations in the Darwin region were dispatched from Sydney in secret. The *Zealandia* had been requisitioned for war work once before, during a mission also involving camouflage. In WWI, as a converted troopship painted in what British and American camouflage experts called 'dazzle' (fig. 1.1), it transported troops from New York to Europe patterned in its startling array of black and white stripes, a design that, to the uninitiated, looked more appropriate for a target than a ship evading enemy attention.[2]

Yet in WWI, dazzle was a new optical weapon. Who invented it was a matter of dispute between those who believed it was British artist and marine painter Norman Wilkinson and those who thought it was United States artist and naturalist Abbott H. Thayer.[3] But what is certain is that the theory of dazzle patterning, like other forms of military camouflage, drew on ideas and aesthetics from natural history, perceptual psychology and modern art. It was linked with experiments in the perception of optical distortions of the type where 'seeing is deceiving', to use the words of Matthew Luckiesh who was a camouflage artist in WWI and later wrote a book on optical illusions.[4] Likewise dazzle was closely related to cubism, that style of painting made famous by Pablo Picasso and Georges Braque who took a conceptual approach to form and space and depicted objects as a series of facets, with the angles of each blending into the next to form a confusing, broken, fragmented pattern. Dazzle was also connected with natural history and functionalist theories of the patterns and colourings in animals, which, it is widely claimed, assist animals to conceal themselves and deceive predators. A case in point is the stripes on zebras that make a moving herd appear a confusing mass of black and white patterns.

1 Mellor 1958, p. 539.

2 For photographs and a history of *Zealandia* see Plowman 2007, pp. 66–67.

3 Roy R. Behrens discusses the debates surrounding the originator of dazzle camouflage. See Behrens 1999, pp. 53–59.

4 Luckiesh 2009 [1922], p. 1.

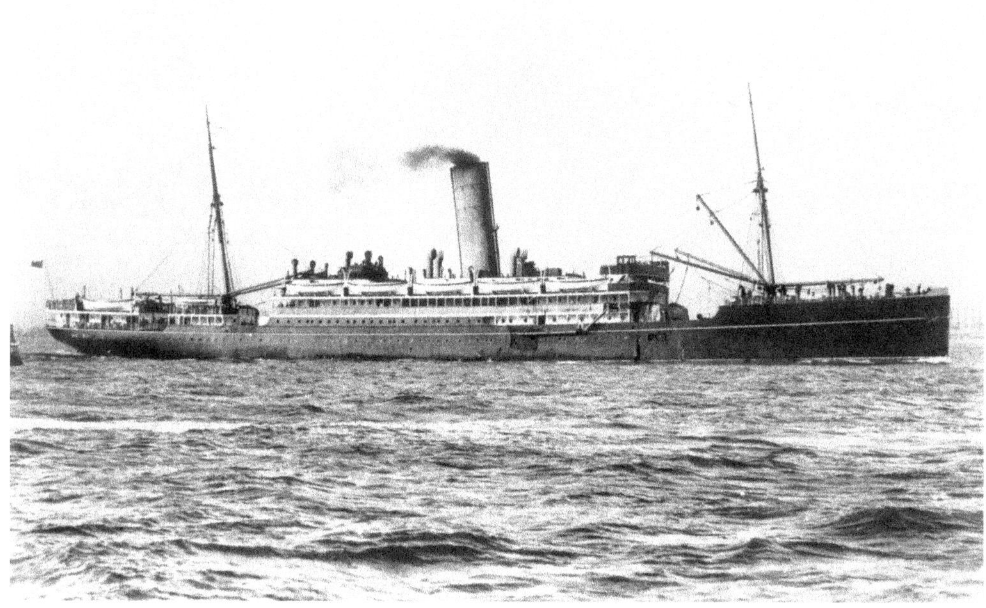

Fig 1.2. Starboard side view of the SS *Zealandia*, c. 1942. AWM, 043177.

The zebra's stripes, cubism, optical distortions and military dazzle have in common the creation of perceptual uncertainty. Otherwise known as 'disruptive patterning', the array of black and white designs on warships like *Zealandia* were intended to confuse and perplex the distant eye of submarine periscopes and transmit misinformation about the ship's shape and form, as well as speed and direction. But disruptive patterning was just one of many strategies in the broader field of military camouflage intended to confuse or negate optical clarity and to trick and deceive the enemy through the interplay of dazzling, to hide weakness, and concealing, to hide strength. It was in WWI, in France and later Britain, Germany and the US, that innovations in military camouflage developed, and when camouflage units and camouflage specialists were first made officially part of military organisations. At this time, too, artists formed the majority of recruits and were known by the title 'camoufleur'. Indeed artists and designers in both world wars dominated camouflage, a field that crossed discipline boundaries of biological science, military science and aesthetics.[5]

The *Zealandia* survived WWI which suggests that dazzle camouflage gave the ship an advantage, because in ordinary civilian guise (fig. 1.2) in WWII, it was not so lucky. Japanese air raids on Darwin sank the *Zealandia* after a spectacular explosion and sent its cargo of camouflage materials to the bottom of the harbour. With the bombing of Darwin—at the 'front door' to Australia—war at home became a reality, and with Japanese pilots having

5 Kahn 1984, p. 1.

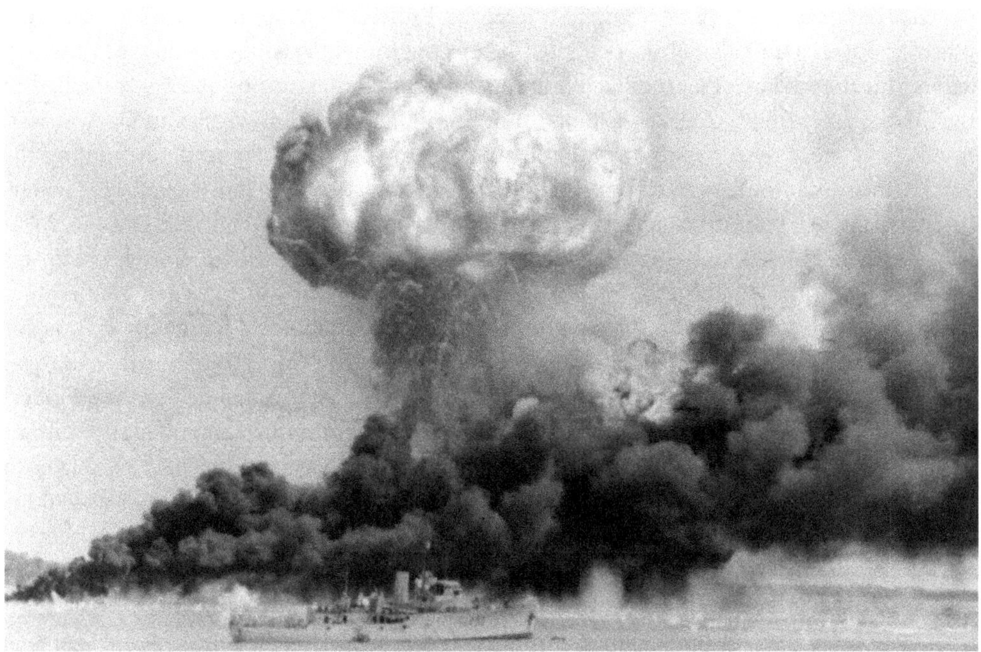

Fig. 1.3. Oil tanks hit in Darwin during the first Japanese air raid, 19 February 1942. RAN Historical Collection, Canberra, AWM, AWM 128108.

the measure of Darwin from the air, the protection of the town's large Allied military base became imperative (fig. 1.3).[6]

Many wondered if the outcome would have been different had the *Zealandia* been camouflaged, if the camouflage cargo had reached Darwin sooner and, if by camouflaging the town itself, the heavy loss of life, extensive physical damage and significant depletion of military strength could have been avoided. Frank Hinder, who was an innovative young modern artist and designer seconded from the Australian militia to work as a camouflage artist for the federal government's DHS, argued that the absence of camouflage on the *Zealandia* had unquestionably increased the ship's visibility and vulnerability.[7] However he conceded that no deception scheme, or preplanning of aerodrome concealment, could possibly have protected Darwin from the bombing by Japanese or from the ensuing devastation.[8] In the flat landscape of the Darwin region, oil tanks and military installations stood out prominently from the air (fig. 1.3). Anyway, Japanese pearl divers already knew the area intimately; the locations of military sites and fuel tanks were no secret.[9]

---

6 For a first-hand record see *Zealandia* Official Log Book 29 October 1941–19 February 1942, Sydney, NAA, Series SP 290/2, Item 1941/Zealandia/4.

7 Frank Hinder, 'Garden island', 5 June 1942, Navy Camouflage, Sydney, NAA, C1707, Item 47.

8 Frank Hinder, 'Records of Lt F.C. Hinder Camoufleur', Canberra, AWM, PR 88/133, File 895/4/182, Item 7 of 12.

9 Mellor 1958, p. 539.

Yet Hinder's supervisor argued differently. Professor William Dakin, an eminent marine zoologist and academic, occupied a key government position throughout the war as Technical Director of Camouflage for Australia and its territories and was attached to the DHS. Preventing the aerial visibility of both military and civilian sites in Darwin was, in his view, more than possible and he blamed the armed forces for underestimating the significance of camouflage to warfare, of ignoring the advice of what he believed were better-informed civilian scientists and artists, and of hindering the camouflaging of Darwin. Most irksome to Dakin was knowing that in early February 1942, two weeks before Darwin was bombed, a secret war cabinet minute indicated that there were difficulties in cooperation between the army and the airforce and that these had impinged on arrangements for transportation of camouflage materials to Darwin and Port Moresby.[10] Further, he was aware that in June of 1942, well after the first raids, the Minister of Home Security, Hubert P. Lazzarini, had complained to the Prime Minister John Curtin that camouflage operations in Darwin were being hindered because 'certain materials were being held by Army to the grave disadvantage of the R.A.A.F.'[11] The cumulative effect of these events was Dakin's readiness to associate every major disaster that befell Australians during the war in the SW Pacific region, including the bombing of Darwin and the bloody Kokoda campaign in July 1942, on lack of preparation with camouflage.

There is no need to press the point that William Dakin was a controversial figure in Australian military politics in WWII. In short, while artists (including architects, painters, designers and photographers) as well as scientists, served a vital role in camouflage defence for Australia, in Dakin's view they were victims of a campaign by the military to discredit civilians and specifically civilian advice, knowledge and contribution to military matters.[12] One clarification that must be made at this point is that artists and designers deployed by the DHS to work in camouflage—a list comprising just over one hundred men and one woman—were not enlisted soldiers. Unlike their colleagues who signed up or were called up for the services, including such distinguished figures in Australian art history as Sali Herman and Donald Friend, those seconded to work for the DHS between 1941 and 1945 were non-combatants. Not soldiers, but intimately involved with military work, Frank Hinder and the camoufleurs who worked under Dakin in the DHS had ambiguous status on military sites. In the view of D.P. Mellor, Official Historian of WWII, it was inevitable that civilian authorities and military authorities would clash over camouflage.[13] And this is exactly what happened, for the entire duration of the war.

At various intervals between 1941 and 1944, William Dakin, Roy Dalgarno, Max Dupain, Frank Hinder, Ronald Rigg and Eric Thompson were among camoufleurs with the DHS who lived in Darwin. It was a grim prospect and an even worse reality: assignment to Darwin during the war was not just unpopular, it was known as 'banishment and even

10 F. Shedden, 'War cabinet minute', 1 February 1942, Establishment of camouflage organisation and procedures, Canberra, NAA, Series A5954, Item 396/2.

11 Minister of Home Security, H.P. Lazzarini to Prime Minister John Curtin, 16 June 1942, Establishment of camouflage organisation and procedures, Canberra, NAA, Series A5954, Item 396/2.

12 See appendix 2 for a full list of camoufleurs with the DHS.

13 In Mellor 1958, p. 539.

as punishment'.[14] The heat, the flatness of the landscape, the monotony of war service punctuated by deep anxiety and fear, and the unfamiliarity of the tropical flora and fauna for Australians used to the temperate zones of south-east Australia, all contributed to a psychological disintegration known as 'going troppo'. The four weeks Frank Hinder spent there in 1944 felt like 'months' and it was an experience, he said, that gave him 'an inkling of the "troppo" danger and sympathy for those affected'.[15] But for a variety of reasons it was important for the DHS to have a presence in Darwin and show leadership in camouflage construction and teaching, and to 'sell' the significance of modern methods of deception and concealment to air, navy and army staff who were otherwise uninterested and unconvinced.

The year before the first Japanese raids on Darwin, Dakin assessed an urgent need for camouflage for the town but was frustrated to find indifference and apathy among military representatives. After the Japanese attack on Pearl Harbour (7 December 1941) and with his relationship with the army becoming increasingly estranged, he complained to a reporter at the *SMH* that the need to 'speed up' camouflage defence in the north of Australia was urgent.[16] The Advisory War Council became concerned about Dakin's behaviour since only the day before, he had also spoken to a reporter with the *Daily Telegraph*. The headline had read: 'War work delay angers expert' and claimed 'Government officials [have] fallen down on the job of camouflaging Australia'.[17] Dakin finally got the outcome he wanted when funds were released to undertake the concealment of the Royal Australian Air Force (RAAF) aerodrome in Darwin.

On 21 January 1942, with a worsening political situation in the Pacific, Dakin sent his 'best man' to the northern town, camoufleur and architect Eric Thompson. Before the war Thompson was a pioneer of modern film production in Australia and had spent time in the United States with Metro-Goldwyn-Mayer working as a set designer.[18] But with the outbreak of war he was deployed to disguise aerodromes in Sydney using elaborate, deceptive facades, and it was this type of camouflage that Dakin wanted applied to airforce bases in Darwin to make them look like sports grounds, racetracks, farm houses, factories—anything but aerodromes. Thompson's role was also to advise the airforce, army and navy.[19] But this was his second trip to the north. In October 1941 he surveyed Darwin by plane and wrote bluntly that if the port and town 'is at present the front door of Australia

14 Lockwood 2005, p. 139.

15 Frank Hinder, 'Records of Lt F.C. Hinder Camoufleur', Canberra, AWM, PR 88/133, File 895/4/182, Item 9 of 12.

16 'Camouflage speed-up; Sydney conference', *SMH*, 24 December 1941, p. 9.

17 'War work delay angers expert', *Daily Telegraph*, 23 December 1941, cited in Establishment of camouflage organisation and procedures, Canberra, NAA, Series A5954, Item 396/2.

18 'Film pioneer dies at 76', 5 January 1978, *SMH*, cited in Frank Hinder, 'Records of Lt F.C. Hinder Camoufleur', Canberra, AWM, PR 88/133, File 895/4/182, Item 7 of 12.

19 W.J. Dakin letter to Officers Commanding RAAF, Army, and Navy, 21 January 1942, in Department of Home Security Staff register—camouflage section, Canberra, NAA, Series A691, Item 1, accumulation dates 1 January 1942–31 December 1945.

as far as defence is concerned, it seems more like the back door as far as camouflage is concerned'.[20]

Despite Dakin's best efforts, the national organisation of camouflage in Australia came too late for the town of Darwin. Despite the lengthy anticipation of war descending on the region, when '17 silver Japanese bombers appeared flying in formation at nearly 20,000 feet', that first raid came as a shock. How strange, one Australian soldier later reflected, that the first Japanese planes were so conspicuous as they approached. He had inspected the wreckage of a Japanese bomber in Darwin harbour and found it carefully painted and 'camouflaged in a green and buff combination'.[21] Wasn't camouflage meant to make objects invisible? Yet the planes could be seen approaching from a long way off by silver flashes in the sky.[22] His question highlighted a popular misconception of camouflage, namely that any pattern of colours aids the concealment of an object. In reality, if the wrong colours and designs are used, camouflage creates more harm than good by making objects stand out rather than become invisible or blend in.

Following the first raids on Darwin, civilian and military residents suddenly understood what it meant to be vulnerable to aerial visibility. The navy, now conscious of the conspicuousness of its white uniforms, instructed personnel to boil their 'tropical whites' in coffee grounds to turn them khaki.[23] But khaki was not the answer either for the tropical environments of the north, because it cut a sharp figure against darker backgrounds of rainforest and thick vines (fig. 1.4). The general problem with khaki as a military colour came to notice in WWI when artist and naturalist Abbott Thayer appealed to the British army to discard it as a solid colour and adopt a patterned uniform to break up, or disrupt, the silhouettes of soldiers.[24] The DHS in Australia in WWII similarly argued that while khaki once served its purpose on dusty earth and rocky kopjes during the Boer War, it failed to meet its objectives in lush, tropical vegetation.[25] It investigated a range of modifications to military uniforms, including a loosely fitted cover of netting to assist the body to blend with its background—something like the diffusion of form in an impressionist painting (fig. 1.5). Finally, the DHS urged the Armed Forces to discard khaki altogether and wear green. The optics of war had changed. Survival and victory now depended on an ability to blend with the jungle in a way that Japanese troops in the Pacific had mastered earlier.

The region of Darwin has a variety of interesting connections with camouflage besides WWII. When John Lort Stokes, surveyor on the third voyage of the HMS *Beagle*, sailed into

---

20 Eric Thompson, 'Camouflage report', 20 October 1941, in Reorganisation of Camouflage Section, 1942, Department of Home Security, Sydney, NAA, Series C1707, Item 29.

21 Baz 1942, p. 172.

22 Macartney 1942, p. 159.

23 Forster, 1992 p. 28.

24 The word 'khaki', which means a dusty colour in Urdu, became part of the English language when the British Army adopted khaki uniforms in the 1880s to assist soldiers to merge with the landscapes of India and South Africa. At this time army units relinquished their more brightly coloured blue and red for the colours of sand, dirt and rocks so they could blend rather than dazzle. See Behrens 2009a, pp. 59–60.

25 Dakin 1947, p. 5.

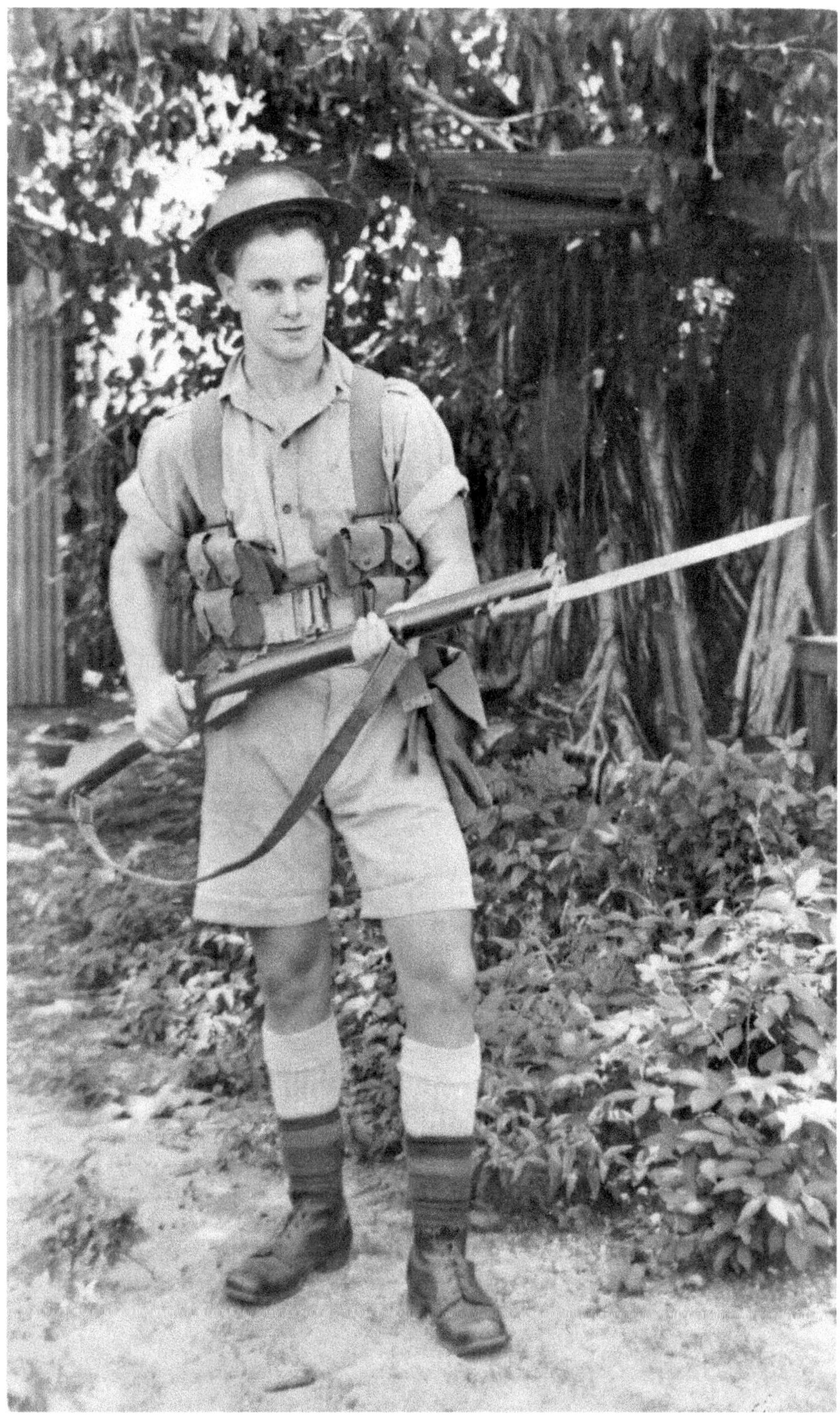

Fig. 1.4. Stoker Richard Lawrence Johnson in Darwin, February 1943. Canberra, AWM, AWM P05291.001.

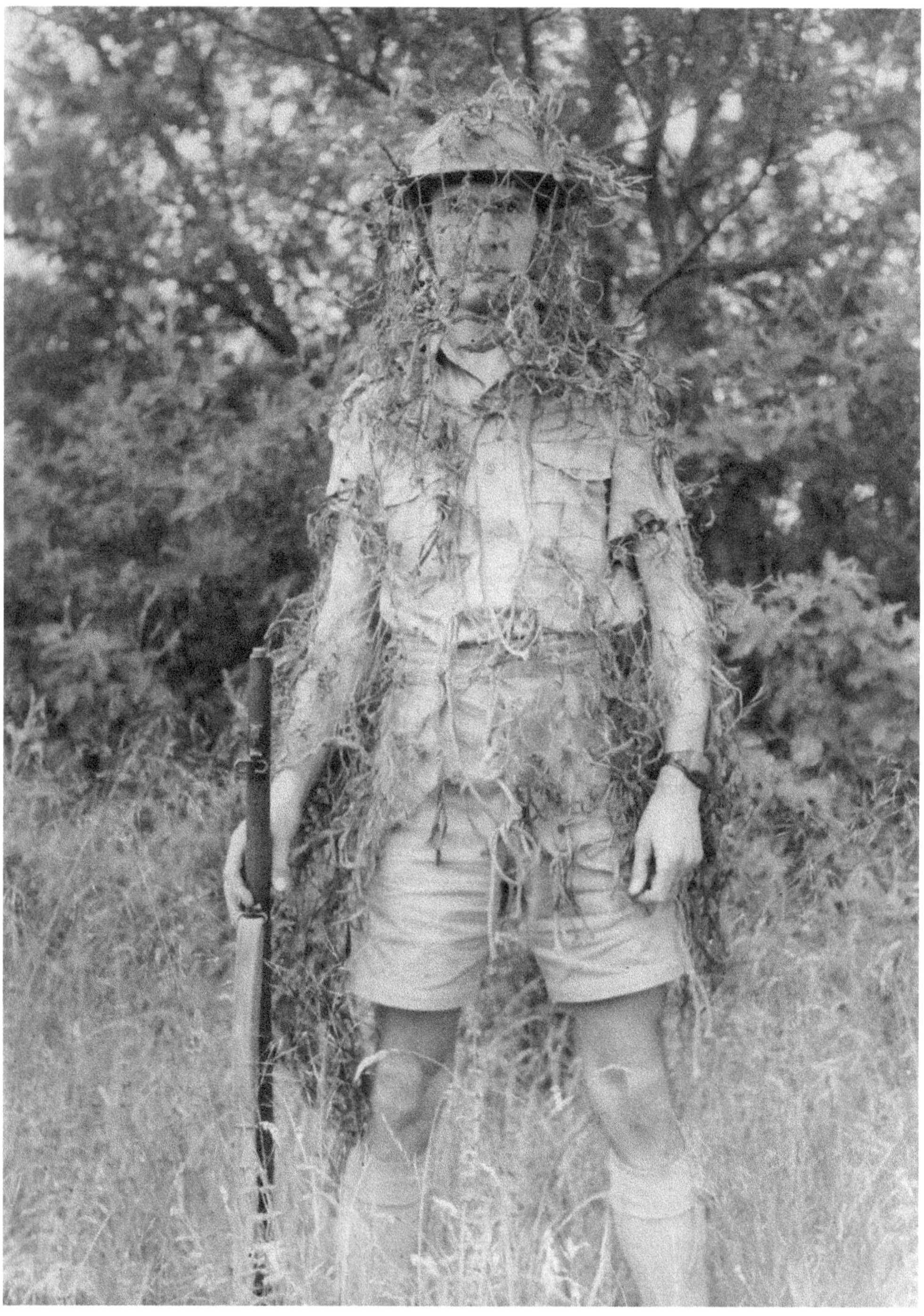

Fig. 1.5. Experiment for individual concealment. Frank Hinder Papers: Collection Archive of the AGNSW. Photograph, Gervaise C. Purcell.

the vast uncharted harbour on 10 September 1839, he named it 'Darwin' in honour of his friend Charles Darwin the naturalist whose reflections on camouflage in nature provided the basis of all modern camouflage studies in biological science and also military science. Charles Darwin and Alfred Wallace, who both wrote about mimicry, influenced every future scientist of camouflage, including those discussed in later chapters: Abbott Thayer, Hugh Cott, Edward Poulton, William Pycraft and William Dakin. Little was understood about the colours and marking of animals and plants before the dissemination of their work, and 20th-century theories on military camouflage are unimaginable without the biological hypothesis that animals evolved colourings and markings to conceal themselves and deceive predators as a way of ensuring species survival.[26] And there is another way that camouflage is relevant to the geographical location of Darwin. Stokes was very much struck not only by the heat in Darwin but by its distinctive flora and fauna.[27] Indeed the faunal range in the region includes one of the world's largest stick insects, a master of camouflage. By virtue of its habitat it is known as the 'Darwin stick insect' (*Eurycnema osiris*) thereby bearing the name of the eminent scientist as well as the region in its own.

The Darwin stick insect is the perfect model for a lesson on military camouflage because it enacts the dual principles of blending for concealment and dazzling for deception, depending on what the situation demands. Out of context, on a hospital sheet observed by a military nurse in WWII, the insect is revealed in all its vulnerability (fig. 1.6). But at home in the bush it blends perfectly with its background by mimicking the morphology of twigs and dissimulating in a way Dakin hoped Australian soldiers would in the green vegetation of the tropics. When threatened, however, it dazzles to deceive its enemy. By flashing the red underside of its wings, it bluffs by proclaiming itself poisonous and dangerous (fig. 1.7).

Simulation, sometimes known in military camouflage as 'bluff', was practised to great effect by both Japanese and Australian militaries in Papua and New Guinea in WWII.[28] Where Darwin and Wallace were enthralled with mimicry as evidence of the processes of evolution, military strategists in WWII were interested in the way it could bring ruination to the enemy.[29] Bluff was an approach to camouflage that made William Dakin nervous, but reports about successful mimicry strategies by Japanese troops in the SW Pacific proved that camouflaging to hide was not enough; it had to be followed up by surprise attack. Increasingly coming to Dakin's notice were Japanese camouflage manuals that made the enemy's philosophy all too plain: 'make the distinction (of objects) difficult, or … cause them to be mistaken when seen from the air or from a long distance'.[30] With this as evidence he continued to push for greater involvement of his staff in Darwin and in all military matters concerning operational camouflage in the north of Australia and the SW Pacific.

While enemy and ally in WWII practised camouflage schemes involving bluff on an heroic scale—and none more daring than an installation on Goodenough Island in

26 Dakin highlighted the importance of zoologists to camouflage studies in Dakin 1941, p. 4.

27 Stokes 1846, pp. 8–10.

28 For photographs and descriptions of the Darwin stick insect see Brock & Hasenpusch 2009, pp. 110–11.

29 See Darwin on mimicry in Darwin 1872, p. 181.

30 'Japanese publication on essentials of camouflage', in Camouflage—General—Camouflage Organisation in New Guinea, Canberra, NAA, Series A453, Item 1943/17/1510 Part 1.

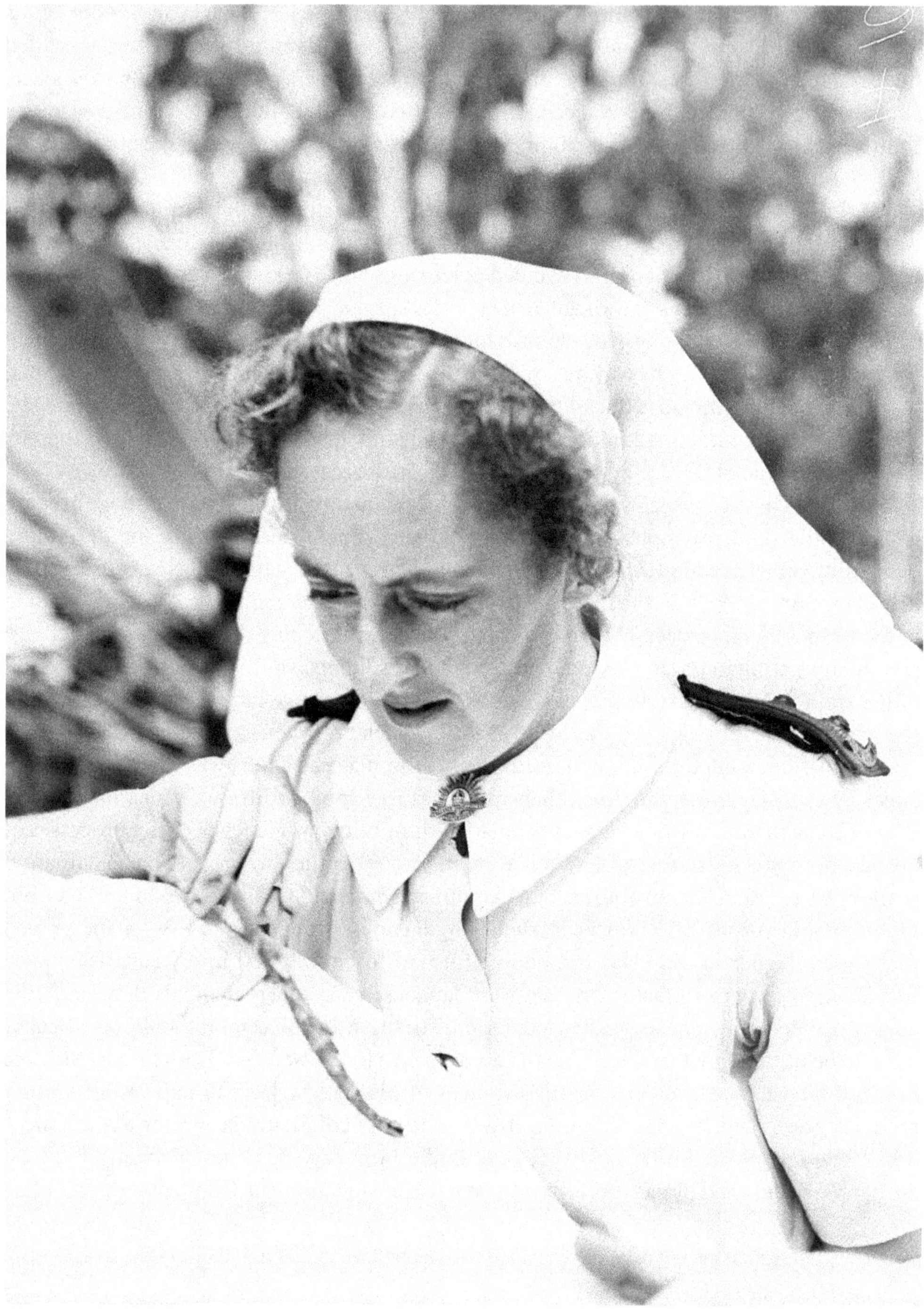

Fig. 1.6. Sister M. Cliff examines a stick insect, Ravenshoe area, Queensland, 25 December 1944. Canberra, AWM, AWM 085155.

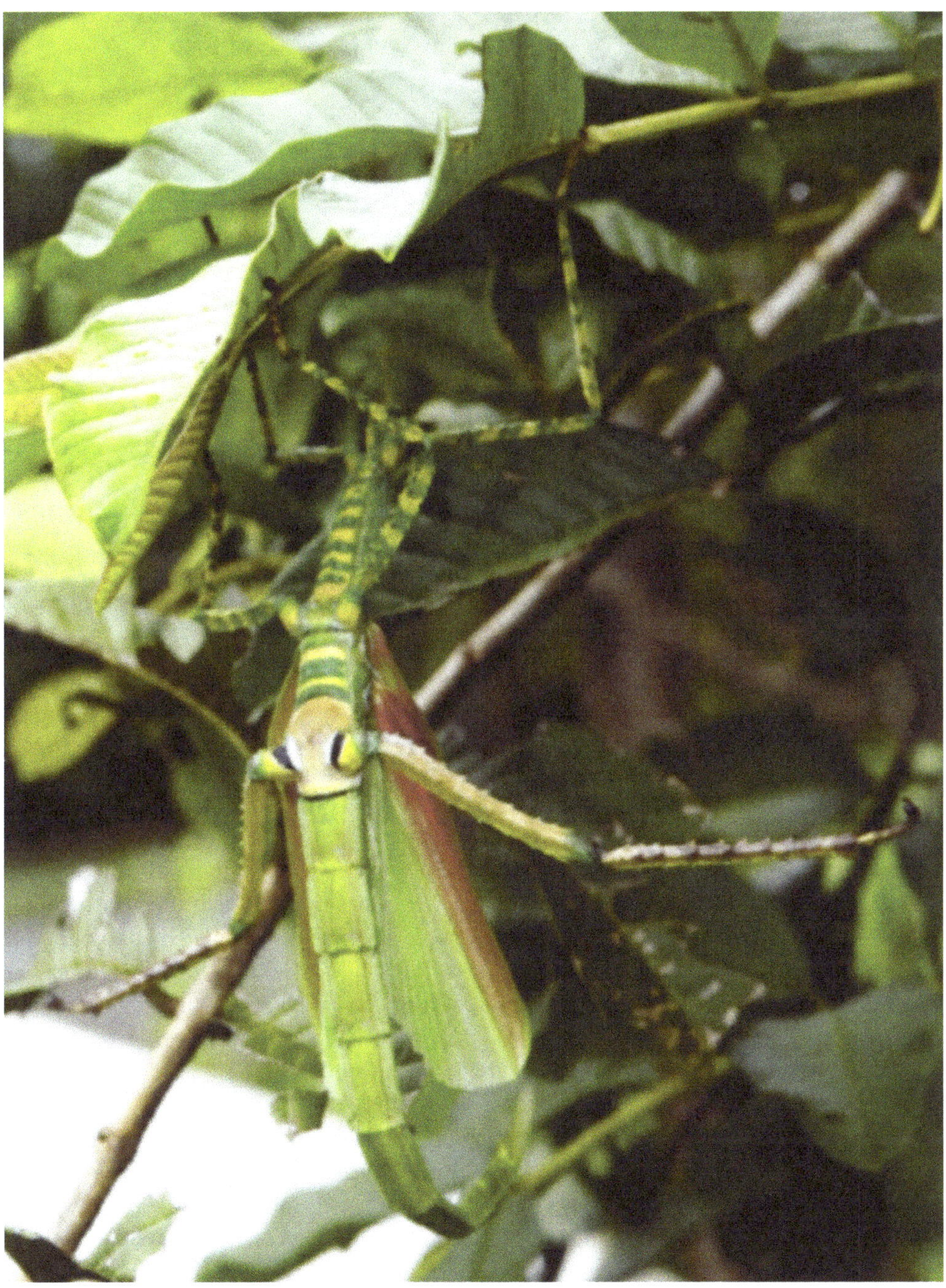

Fig. 1.7. Darwin stick insect female startle display. Photograph in Brock & Hasenpusch 2009. p. 111.

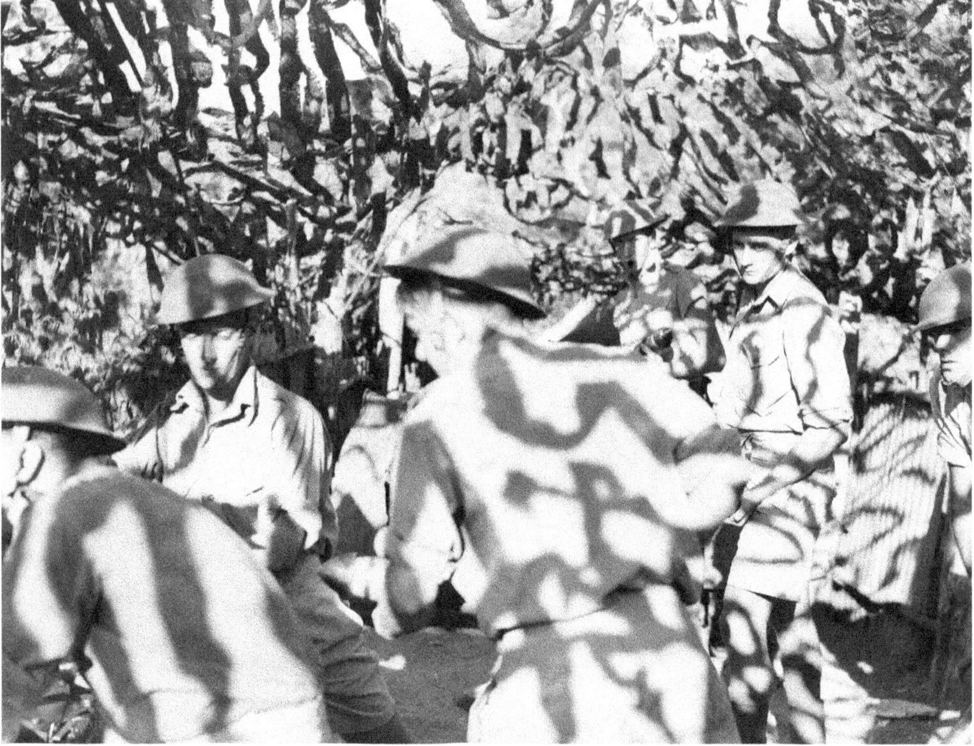

Fig. 1.8. East Point Coast artillery, fixed defences, Darwin 23 June 1943. Canberra, AWM, AWM 052909. Photograph, Geoffrey McInnes.

Papua by the Australian Army—the DHS's main objective was individual concealment. The protection of Australian troops fighting in war zones required that they learn simple ways to use shadows for concealment. The war record is well represented by photographic images of coastal defence sites in Darwin showing shadows falling through camouflage nets and throwing disruptive patterns across solid objects and bodies which appear almost flat, transparent and invisible (fig. 1.8). In the absence of nets, troops were encouraged to utilise tree-cover and let the striations of light falling through branches provide the effect of optical dissolution.

Dakin's view was that observing nature closely, indeed of returning to a state of nature and being attuned to the smells, sounds and appearances of the tropical north, was the best way to use camouflage to survive. But his call for a return to primitivistic behaviours, coupled with primitivising anecdotes used as teaching aids to set free the warrior instincts of troops in the Pacific area, were racially insensitive. For example, a publication on the principles behind individual concealment claimed that when the first bombs fell on Darwin: 'a number of aborigines quickly removed their civilized, brightly coloured clothing, dashed for tree cover and then "froze". The natural instincts of ages enforced the correct action in this case.'[31]

31 W.J. Dakin and the Camouflage Directorate, 'Concealment, and camouflage of the individual in

In fact, when bombs fell on Darwin, Indigenous Australians were among the war's many casualties, but such was the apartheid in the north, they did not have access to the same wards as whites.[32] Dakin's anecdote of Aborigines returning naked to trees for natural concealment was symptomatic of a more widespread wartime myth about camouflage, namely that to be effective, individual concealment in jungle areas required black skin, a tribal mind, and a state of devolution to instinctual behaviour of a type believed to have once belonged to the Anglo-Saxon race, but one that had disappeared with civilisation.[33] Accordingly, in Dakin's view only good could come of soldiers modelling themselves—for the duration of the war—on Aborigines, people who Australian anthropologists of his generation, including Walter Baldwin Spencer, believed were living examples of the early history of humankind.[34]

Frank Hinder, on the other hand, found Aboriginal body painting of great insight when considering camouflage.[35] The war offered many white Australians their first encounter with Aborigines and, during Hinder's spell in Darwin, he observed how painted marks on Indigenous Australians broke up the surface of the body. He speculated on the advantages these might bring to hunting and combat:

> The tone and colour of the aboriginals skin, plus the breaking up of body areas with paint would obviously assist them to 'vanish' so far as the average European was concerned. And these painted areas were probably connected with totem or magic (as we would call it) rather than with the intention of camouflaging the individual in the modern sense of the word. But it must have been extremely effective, especially at dusk or early morning.[36]

Hinder's interest in Aboriginal culture, as a model for camouflage, was based on the perceptual impact of Indigenous designs on the eye. William Dakin's interest in the example of Aborigines, however, was based in social Darwinism. He looked upon human progress as a matter of evolution, but in times of war he felt progress and civilisation had come at a cost.[37] In *The descent of man* (1882) Charles Darwin wrote that the most successful

---

warfare', Canberra, Department of Home Security, 1944, p. 31 in William J. Dakin, Camouflage report 1939–1945 (with appendices K–N), Canberra, AWM, Series 81 [77 Part 3], 1947, unpaginated.

32 Hall 1980, p. 122.

33 While in Dakin's view soldiers in the jungle did not behave instinctively enough, accounts from the SW Pacific focus on going to ground and 'snaking'. See Armstrong and Kerr 1945, p. 109.

34 Frame 2009, p. 240.

35 Hinder was most likely familiar with *The Australian Aboriginal* (1925), an anthropological study by Herbert Basedow that discussed the practice of Aboriginal body painting as a 'cryptic covering' for hunters trying to approach their prey unseen. See Basedow 1925, pp. 141–42. Basedow's investigations of the camouflage techniques of Indigenous Australians were in turn influential on British military camouflage theorist and zoologist Hugh Cott. Herbert Basedow is cited in Cott 1940, p. 361.

36 Frank Hinder, 'Records of Lt F.C. Hinder Camoufleur', Canberra, AWM, PR 88/133, File 895/4/182, Item 9 of 12.

37 This attitude was shared by the German military. A German document read: 'As the natural instinct for concealment has died out in present day man, the young soldier must be taught the art of camouflage', see W.J. Dakin, 'The modern position and value of camouflage in regard to war in forward

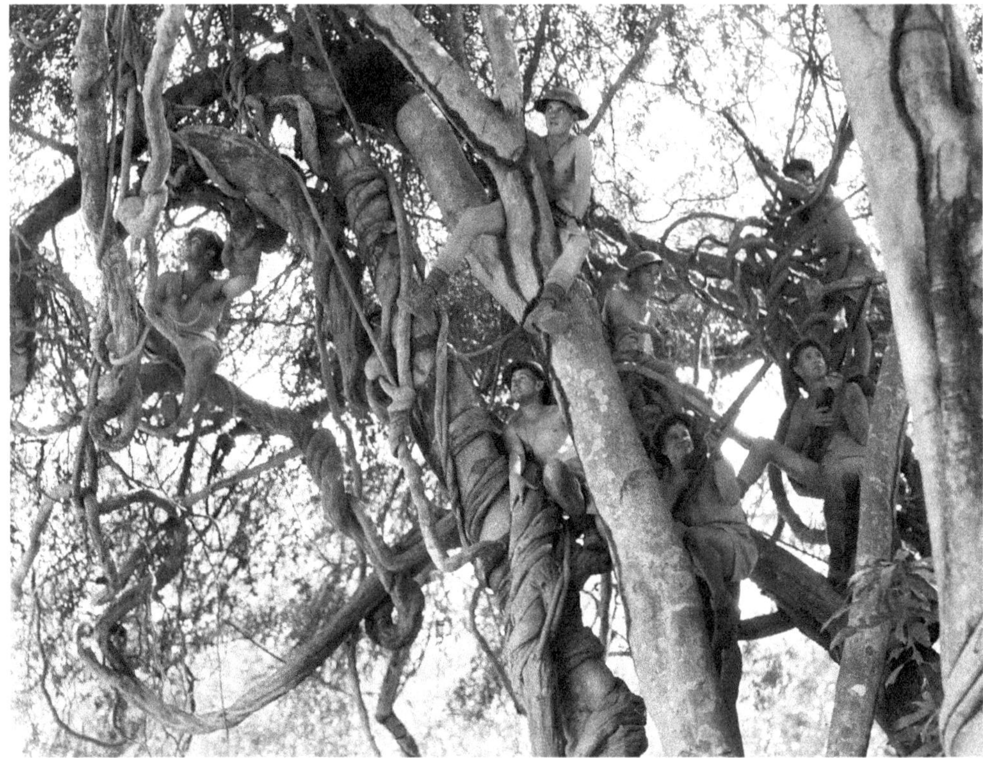

Fig. 1.9. Concealed troops in tropical vegetation, Darwin, 7 September 1942. Canberra, AWM, AWM 013187. Negative by McNeil.

primitive tribes 'in the rudest state of society' had increased in greatest numbers through their use of cunning.[38] And this was precisely why Dakin dedicated his wartime life to camouflage; it was his way of ensuring the survival of the Anglo-Saxon race. One of his great concerns was the pinkness of Australian skin in the jungle. In October 1942 he noted that soldiers in Darwin were working 'with nothing on but shorts, boots and helmets' and neglecting the first principles of camouflage in tropical warfare, namely concealment of conspicuous European skin from the aerial eye of the enemy (fig. 1.9).[39] His solution was to design black camouflage paint manufactured under the name Skin Tone Commando Cream. It was intended as a skin covering for jungle-fighting troops, and during a later period of the war when the DHS established a centre on Goodenough Island in Papua, Dakin and his staff embarked on a campaign to encourage Australian soldiers to blacken their skin, one that ultimately failed.

The links between jungle and tropical camouflage and primitivism—the western nostalgia and fantasy for origins, and for instinctual, primal states—become increasingly

---

and base areas', Air intelligence reports, Sydney, NAA, Series C 1707, Item File 36, p. 5.

38 Darwin 1882, p. 128.

39 Dakin 1947 [1942], p. 6.

important to the story of camouflage in the SW Pacific as will be discussed in later chapters. What has been established in this chapter is that the modern history of camouflage begins in WWI, that this history involves an exchange of ideas between artists, scientists and military tacticians, and that the bombing of Darwin brought urgency to the organisation of camouflage in Australia (including preparations for war in the SW Pacific). Two key figures in the story of camouflage in Australia are William Dakin and Frank Hinder, and their roles in camouflage precipitated conflict between military and civilian bodies. It is opportune now to turn to prewar years when the seed of an idea to mobilise artists and designers to work in this specialised field began to germinate in Sydney.

In 1939 representatives of Australia's artist community in Sydney, consisting of painters and sculptors, photographers, draughtsmen, display artists, designers and architects, people who otherwise belonged to a peace-seeking culture, were proactive in forming an alliance to work in camouflage, believing it a positive way in which they could help resist the spread of fascism and a Japanese invasion of Australia. But once WWII commenced, those who put up their hands to work in camouflage, whether for military or civilian organisations, encountered a harsh military environment where an already existing prejudice to the contributions of artists to society was compounded. In Australia in the 1930s, according to expressionist painter and pacifist Albert Tucker, 'even to be categorized as an artist immediately put one in queer street; artists were people trying to avoid work'.[40] Stereotypes of artists became exaggerated in the war and camouflage became known as the soft option for war service. Those with a commitment to winning the war through face-to-face combat rather than the indirect means of concealment and deception were suspicious that camoufleurs were people trying to avoid the proper work of wartime. Yet the history of advancements in camouflage warfare in Australia in WWII is unthinkable without the participation of artists.

40 Mollison & Bonham 1982, p. 23.

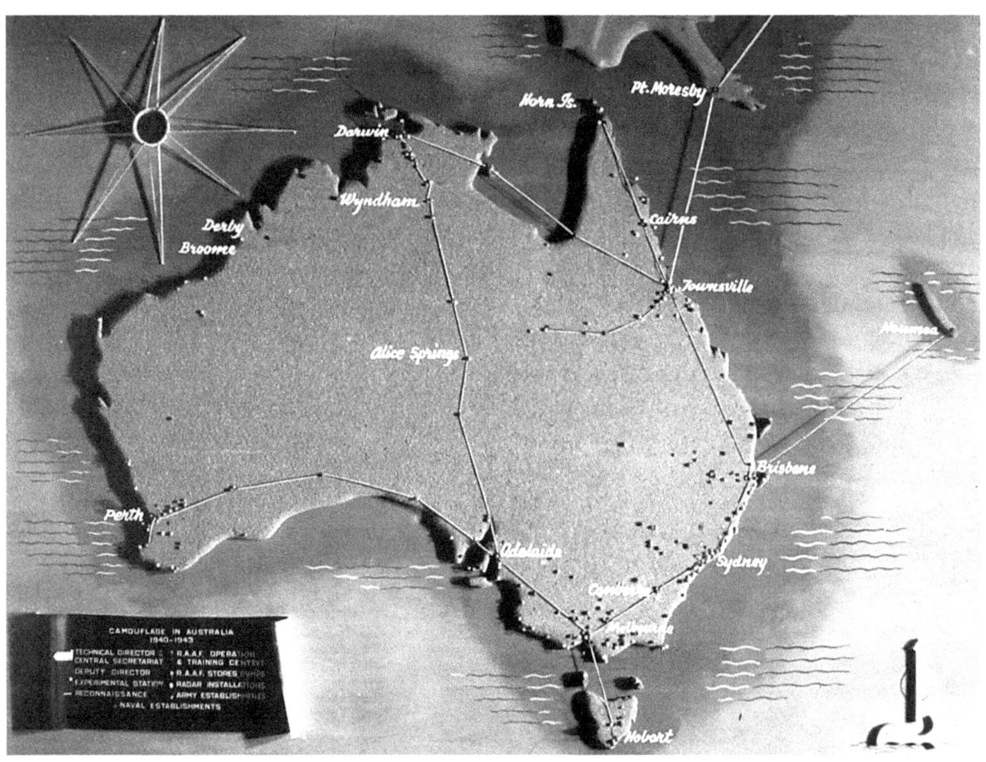

Fig. 2.1. 'Camouflage in Australia 1940–1943'; Department of Home Security map, c. 1943. The collection of the NAA, C1908, 3[24].

Chapter 2

Sydney

Whereas Darwin, as a forward area of the war, was important for operational camouflage, Sydney was the primary centre for research and experimentation (fig. 2.1). But as the biggest city in Australia with the largest population, the largest commercial and industrial infrastructure, the most aerodromes and the greatest number of oil storage installations, munitions establishments and essential public utilities such as power and water supplies, Sydney also presented the biggest challenge for civil defence.

The story of Sydney as a camouflage centre begins in 1939, before war was declared, with the proactive formation of the Sydney Camouflage Group, an informal committee of scientists, military representatives, engineers and architects and a list of artists and photographers that today reads like a Who's Who of Australian art history. The group's mission was to advance the state of camouflage knowledge in Australia and adapt northern hemisphere methods to Australian conditions. Its members were concerned about Australia's lack of preparation for what then seemed like an inevitable Japanese attack and so, equipped with access to and cooperation from a major arts community, state military organisations, and local scientific and industrial research institutions, they applied their professional knowledge of concealment and deception in nature, art and war, to experiments in the Sydney region, and produced a book for use by civilians, and later used by the Australian military, *The art of camouflage* (1941).[1]

The group had 30 members, and in addition to Dakin, its Chairman, included architect John D. Moore, architect Professor Leslie Wilkinson, four high-ranking members of the airforce, army and navy, arts administrator Sydney Ure Smith, Victor Tadgell of the Orient Steam Company, artists and designers Frank Hinder, Adrian Feint, Douglas Annand, Peter Dodd, Robert E. Curtis, Frank Medworth and Arthur Murch, photographers Russell Roberts and Max Dupain, a representative of the chemical paint industry and additional representatives of the professions of architecture, engineering, model-making and commercial art.[2] As a patron of the arts, Ure Smith was close to Adrian Feint and Max Dupain but in general, members of the group had little personal acquaintance, sometimes polarised views on modern art as in the case of Ure Smith and Hinder, and seemingly

1 The Sydney Camouflage Group was also known as the 'Associated Sydney Camouflage Group of Camoufleurs'. See John D. Moore to A.W. Welch, 6 March 1942, Camouflage personnel NSW Camoufleurs staff appointments, Canberra, NAA, Series A453, Item 1942/28/2298.

2 See appendix 1 for a full list of members of the Sydney Camouflage Group.

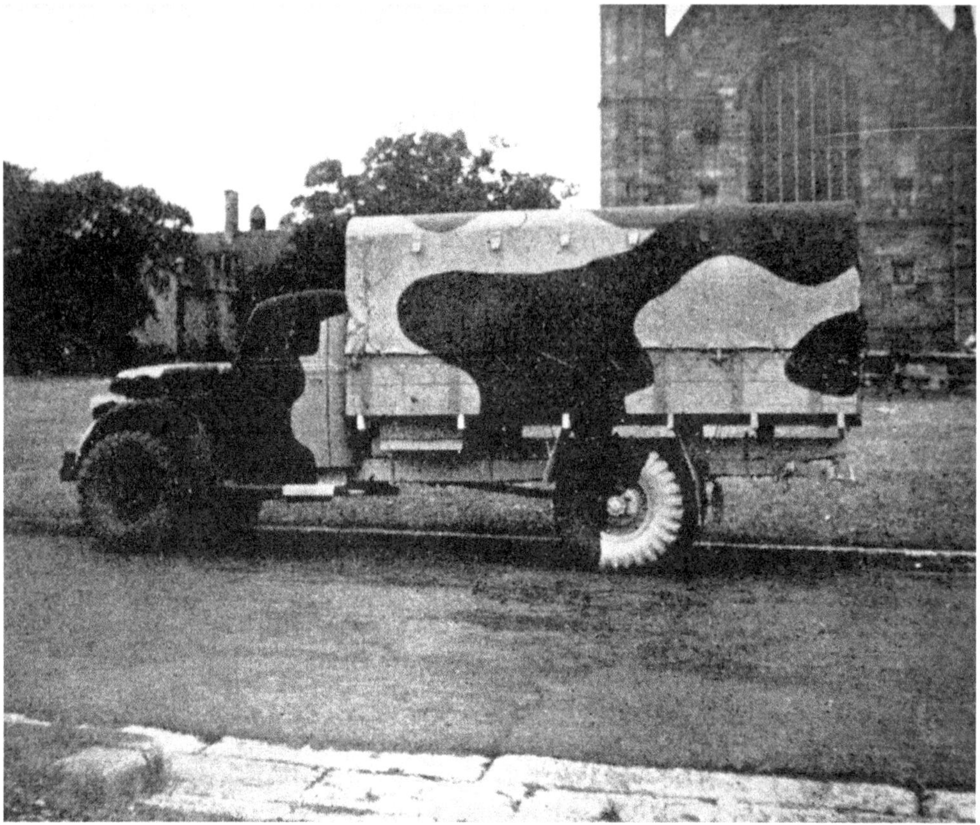

Fig. 2.2. Camouflaged army truck at the University of Sydney. Dakin 1942, p. 39. Photograph W.J. Dakin.

nothing in common but a concern about the imminent outbreak of WWII and the increasing threat of Japanese invasion.[3] They conducted experiments in and around the city of Sydney, and since Dakin and Wilkinson were both staff at the University of Sydney, the grounds of the University were used to test concealment and deception methods (fig. 2.2).

Of the older members in the Sydney Camouflage Group, John Moore and Adrian Feint both served in WWI, and both in France where camouflage was first formalised into a separate military unit by the French army known as *La Section Camouflage*.[4] Among the camouflage designs made infamous on the battlefields in France in WWI were trenches, narrow slits in the ground where strategic concealment and exposure in the form of advancement and retreat became embodied experience for soldiers confined underground. In the French war landscape Feint and Moore were witness to new experimental camouflage designs sporting cubist geometric patterns that Salvador Dalí attributed to the influence of

3 For associations between Ure Smith, Feint, Dupain and Moore see Underhill, 1991, pp. 41–51. For professional animosity with Frank Hinder and Margel Hinder in relation to modern art, see Underhill 1991, pp. 21–23.

4 Kahn 1984, p. 11.

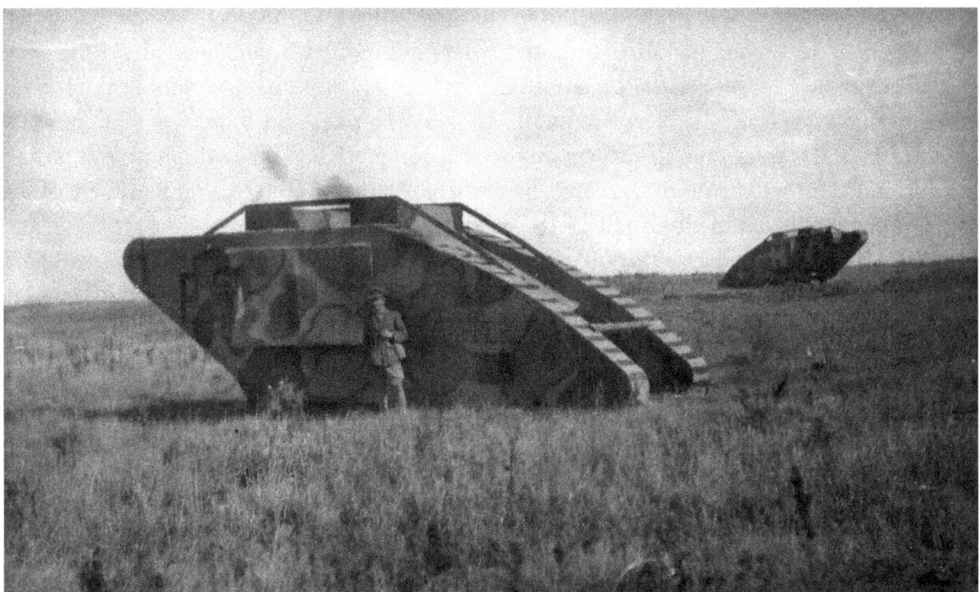

Fig. 2.3. Australian dummy tanks made of canvas and wood, 4th Australian Division, France, c. September 1918. Canberra, AWM, AWM C04505, original half plate glass negative.

Picasso and specifically the patterning of Picasso's harlequin figures.[5] And among the many strange sights they would have seen on the battlefields were military decoys, including fake tanks designed by Australians, that were light enough to be carried by small groups of men. They were far from convincing on the ground but were likely to fool the aerial eye (fig. 2.3).

Louis McCubbin, Frank Crozier, Will Longstaff, James F. Scott, J.S. Macdonald and George Benson were among the now eminent Australians who worked in camouflage in Europe in WWI.[6] But it was John Moore and Adrian Feint, two artists who were exposed to a myriad of camouflage innovations, abstractions, and trompe l'oeil illusions in France, that were likely conduits for relaying insights and information on camouflage to the group in Sydney in 1939. Their encounters with WWI camouflage, coupled with Dakin's extensive knowledge of animal camouflage, in combination with the group's collective expertise with military matters, aerial and ground photography, architectural space, realistic painting, modernist abstraction, perspectival drawing and scale modelling, allowed the profile of the Sydney Camouflage Group to take shape as a well-rounded team.

Among the historical precedents for the formation of the Sydney Camouflage Group was an American alliance of artists in WWI named the American Camouflage Corps which, likewise, set about independently preparing itself for future service, in its case to the

5 Dalí 1998 [1942], p. 342.

6 For a useful source of information on Louis McCubbin, Frank Crozier, Will Longstaff, James F. Scott, J.S. Macdonal, and George Benson in WWI see the collection and biography databases of the AWM, www.awm.gov.au.

US Army.[7] While there is every likelihood that the Sydney Camouflage Group knew that its own organisation had historical links to earlier ones in the US, such were the censorship and secrecy laws in WWII that it is unlikely that the group knew of contemporaneous developments overseas. In New South Wales (NSW) alone, and following the bombing of Darwin, the State Publicity Censor imposed 'a complete ban on both photographs and letterpress relating to camouflage'.[8] The chances are remote, therefore, that members of the group were aware that in Chicago László Moholy-Nagy, the preeminent modern artist and designer, had developed a course on civilian camouflage,[9] that in Britain surrealist Roland Penrose had produced a book on camouflage for the Home Guard,[10] or that in the US abstract painter Ellsworth Kelly had enlisted in a secret camouflage regiment known as the Ghost Army.[11]

However, a store of wartime newspaper clippings collected by Frank Hinder shows that, by 1941, news had reached the Sydney group that artists in the US were pressuring their government to make more effective use of the art sector's expertise for the war effort.[12] The Australian group of camoufleurs almost certainly understood that this was a unique moment in history, and that they were part of an international alliance of western artists. When Max Dupain later talked with Hazel de Berg about the war years, he spoke of his participation in 'the camouflage movement', a comment that strongly supports the claim that the Sydney Camouflage Group saw itself as vanguard, future looking, progressive and political.[13]

In the early months of 1939 the group met fortnightly and, if Frank Hinder's memory was right, some meetings were held in the offices of Sydney Ure Smith.[14] Of all the members, Ure Smith (1887–1949) is the most surprising because, while so much is known today about this powerful publisher, editor and administrator, nothing is written about his role in camouflage defence, a situation that reflects the near-invisibility in Australian art history of artists' involvement with WWII camouflage.[15] Therefore it is intriguing to read Dakin's

---

7 See 'Faking as an art in conducting war', *The New York Times,* 24 June 1917, p. 35.

8 The ban did not stop Dakin talking to the press about the government's slowness to act on matters of camouflage, but it prevented him from discussing the details of his work. See edict by H.H. Mansell, 'Publicity censors', 1 April 1942 in Camouflage, Sydney, NAA, Series SP106/1, Item PC 490, p. 32.

9 Moholy-Nagy 1969, pp. 182–83.

10 Penrose 1941.

11 Behrens 2009a, pp. 157–58.

12 'Camouflage', *The New York Times Magazine,* 25 May 1941, p. 14, cited in Frank Hinder, 'Records of Lt F.C. Hinder Camoufleur', Canberra, AWM, PR 88/133, file 895/4/182, Item 4 of 12.

13 Max Dupain, 'Max Dupain interviewed by Hazel de Berg', Canberra, National Library of Australia, 3 November 1975, Oral DeB 1/874, unpaginated transcript.

14 Frank Hinder, 'Records of Lt F.C. Hinder Camoufleur', Canberra, AWM, PR 88/133, file 895/4/182, Item 7 of 12.

15 For the minutes of a meeting of the Sydney Camouflage Group see Lt David Barker, 'Report on the Group of Camoufleurs', 29 November 1939, in Camouflage organisation, Sydney, NAA, Series SP 1008/1, Item 469/2/28. The address of the meeting is given as '116 Phillip Street, Sydney'. As Ure Smith's address was 166 Phillip Street, the street number is probably a typographical error since Ure Smith was the only

account of the formation of the Sydney Camouflage Group and learn it was Sydney Ure Smith and Victor Tadgell who initiated the formation of the group:

> In 1938–39, some citizens at least were beginning to be very concerned about the possibility of war in Europe and their own spheres of usefulness in the event of such a catastrophe. The writer of this report had stated his interest in camouflage to brother scientists and written on the same subject to the Commonwealth Government. He also gave a broadcast talk on the subject. About the same time (April, 1939) Mr. V.E. Tadgell of the Orient Line and Mr. Sydney Ure Smith, a member of the Council of the Society of Artists visited the Navy Department at Garden Island, Sydney, to inquire whether there was any part that could be played by artists in the event of war. It was as a result of this that Mr. Tadgell and Mr. Ure Smith contacted the writer and finding that he had been interested in camouflage for many years, suggested that a small Sydney Camouflage Group of interested Sydney gentlemen (chiefly artists) should be brought together for camouflage study with a view to giving aid in the event of hostilities.[16]

It makes perfect sense that Ure Smith had a central role in the group's formation because, such was his influence on Australian cultural life, that the period 1920 to 1945 was referred to by artist Lloyd Rees as 'The Ure Smith Era'.[17] It fits his reputation for activism, and his aspirations for industry, design and architecture to work together in developing a progressive and successful, if somewhat conservative, nation.[18] In addition, Ure Smith was already a close associate of many artists in the group and in 1939 administered two major modernist design projects, one for the Australian Pavilion at the New York World's Fair and the other for the Australian Pavilion at the New Zealand Centennial Exhibition in Wellington. Both events employed Douglas Annand, Russell Roberts, Max Dupain, Adrian Feint, Frank Hinder and Bob Curtis on their design teams.[19] He was president of the New South Wales Society of Artists where many members of the Sydney Camouflage Group exhibited, and was publisher of the *Home*, *Art in Australia* and *Australia, National Journal,* three periodicals that commissioned graphic art and photography from the same list of artists. And lastly, Ure Smith was closely associated through the Australian Academy of Art with Robert G. Menzies who, as Attorney-General, established the Academy of Art in 1937.[20]

The connection between members of the Sydney Camouflage Group and Robert Menzies—later prime minister of Australia, leading the conservative United Australia Party from 1939 to October 1941 until defeat by Labor's John Curtin—proved significant. In November 1940, in his capacity as prime minister, and after a period of lobbying by

---

member of the group with an address at Phillip Street (See appendix 1).

16 W.J. Dakin, 'The first Australian steps in camouflage instruction and the organisation of camouflage research before war broke out with Japan', in W.J. Dakin, Camouflage report 1939–1945, Canberra, AWM, Series 81 [77 part 1]. 1947, p. 7.

17 The quote is from Lloyd Rees in conversation with Nancy Underhill, quoted in Underhill 1991, p. 10.

18 These and other aspects of Ure Smith's attitudes discussed in Underhill 1991, p. 169.

19 Names of designers of Australian Pavilion in New York and Wellington sourced from Goad 1999, pp. 26–27, and from Stephen 2006, p. 36.

20 Underhill 1991, pp. 38–42.

Dakin, Menzies called a meeting of civilian and military advisors. The outcome was official government recognition of research into camouflage for military purposes by the Sydney Camouflage Group. This brought to an end the unofficial organisation of camouflage by civilians in the group and led to the transformation of the roles of its members. Dakin's plan to set up a civilian advisory committee on camouflage for the armed forces had succeeded.

In 1941 the DHS was formed, a federal organisation that mirrored the British Ministry of Home Security. It appealed nationwide to the arts sector to step forward and work for its innovative new camouflage section. In response, and after July 1941, the majority of the Sydney Camouflage Group's members became official camoufleurs with the DHS.[21] Over time the number of artists, designers, architects and photographers listed with the DHS swelled to over one hundred and on the final register included well-known names in Australian art such as George Bell and Louis McCubbin, in addition to other lesser known individuals whose names reappear frequently in the following chapters, among them Victor Bragg, Charles O'Harte, Gervaise Purcell, Ronald Rigg and Clement Seale.

That meeting between Dakin and Prime Minister Robert Menzies in 1940 was also the beginning of Dakin's future government role as Technical Director of Camouflage for Australia, serving the DHS and reporting through the Defence Central Camouflage Committee (DCCC) which he chaired from 1942 to 1945.[22] And it was the symbolic start of Dakin's embattled wartime life and disappointment that civilian experts and military personnel found it almost impossible to work easily together. His utopian vision was a harmonious collaboration to perfect the art of camouflage, guided by scientific research for the salvation of nation and empire.

Perhaps it was the fervour of those times, or just blind faith in the power of camouflage, but in 1941 there were at least three organisations in Sydney, and nationwide, that handled this new art of war. First there was the DHS, an organisation that was itself divided into two scientific sections, one 'Research and Experimentation', the other 'Camouflage' and often the work of both sections overlapped on matters of concealment and deception, not least because the first advised on protection of civil populations from air raids, and the second was concerned with concealment from air attack.[23] Second, there was the Allied Works Council in the Department of the Interior and its members, who were conscripted for war service, were collectively known as the 'Labour Squad'.[24] Through the Manpower Directorate the Australian government introduced civil conscription; men and women were to be placed in any form of work, including camouflage. Occasionally Dakin seconded talented members of the Labour Squad to work as camoufleurs for the DHS, as in the case of Eric Thompson, Ronald Rigg, and William Dobell.[25] Camouflage artists with the

21 W.J. Dakin, 'The outbreak of war with Japan', in Draft Camouflage Report, Sydney, NAA, Series C 1908 Item 5.

22 'Early history in Australia', in History of the development of camouflage organisation in Australia, September 1941, Sydney, NAA, Series C1707, Item 55, p. 2.

23 Mellor 1958, p. 531.

24 For clarification on both camouflage groups see W.J. Dakin, Camouflage report 1939–1945 (with appendices A–D), Canberra, AWM, Series 81 [77 part 1], 1947, p. 39.

25 Among artists and designers deployed to work as builders or labourers for the Allied Works Council

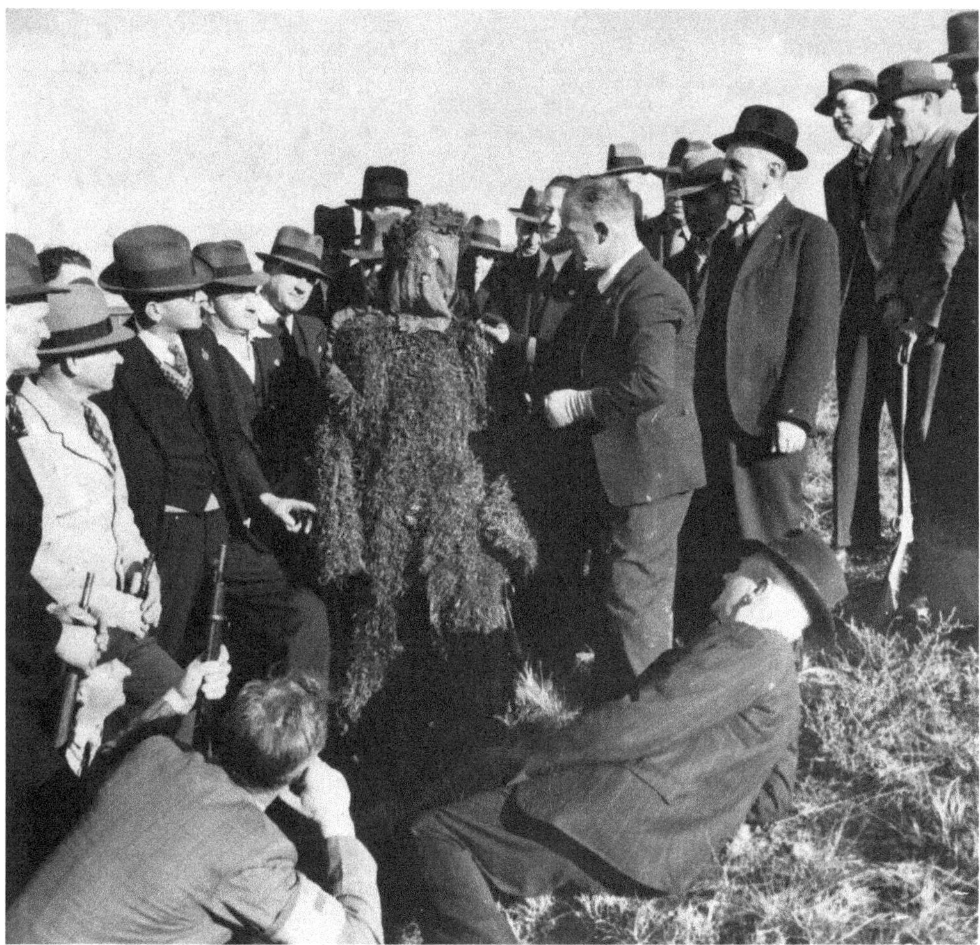

Fig. 2.4. Camouflage training of Volunteer Defence Corps members, c. 1943. State Library of Victoria: H99.201/1833. Argus Newspaper Collection of Photographs.

DHS and the Department of the Interior worked alongside the airforce, army and navy and a primary objective of their work was to disguise local Sydney aerodromes, including those at Richmond and Bankstown, and radar stations, in preparation for a Japanese aerial attack. Third, there was the Volunteer Defence Corps (VDC) an organisation comprising ex-servicemen and other civilians, that assembled in 1940 to assist the war effort and that not only claimed an interest in the intriguing world of camouflage but, as the photographic record of the war shows also devised some unlikely solutions for defence (fig. 2.4).

---

but transferred to the DHS in June 1942 were George Howard Adams, Victor Corlett, Phillip Handfield, John Ranshaw, Ronald Rigg, Clement Seale and Charles O'Harte. See V.C. Yeates to A.W. Welch, 'Transfer of employees', 4 April 1945, Camouflage personnel NSW Camoufleurs staff appointments, Canberra, NAA, Series A453, Item 1942/28/2298. For a complete roll of DHS camoufleurs see Department of Home Security Staff register—camouflage section, Canberra, NAA, Series A691, Item 1, accumulation dates 1 January 1942–31 December 1945.

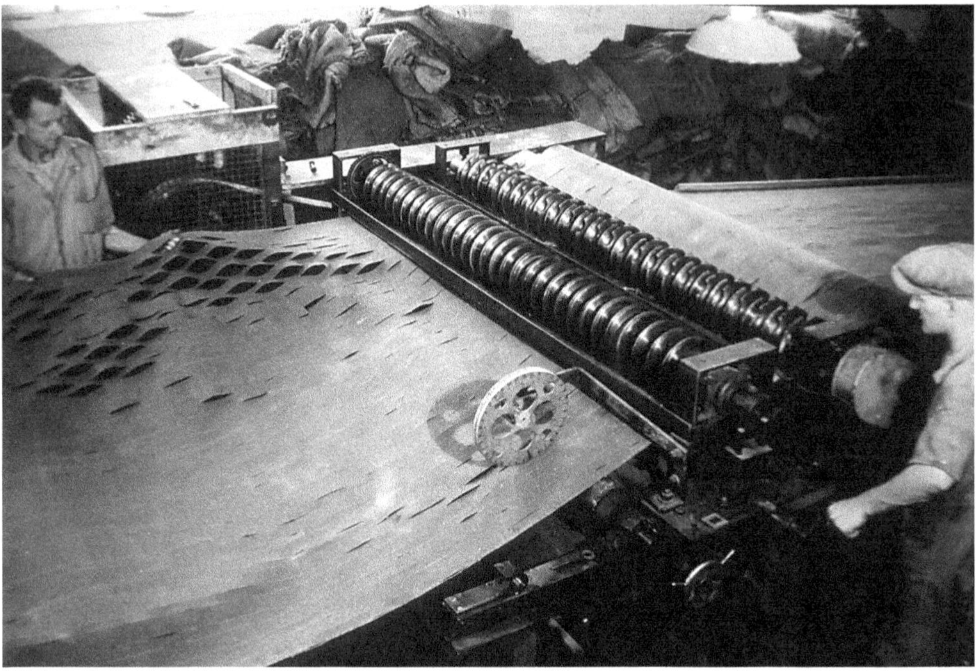

Fig. 2.5. Linoflage factory, Sydney, NSW, c. 1942. The collection of the NAA: C1905 T1, 1[94]. Photograph, Max Dupain.

Sydney became the main centre in Australia for mass-produced camouflage materials.[26] Hessian used in peacetime for wheat sacks and wool bales was cut into strips, dyed with colours that did not fade in the Australian sun, and baked in five-mile lengths in industrial ovens (fig. 2.5). The result was 'linoflage' also known as 'camsheen', a multi-purpose fire-resistant material used in perforated form as camouflage netting (especially for aerodromes), in sheet form to construct dummy buildings used as aeroplane hideouts and munitions stores, and in strip form for painted disguises attached to the outer walls of critical civic utilities such as water and fuel tanks (fig. 2.6).[27]

Sydney's most important sites for military defence were camouflaged with an amazing array of concealments and deceptions. Artificial trees and painted hessian backdrops helped redesign the landscape into illusory scenes that, from a distance, were difficult to differentiate from reality. In transforming the urban landscape into artificial scenes of nature, concrete fuel tanks were turned into sandstone cliffs, airfields into countryside. As new radar stations and wireless stations were installed in the city, so the landscape transformed into parkland that wasn't there the day before (fig. 2.7). These visual transformations were

26 W.J. Dakin, 'Notes of conference held at Premier's Department', 9 July 1940, in Notes of Camouflage, Sydney, NAA, Series SP 1048/7 Item s10/1/329, p. 8.

27 For images of linoflage cutting machines and ovens, see 'Department of Home Security circulars & bulletins', in W.J. Dakin, Camouflage report 1939–1945 (with appendices P–R), Canberra, AWM, Series 81 [77 part 5], 1947, pp. 1–4.

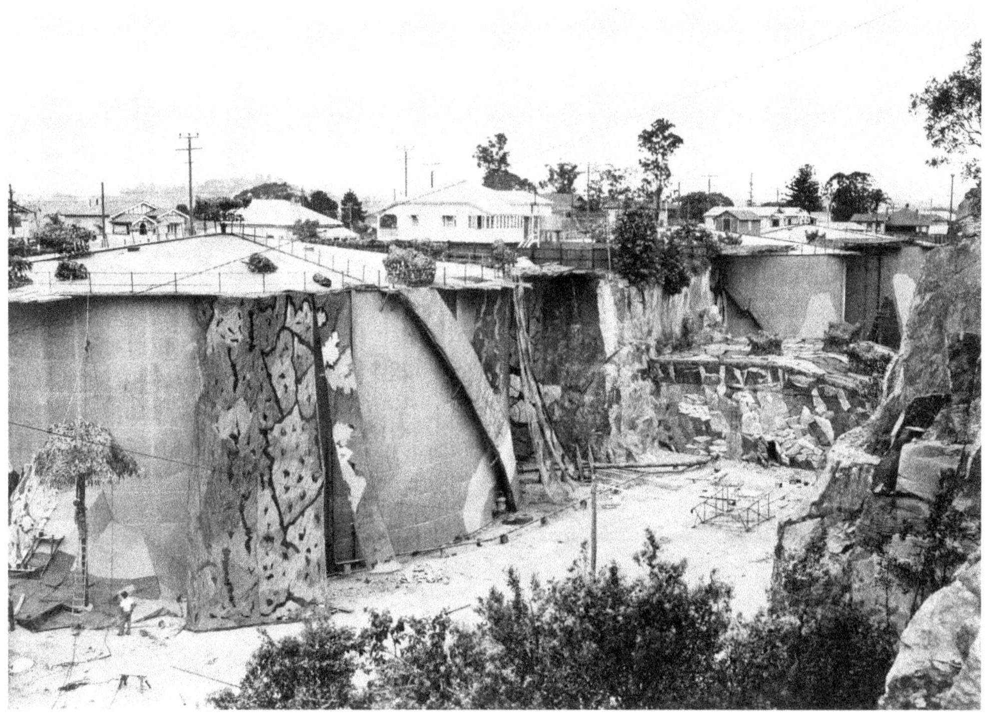

Fig. 2.6. Hessian strips to camouflage tanks in Sydney, NSW, c. 1942. The collection of the NAA: C1907, item 00999.

fairly standard stuff for civic camouflage in Allied and Axis countries in WWII, but a first for the Australian urban experience.

The public in Sydney felt it had a rightful stake in camouflage as well. After all, wasn't civil defence the public's business? Camouflage was a passionate subject and captured everyone's imagination. Here was an aspect of the war that was intriguing, even glamorous, and seemed to belong as much to mass culture and entertainment, humour and artifice—like magic, Disneyland, dioramas, fashion and exhibitions—as it did to political conflict. And it had novelty value. The editor of the *SMH,* for example, was eager for Donald Friend—prominent artist turned army recruit—to paint his tin safety-helmet in camouflage colours for what Friend believed were reasons of beautification and fashion rather than protection.[28]

At the same time the press, including the *SMH,* kept an interested public informed of camouflage developments in Europe:

> Germany and Russia have been developing camouflage for some time, evidence of which is Russia's claim, as a counter to Germany's boast that the Red Air Fleet had been destroyed, that papier mache machines had been deliberately placed on the aerodromes to mislead the Germans.[29]

28 Friend 2001, pp. 97–98.

29 'German Fears of R.A.F.', *SMH,* 22 July 1941, p. 5.

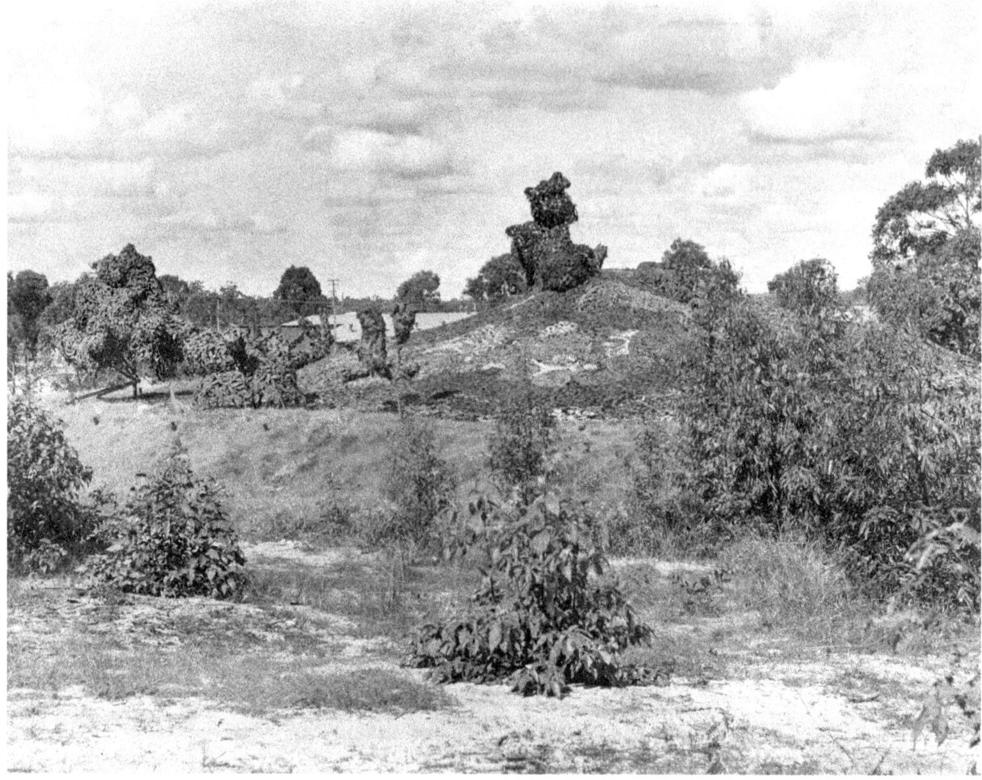

Fig. 2.7. Camouflaged RAAF site in the Bankstown area, c. 1943. The collection of the NAA: C1907, item 01007.

More indicative of public concern and desire to be involved in civic methods of concealment and deception were letters written by concerned citizens to Dakin in his capacity as Technical Director of Camouflage. They urged him to take up their ideas for camouflage defence, some of which included well-used ploys in WWI, while others exhibited wild imagination.[30] One was a recommendation to attach incendiary bombs to birds migrating to Japan to burn Japanese houses; another suggested that a false network of lights mimicking the urban city grid be installed in Sydney's empty parkland to lure enemy bombers away from dense population. The person who suggested attaching mirrors to the sides of warships to make them invisible was clearly not aware that in the US in WWI an earlier group of 'patriotic amateurs' proposed the same idea.[31]

But public misconceptions about camouflage as a form of magic plagued Dakin throughout the war. He was overwhelmed by the sudden curiosity that emerged in the months surrounding the bombing of Darwin, but he also found it annoyingly full of

30 'Suggested means of deceptive camouflage for aerodromes, etc', Letters seeking employment, offering help, or suggestions, Sydney, NAA, Series C1707, Item 5, unpaginated.

31 Behrens 1999, p. 55.

ignorance.[32] It was the unscientific mystique surrounding the public's conception of camouflage that bothered him most:

> The general public has an erroneous idea that they should be seeing all round them, church towers, large public buildings, military camps etc., being made invisible in some highly mysterious manner.[33]

True, he argued, illusions play an important role in camouflage, but not because camouflage is a matter of magic and fetish, but rather of science and modern research.

In Dakin's way of looking at it, science and art in combination provided the ideas and camouflage was nothing other than a rational enterprise. When he seconded artists to work for the DHS they were chosen for their scientific approach to camouflage problems. For this reason, while the organisation of camouflage in Sydney in WWII was complicated by the number of departments working on defence solutions, to William Dakin it was quite simple; the department he headed, the DHS, controlled the conceptual side of camouflage while those in the Department of the Interior in the 'labour squads' specialised in technical knowledge.[34] It was a division of labour that reflected a traditional dichotomy between mind and body, art and craft:

> And so it came about that commercial and landscape artists, preferring manual work of an interesting kind to some job, probably in an Army camp, for which they were totally unsuited, took up the work of labour squads. These men were paid by the Department of the Interior. Most of the men so recruited were sent to aerodromes near Sydney where both emergency camouflage and also experimentation were taking place and there they came into close contact with the best camouflage officers of the Department of Home Security supervising the work on which they, themselves, were engaged.[35]

The DHS employed a selective team; indeed they were the only camouflage artists entitled to call themselves 'camoufleurs'. Some were original members of the Sydney Camouflage Group, including Max Dupain, Douglas Annand, Bob Curtis and Frank Hinder. Of these it was Max Dupain, Australia's leading modernist photographer, who was singled out by Dakin as the most talented camoufleur in the department and whom he described as a:

> Special photographer. Possesses a strong creative imaginative mind & makes excellent contacts with all men. Is energetic & keen on his work, expresses himself clearly at all times. Has good construction sense. Capable & qualified to undertake general camouflage design & supervision.[36]

---

32 See W.J. Dakin to Captain C.P.T. Throsby, 29 January 1942, in Letters Seeking Employment, Offering help, or suggestions, Sydney, NAA, Series C1707, Item 5, unpaginated.

33 W.J. Dakin, 'Camouflage and censorship', 12 January 1942, Establishment of camouflage organisation and procedures, Canberra, NAA, Series A5954, Item 396/2.

34 Dakin insisted upon this division. See W.J. Dakin, 'Instructions to camoufleurs', 19 May 1942, Department of Home Security, Sydney, NAA, Series C1707, Item 3.

35 W.J. Dakin, Camouflage report 1939–1945, Canberra, AWM, Series 81 [77 part 1], 1947, p. 39.

36 'Dupain, Max', in Department of Home Security Staff register—camouflage section, Canberra, NAA,

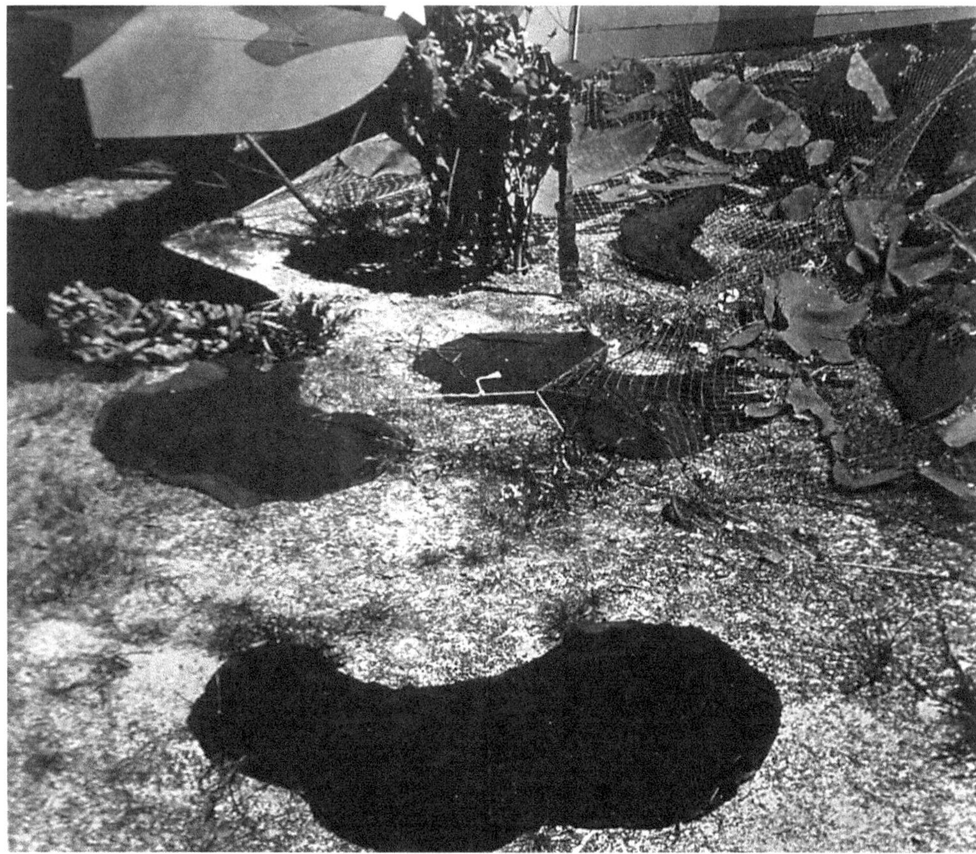

Fig. 2.8. Bankstown aerodrome 25 February 1943; 'areas surrounding aircraft were disrupted by large patches of dark earth hessian'. The collection of the NAA: C1905 T1, 3[2]. Photograph and notes, Max Dupain.

Dupain was retrained in the war at Bankstown aerodrome before spending a year or more stationed in Darwin and Papua New Guinea, making and interpreting aerial photographs, getting to know exactly what objects looked like on the ground from the air, and experimenting with new methods for more effective ground concealment. He researched and tested a wide range of colours and textures to use as backgrounds for planes. In one experiment he arranged patches of hessian painted in the standard Australian wartime colour 'dark earth' to:

> create a disruptive scheme immediately underneath two wings and part of fuselage to destroy the shadow cast thereby and at the same time to use similar disruptive pattern on the aircraft itself so as to fuse both patterns into each other. (fig. 2.8)[37]

Series A691, Item 1, accumulation dates 1 January 1942–31 December 1945.

37 Max Dupain, 'Report on mobile camouflage experiment No. 4 Bankstown', in Bankstown experiments, Sydney, NAA, 29 January 1943, Series C 1905 T1, Item 3.

Coloured patches helped make the contours of planes blend with their background and helped to confuse the shadows cast by the plane's body. The resulting effect was a complex, visually ambiguous semiotics of tones, shapes, edges, shadows and forms comparable to the way patterns on the skins of frogs and snakes function to conceal them. Disruptive patterns on reptiles interrupt their shape while coincident disruptive patterns reconfigure their true shape into different forms. In the words of zoologist Hugh Cott, who coined the term 'coincident disruptive colouration' for the visual impact of animal skins, illusions are created by 'the optical destruction of what is present' on one hand, and 'the optical construction of what is not present' on the other.[38]

While Max Dupain's work for the DHS stood out, so too did William Dobell's for the Department of the Interior—so much so that Dakin tried to second him to the DHS.[39] Instead, Dobell became an Official War Artist. However, before he did, in 1942 this well-respected artist living in Sydney built papier-mâché farm animals to act as decoys for camouflaged airfields at Bankstown aerodrome.[40] A photograph of a papier-mâché cow that the Women's Land Army used to demonstrate milking techniques during the war provides an approximation of what Dobell's animals looked like (fig. 2.9). They were part of a large deception scheme at Bankstown to disguise the area as a farm property and keep its true military function concealed in the event of enemy aerial reconnaissance. Paint, soot and oil were applied to the ground to eradicate signs of disturbed earth made by aircraft traffic and to create a seemless impression, from 2000 feet, of a typical Australian rural scene.[41] To strengthen the illusion, fake trees were 'planted' and real trees transplanted, and where their foliage was sparse, artificial foliage was used to enhance the cover.[42]

As history has shown, Dobell was implicated in matters of deception and concealment in more ways than one during the war. Over the course of 1943 and 1944 the 'Dobell scandal' was a controversy involving the artist's painting, *Portrait of Joshua Smith* (1943), a modernist portrait of a fellow camouflage artist in the Labour Squad.[43] The question was whether the painting, which won the coveted Archibald Portrait Prize, was an authentic portrait and a true likeness, or a 'caricature' and by implication a sham. The controversy devastated Dobell, but ironically, his critics' frustrations at trying to distinguish between a true likeness and a fictitious caricature replicated the very confusions that Dobell's work in camouflage aimed to generate. The 'Dobell scandal' is extensively discussed in Australia, but not in relation to general public anxieties during WWII about the problematising of reality in a war culture that fetishised deceptions, bluffs and dummies. There is a good case

---

38 Cott 1957 [1940], p. 70.

39 Bob Curtis to William Dakin, 10 August 1942, Camouflage personnel NSW Camoufleurs staff appointments, Canberra, NAA, Series A453, Item 1942/28/2298.

40 Wilkins 1999, p. 44.

41 'Guide to camoufleurs on framing a report and recommendations for the camouflage of an Australian aerodrome', Department of Home Security, Sydney, NAA, Series C1707, Item 3.

42 Information on false trees published by the DHS in pamphlet Camouflage, 1943, pp. 24–25, cited in Frank Hinder, 'Records of Lt F.C. Hinder Camoufleur', Canberra, AWM, PR 88/133, File 895/4/182, Item 11 of 12.

43 For an account of the 'Dobell scandal' see Haese 1982, pp. 245–46.

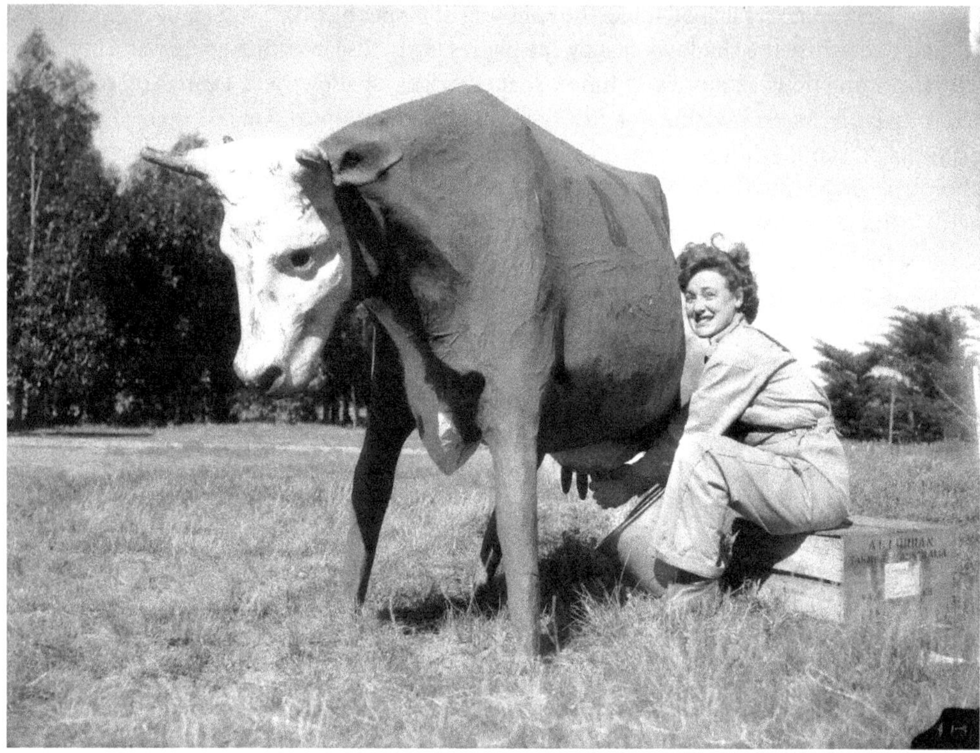

Fig. 2.9. Darley, Victoria, 1944. A member of the Australian Women's Land Army practises milking technique on a dummy cow made of papier mâché, 1944. AWM, AWM 141684.

for the argument that Dobell was as much a victim of the period's exaggerated concern for certain identity, brought about by the ubiquitous deceptions of WWII camouflage, as he was a victim of conservative tastes in art.[44] Certainly arguments around the immorality of deception were a major point of discussion throughout the war, often precipitated in Australia by debates on the ethics of fakery in war photography.[45]

All of this begs the question of what the Bankstown deceptions that William Dobell was involved in looked like. With the historical background to artists' involvement in camouflage research in Sydney mapped out, it is now to the visual nature of the work that the discussion turns. An exciting array of visual examples of disguise, decoy, concealment and deception, as practised and applied in the Sydney region, survives in the nation's archives. It provides insight into camouflage as practised at Bankstown aerodrome where Dobell worked and where the DHS and the Department of the Interior worked side by side. It also provides insight into work conducted at Georges Heights where a Camouflage Research Station was shared between the army and the DHS (until the two organisations

---

44 A general argument about anxieties of identity, truth and reality in relation to camouflage has been put forward in Schwartz 1996, pp. 175–211.

45 See 'Dispute over air picture', in *SMH*, 13 August 1940, p. 10; and discussion on the ethics of war photography in 'War photography', *Australasian Photo-Review*, August 1944, pp. 253–54.

split under acrimonious circumstances), as well as into experiments at the University of Sydney where William Dakin's office in the Zoology Department was also a camouflage laboratory. Whether working under the aegis of the Sydney Camouflage Group, the DHS or the Labour Squad, camouflage workers in Sydney were involved with the ambiguous visual and psychological edge between reality and illusion, and with paradox, crypsis and visual puzzles, all exploited for military gain.

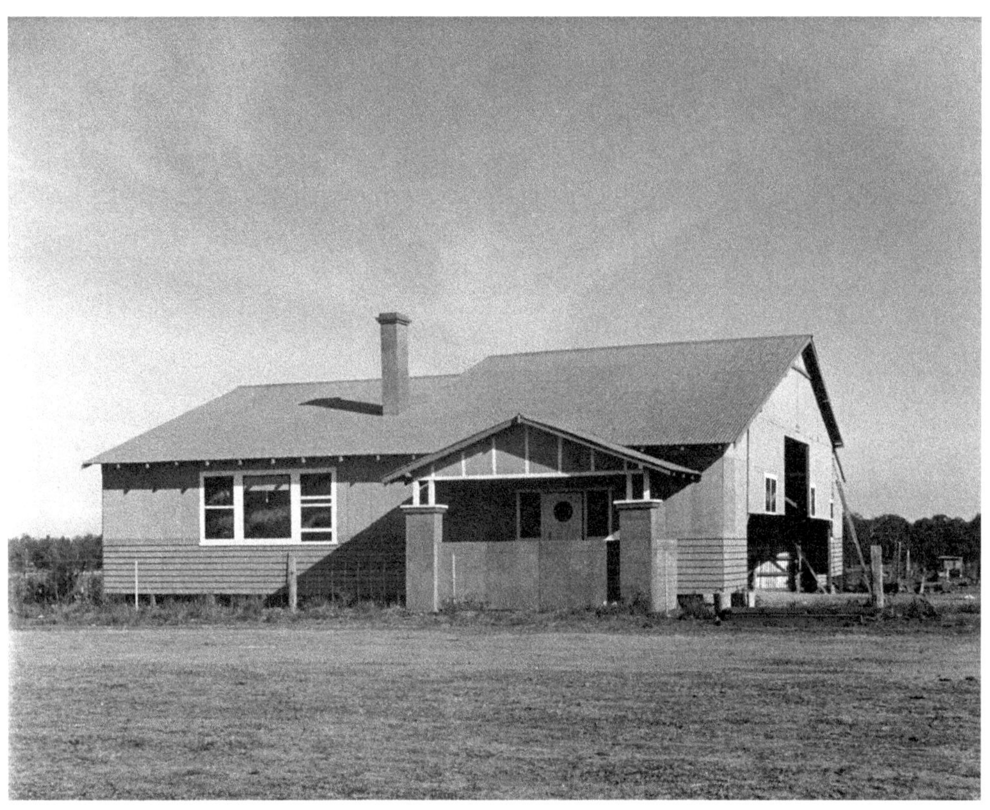

Fig. 3.1. Bomber hideout no. 6 at Bankstown aerodrome, c. 1943.
The collection of the NAA: C1905, 19[28].

Chapter 3

Sydney experiments

During the Great Depression Julian Ashton, an influential art teacher in Sydney, felt that in periods of national trouble the artist's duty is to 'uplift the minds of the people by some excursion into the Realm of Beauty'.[1] Such romantic visions of the artist's social role all but disappeared in WWII when, in January 1942, the Manpower Directorate conscripted thousands of citizens into the labour force to work in industries such as munitions factories, and for the Allied Works Council to build war infrastructure including camouflage materials.[2] Pacifist or otherwise, artists who had not volunteered for the armed forces were put to work on military duties or civil construction and defence.

Those put to work in camouflage were largely responsible for defending Australia from aerial attack. Considering the purpose of camouflage—which is primarily to deceive in the visual domain—their training focused on concealing and transforming airforce bases by ingenious methods of hiding aircraft and disguising telltale signs of their activity, including 'scars' on the ground, from the aerial enemy observer.

Bankstown aerodrome in Sydney was a special case. It was there that William Dakin oversaw an ambitious experimental deception scheme involving artists, designers, architects and photographers from the DHS, including Max Dupain. Others involved were from the Labour Squad, including William Dobell. All were trained at a Camouflage Research Station at nearby Georges Heights, a facility shared with the army.[3] Additional contributors to the Bankstown deception scheme were previous members of the Sydney Camouflage Group who, when that group became defunct, took positions in the NSW state branch of the federal government's DCCC. Staffing this NSW Directorate of Camouflage were John Moore and Russell Roberts. They supervised another camouflage research laboratory set up at the University of Sydney near Dakin's professorial offices in the Zoology Department.

The wartime history of camouflage in Sydney is indicative of the complexity of its organisation on a national scale. Never before had camouflage been systematically integrated into Australian military activity, let alone developed as a military science on the advice of Australian civilians. It is little wonder that the bureaucracy surrounding communication

1 Eagle 1989, p. 132.

2 Penglase & Horner 1992, pp. 136–37.

3 W.J. Dakin, 'Special camouflage methods at Bankstown aerodrome and Laverton' in W.J. Dakin, Camouflage report 1939–1945 (with appendices A–D), Canberra, AWM, Series 81 [77 Part 1], 1947, p. 75.

and collaboration between state, federal, military and civilian bodies became unwieldy. The boundaries between disciplines and professions that were crossed resulted in competition, even jealousy, but that part of the story unfolds in later chapters. The concern here is to introduce the reader to the empirical work and theoretical ideas of the Sydney Camouflage Group and the DHS in relation to operations in Sydney. As the largest city, Sydney had urgent problems for defence and, as the extensive photographic record of the war shows, the deception schemes that were put into practice, especially the disguises at Bankstown aerodrome, were remarkable.

Equipped with aerial knowledge of the Sydney landscape and an intimate knowledge of vernacular architecture, and motivated by the imperative to make military operations on the ground invisible from the air, the camouflage workers assigned to Bankstown worked in secrecy to design an elaborate simulation of a nonexistent rural and agricultural community to disguise the site's true function. Bankstown had become an airforce fighter interception station. The plan was to disguise it as an ordinary rural town. Instructions for the redesign of Bankstown aerodrome read: 'simulate activities that would be normal to what the disguise pretends should be carried on'.[4]

Making the aerodrome look agricultural rather than aviational, and civilian rather than military, entailed a visual transformation of the site that required creativity in construction and talent with the simulation of 'authentic' detail. The results exceeded Dakin's expectations. Out of plywood, hessian and linoflage, camouflage labourers built spectacular structures mimicking the types of domestic and commercial buildings commonly found in the Bankstown region in 1940. But hidden behind their innocent-looking facades, ones that blended so well with the Sydney environs, were fighter and bomber aircraft and munitions dumps. Aircraft hideouts masqueraded as domestic houses (fig. 3.1), other buildings as a sawmill, an ironmonger's store with a quick release door in a simulated wall (fig. 3.2), a grandstand, an advertisement hoarding, and a 'hovel' to fit in with the low socio-economic profile of the area. Most difficult of all to construct was a wireless station masquerading as a farm and orchard, one William Dobell helped decorate with papier-mâché animal decoys.[5]

Camouflage in nature, particularly the concealment methods of the crab, seahorse and octopus, were a constant delight to Dakin. But the deceptions at Bankstown aerodrome in 1942 and 1943 were a revelation that fired his imagination in an altogether different way and brought out his competitive spirit. Britain and Germany were by then famous for ingenuity with large-scale war deceptions, but in Dakin's view, Australia too was in that league:

> fake houses, a fake agricultural show grandstand, a fake timber cutting yard, and even a fake road were built of flimsy material over aircraft dispersal points. They had to be seen to be believed. Two houses and a country store standing at the corners of a road apparently like all others in the district, were nothing but painted shells of ply-wood and 'cardboard' as wooden frames over the solid blast-proof shelters. Whole sides of the fake houses were

4 'Disguise and concealment', in Air intelligence reports, Sydney, National Archives of Australia, Series C 1707, Item 36, p. 1.

5 For additional photographs of the Bankstown scheme see W.J. Dakin, Camouflage report 1939–1945 (with appendices P–R), Canberra, AWM, Series 81, [77 Part 5], 1947, pp. 17–21.

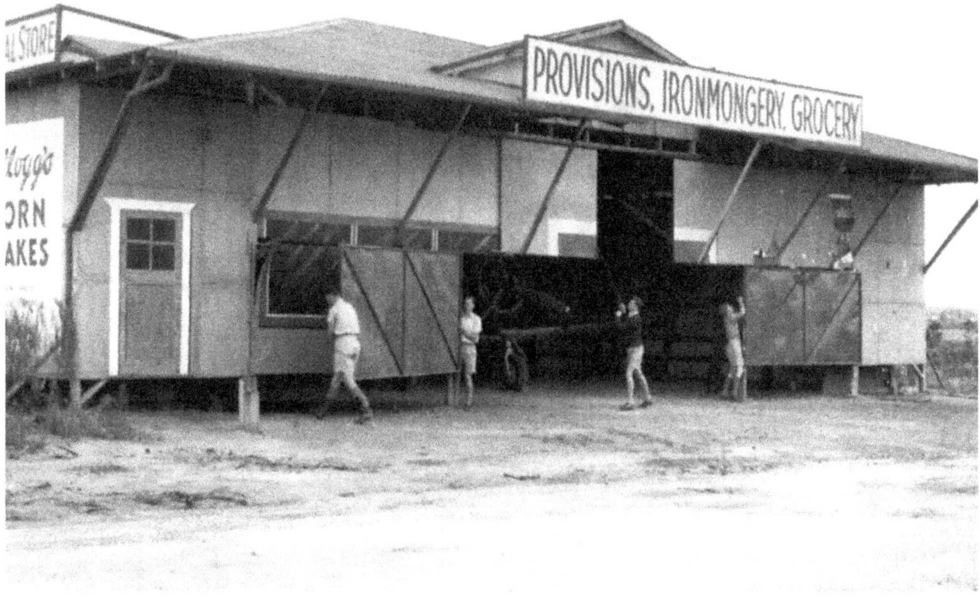

Fig. 3.2. Bomber hideout at Bankstown posing as a general store, c. 1943. The collection of the NAA: C1908 P1, 3[14].

> ingenious folding doors. In a few seconds the side of the house could be open and the aircraft outside.[6]

The scheme went beyond the usual definitions of civil camouflage, which the British Home Defence Service claimed was 'the use of any means of visual deception, direct or indirect, which will make it more difficult for an enemy to find targets'.[7] The purpose of the Bankstown scheme was more than just hiding equipment and targets and misdirecting enemy attention. It was designed for attack as well as defence, and for aggressive surprise. These roughly made large-scale buildings mimicking the vernacular architecture of Sydney concealed bomber and fighter planes ready to take to the air at short notice in defence of Sydney.

To increase the illusion of a rural scene at Bankstown and at other aerodromes around Sydney, fake gumtrees, or 'gumtree suckers', made from poles and wires with garnished

---

6 W.J. Dakin, 'Special camouflage methods at Bankstown aerodrome and Laverton' in W.J. Dakin, Camouflage report 1939–1945 (with appendices A–D), Canberra, AWM, Series 81 [77 Part 1], 1947, p. 75.

7 'Extract from report by a sub-committee on British Home Defence Services, Middle Head (New South Wales): experimental camouflage, Canberra, NAA, Series A663, Item O30/1/192, p. 1.

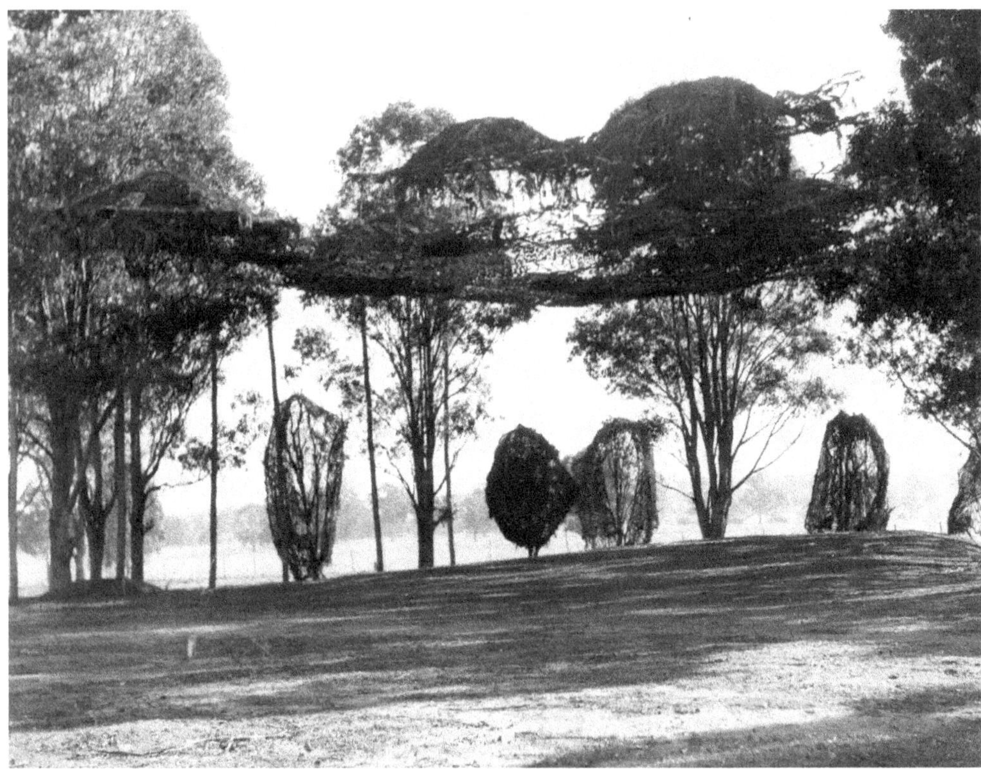

Fig. 3.3. Hideout experiment with artificial trees and nets at Sydney, NSW, c. 1943. The collection of the NAA: C1905, 10.

canopies of camsheen-netting and hessian strips, were 'planted' beside buildings (fig. 3.3). Some fake trees were merely tree shapes hung on wire. They looked odd from below but were convincing enough from the air.[8] Max Dupain kept notes on the mature ti-trees he replanted as cover for hideouts and which he spray-painted green to give a fresh, alive look.[9] Dupain, who was attached to the RAAF at Bankstown, and John Moore who was Deputy Director of the NSW Directorate of Camouflage, wove real ti-tree into fake tree-garnished netting to improve the Australian quality of the shadow.[10]

When Dakin later viewed the outcome through aerial photographs he was convinced 'that the concealment was perfect'.[11] His favourite deception was a 'fake Grandstand—built

8 Information on false trees published by the DHS in pamphlet 'Camouflage', 1943, pp. 24–25, cited in Frank Hinder, 'Records of Lt F.C. Hinder Camoufleur, Canberra, AWM, PR 88/133, File 895/4/182, Item 11 of 12.

9 Max Dupain, 'Mounted photographs of camouflage projects' in Bankstown experiments, Sydney, NAA, Series C 1905 T1, Item 23.

10 Max Dupain, 'Report on mobile camouflage experiment No. 5 (John Moore's Umbrellas), in Bankstown experiments, Sydney, NAA, Series C 1905 T1, Item 3.

11 W.J. Dakin, 'Special camouflage methods at Bankstown aerodrome and Laverton' in W.J. Dakin,

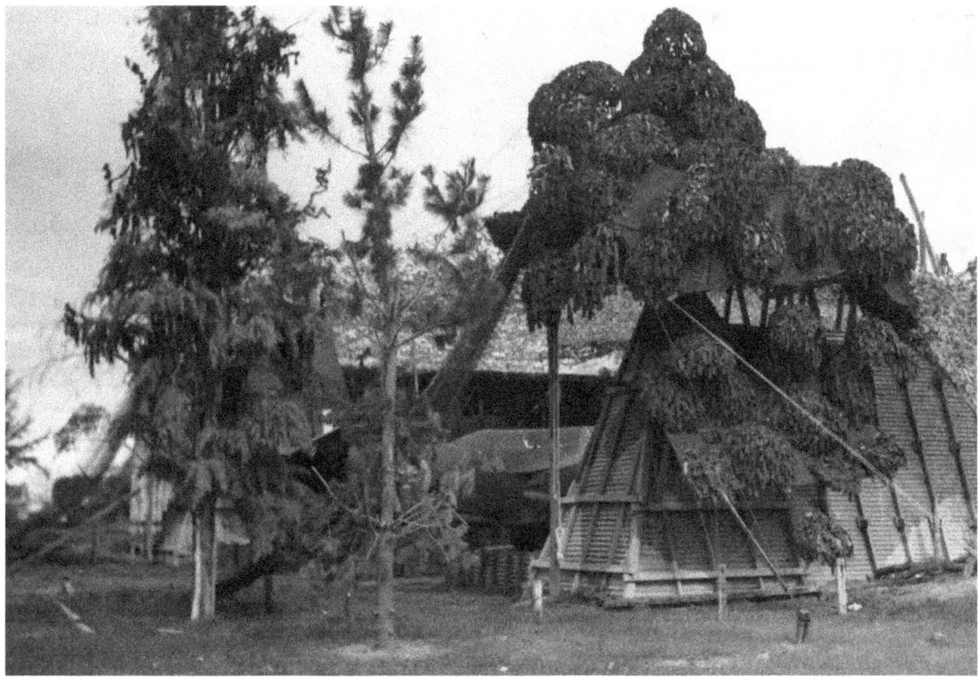

Fig. 3.4. Plane hideout with artificial trees at Bankstown aerodrome, c. 1943. The collection of the NAA: C1905, 10.

with humour and imagination—as a complete wreck' (fig. 3.4).[12] At an early stage of the Bankstown scheme aerial photographs revealed the faint impression of an abandoned sports oval near the landing ground. To Dakin's eye it emerged like the traces of old Roman forts in Britain. With a disused oval dominating aerial views of the site it was decided to make full use of this feature and build a collapsing grandstand with the purpose of concealing a bomber inside. Paradoxically, in order for the bomber to escape unwarranted attention, the collapsed grandstand had to be in full view. The objective of the Bankstown scheme was not to prevent military objects from being seen, but from being recognised. Indeed for the most part the scheme—like many others in WWII—was only detectable through stereoscopic photo-analysis, and from aerial photographs taken at an oblique angle with the camera pointing at the ground but on an angle to the horizon to provide a more detailed view than one taken straight down.

The Bankstown scheme displayed the same vivid imagination as British methods of hiding artillery in fake petrol stations, and compared well to the daring concealment of

Camouflage report 1939–1945 (with appendices A–D), Canberra, AWM, Series 81 [77 Part 1], 1947, p. 75.

12 W.J. Dakin, 'Special camouflage methods at Bankstown aerodrome and Laverton' in W.J. Dakin, Camouflage report 1939–1945 (with appendices A–D), Canberra, AWM, Series 81 [77 Part 1], 1947, p. 76.

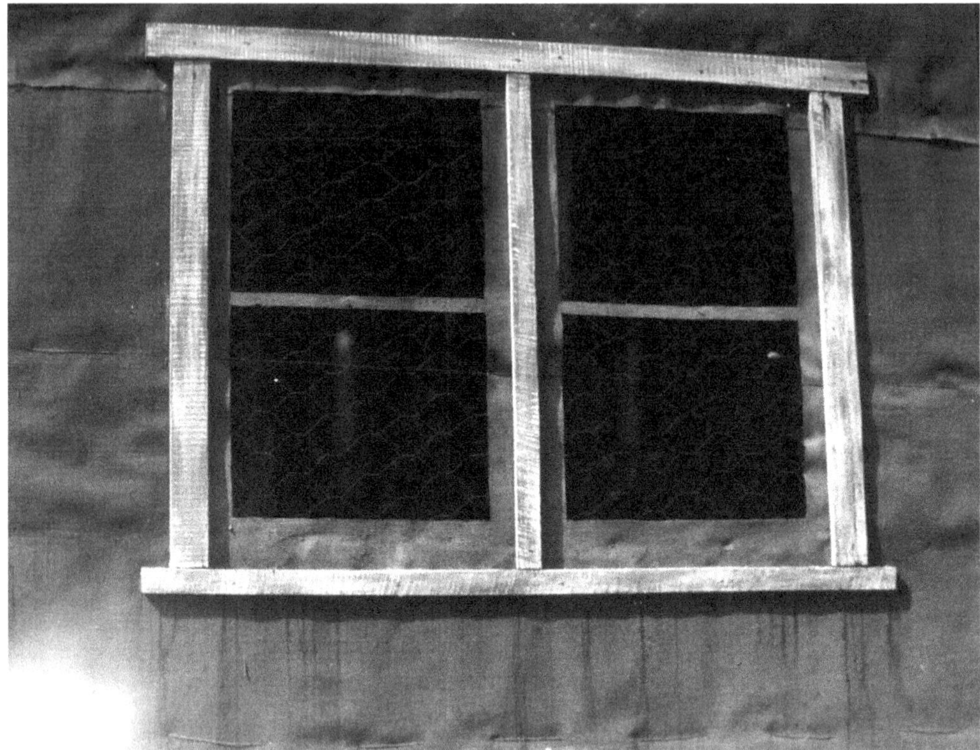

Fig. 3.5. False window in bomber hideout at Bankstown aerodrome, c. 1943.The collection of the NAA: C1905, 19.

the Lockheed–Vega aircraft plant at Burbank California which was hidden beneath netting and canvas and made to look like a 'suburb' of houses, roads and shops.[13] It also had ideas in common with German aerodrome disguise since, as Dakin was keen to point out, by 1942 it was well known that 'anything which resembles an aerodrome in Germany is now regarded as a fake'.[14] By 1942 camouflage schemes by both Allied and Axis powers were becoming more legendary, and more theatrical.

Military historians claim the word 'camouflage' derives from the verb 'camoufler' meaning the act of putting on makeup for the stage.[15] The conceptual links between camouflage, theatre and makeup are clearly illustrated by the Bankstown scheme. Dramatising and obfuscating were its operative methods, similar to the way stage makeup exaggerates but also conceals the faces of actors. And as with theatre, large-scale camouflage schemes in WWII depended on teams of specialists working cooperatively rather than alone, which explains why so many designers, architects and commercial artists were sought after in WWII—they knew how to work creatively and collaboratively. Elizabeth

13 Reit 1978, pp. 87–89.

14 W.J. Dakin, 'Camouflage in Australia', 31 March 1942, in Camouflage Costs, Business Board etc. Sydney, NAA, Series C1707/30, Item 1, p. 3.

15 This definition and translation is found in Dear & Foot 2001, p. 220.

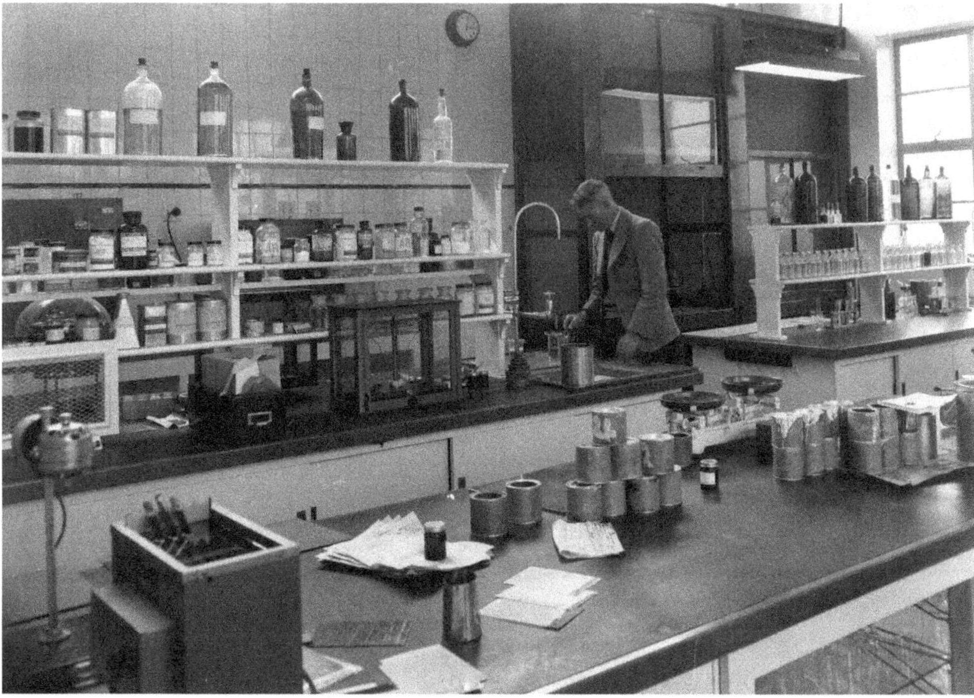

Fig. 3.6. Georges Heights, NSW, camouflage laboratory (chemicals), 1942. The collection of the NAA: C1905, 10.

Kahn made this point in 1984 in relation to European camouflage artists in WWI who, when plucked from their studios and individualist practices, found a foreign world where:

> the long-standing stress on individual genius, the ever-intensifying identification of art work as private and personal—seemingly independent of physical reality or everyday life that still remains the image of the Western artist today—was irrelevant to the work of the camoufleur.[16]

Eric Thompson, the architect and set designer who Dakin sent to Darwin in 1942 to advise the armed forces on concealment of that town from Japanese bombers, had the perfect balance of creative, conceptual, professional and social skills. Like Dobell, he too gained his experience as a camoufleur working for the Labour Squad on Sydney's aerodromes. Given the emphasis on simulation and mimicry at Bankstown, set designers, architects and realistic painters were in demand. Take for example the 'windows' of each fake house and corner store which were, in reality, two-dimensional paintings posing as the glass threshhold to interior space (fig. 3.5). Aesthetically they belonged to the history of trompe l'oeil effects where flat surfaces are painted so illusionistically that, from a certain distance, the viewer is completely fooled into thinking them three-dimensional.

16 Kahn 1984, pp. 2–3.

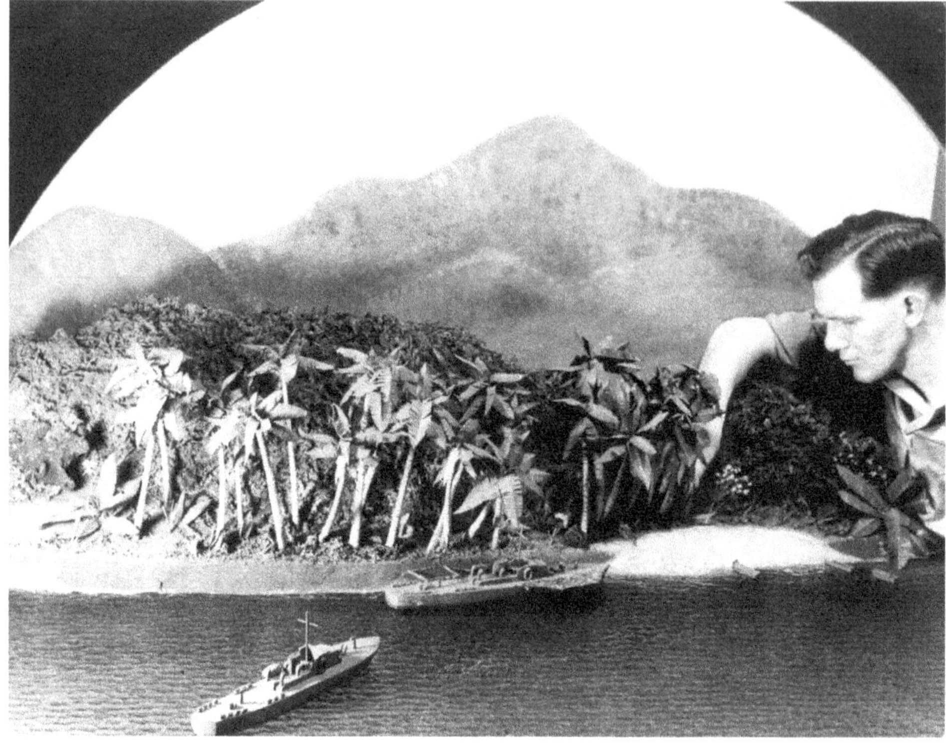

Fig. 3.7. Department of Home Security model for ship and jungle camouflage, 1942. The collection of the NAA: C1905 T1, 10[57]. Photograph, Max Dupain.

Jean Baudrillard likened the experience of being tricked by trompe l'oeil to a momentary loss of self and attributed its hyperreal effect to counteracting 'the privileged position of the gaze' in viewers who normally feel that their powers of seeing constitute them as rational beings.[17] Similarly the trompe l'oeil deceptions at Bankstown were designed to trap the gaze of the enemy for long enough to make them a victim of the deception. Of course with camouflage in military contexts there are different issues at stake. Trompe l'oeil ultimately makes the viewer consider the processes of vision, the question of truth, their own sense of self, and above all the nature of jokes and wit. With war, however, the question of camouflage's comical nature is repressed by war's grave and tragic nature. Yet it struck Frank Hinder's daughter Enid Hawkins who, looking back on WWII when she was just eight years old, noticed that her father's assignment to the field of camouflage suited his fascination with eccentric inventions, and curiosity with the ability of colour, light and form to fool, confuse and deceive the eye.[18]

Bankstown was only one significant site in Sydney for camouflage experimentation. The Camouflage Research Station at Georges Heights on Sydney Harbour was another. It was an army establishment but, by September 1941, Dakin, in cooperation with the army,

17 Baudrillard 1991 [1988], p. 58.

18 Correspondence Enid Hawkins to Ann Elias, 12 September 2003.

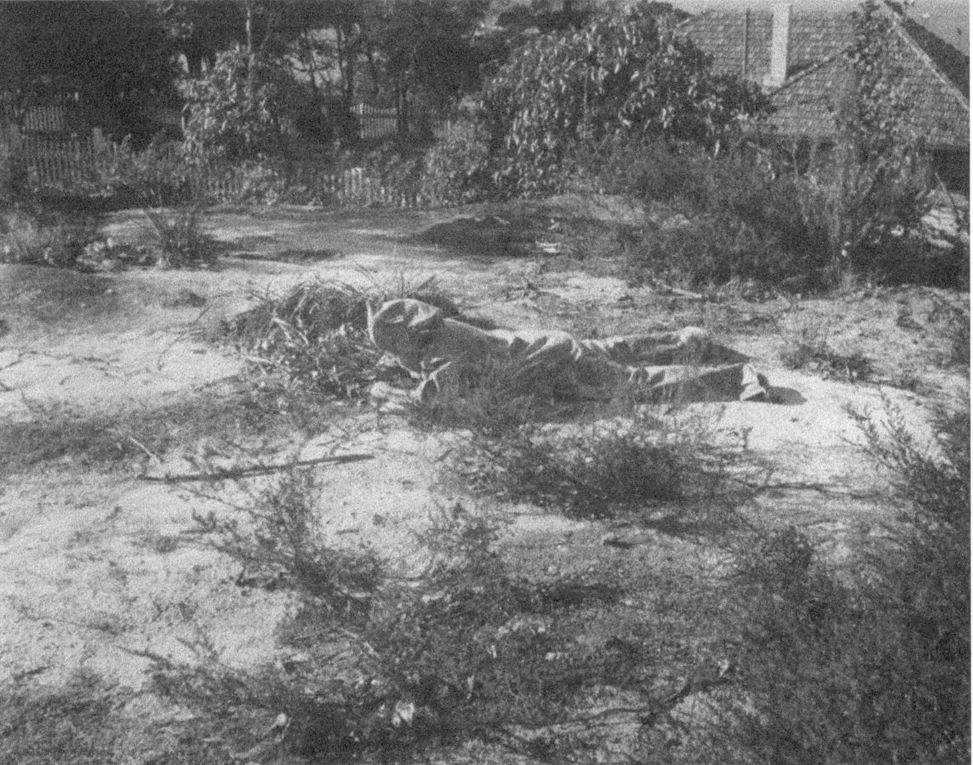

Fig. 3.8. 'Cam. Wing: Full Costume', Sydney, 1942. Frank Hinder Papers: Collection Archive of the AGNSW. Photograph, Gervaise C. Purcell.

turned it into a camouflage research base, complete with chemical laboratory to test a wide range of defence experiments requiring concealment and deception. These ranged from personal cover for soldiers, foliage cover for buildings, colours for ships in the tropics, durable dyes and paints, and paints resistant to infra-red photography (figs. 3.6, 3.7, 3.8). The building at Georges Heights was dual purpose and included one space for model-building and photography and another for lectures.[19] Members of the research team included Sali Herman who was at that time enlisted in the army and who worked in camouflage for the Royal Australian Engineers (RAE). Later he was appointed Official War Artist. Frank Hinder worked there too, but as a member of the Australian Military Forces, prior to secondment to the Department of Home Security to work with Dakin.

The team at Georges Heights was instrumental in devising a series of standard camouflage colours for Australia to add to the palette already in use by the British (fig. 3.9).[20] Among them was 'Darwin stone', invented after studying the dominant colours of different

19 W.J. Dakin, 'Scheme for operation of a Camouflage Research Station', in W.J. Dakin, Camouflage report 1939–1945 (with appendices A–D), Canberra, AWM, Series 81 [77 Part 1], 1947, p. 16.

20 Information from DHS pamphlet Camouflage, 1943, pp. 24–25, cited in Frank Hinder, 'Records of Lt F.C. Hinder Camoufleur Canberra', AWM, PR 88/133, File 895/4/182, Item 11 of 12.

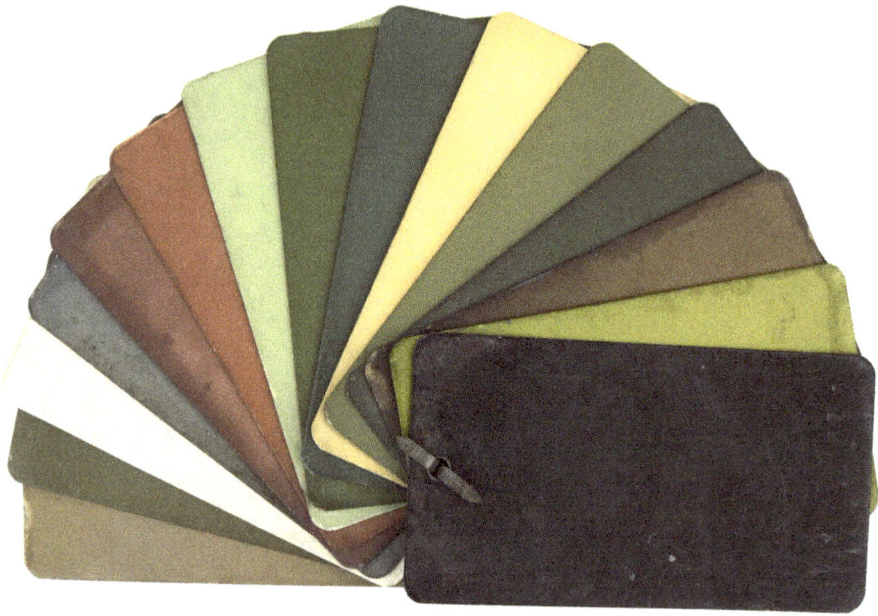

Fig. 3.9. Camouflage colours: set of 17 metal camouflage paint tablets, colours tested and developed by the Camouflage Wing, RAE and Frank Hinder, 1942–1943. AWM REL/16500.

Australian landscapes and observing the impact of changing light and shadow conditions. Dakin was relieved and excited to witness the ingenuity of his team as it developed Australian solutions to problems of concealment. The southern environment, he said, is a place where 'shadows are much darker, and it is the shadows of objects which are the greatest guides to observers in aeroplanes'.[21] He noted that colours in Australia are more visible at a distance than in England. Early on in the war he sought advice from England, but soon came to realise that regional solutions were more important:

> Camouflage in Australia provides considerably different problems from Camouflage in Europe, owing to different conditions of light background. Relatively little information concerning the details of large scale treatments in England has been forthcoming in spite of official requests. This is a handicap, although principles and experimental practice in Australia are sufficiently advanced by virtue of the amount of research done to date to give confidence that all problems in this country can be adequately handled when the organisation is complete, except possibly for difficulties of supply, consideration of which aspect will be an important responsibility of the organisation.[22]

---

21 W.J. Dakin, 'Notes of conference held at Premier's Department', 9 July 1940, in Notes of Camouflage, Sydney, NAA, Series SP 1048/7 Item s10/1/329, p. 7.

22 'Development of camouflage organisation in Australia', September 1941, in History of the

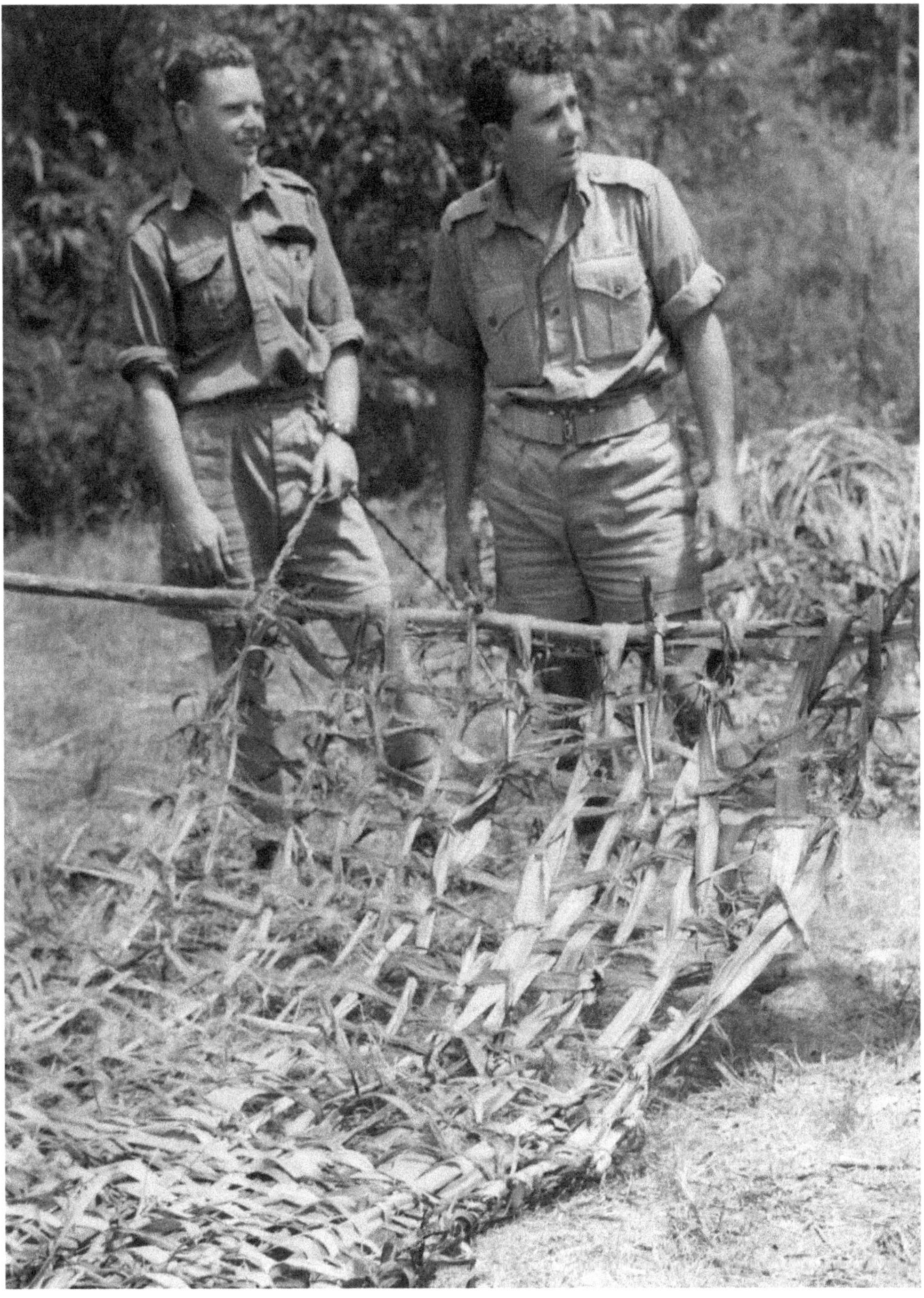

Fig. 3.10. Stringy-bark camouflage net, c. 1943. The collection of the NAA: C1905, 10.

Solving problems specific to the Australian environment, ones that used local materials for camouflage, was Frank Hinder's personal passion. One initiative was the manufacture of camouflage nets from the stringy bark of gum trees (fig. 3.10). Eucalypt bark offered a cheaper alternative to fabric, and was in plentiful supply on the east coast of Australia. But it also occurred to Hinder that, as a covering, bark was inherently more deceptive. It blended naturally with the light and colour of Australian bush, therefore increased the chance of avoiding attention. And whereas in 1839 Charles Darwin described the long shreds of eucalypt bark that hang off trees in NSW as 'desolate & untidy', in WWII the bark's unkempt, rough texture produced useful patterns of reflected light and shadow that contributed to the dissolution of forms being concealed.[23]

Hinder's undertaking with stringy-bark netting was also part of a concerted effort to build up additional supplies of much needed camouflage nets. Dakin initiated a training scheme in net-making. First he taught female colleagues and female students at the University of Sydney how to make nets from fabric. Then he sent them into the community to pass on their knowledge to organisations such as the Women's National Defence League.[24] In no time, women and some men belonging to volunteer groups throughout Australia produced camouflage nets for the homefront. But nets made of stringy bark were different. They were produced by soldiers in the field forced by circumstance to make do:

> Am sending you a net made of stringy bark, plus a sample of heavy rope—partially finished showing the thickness of the slivers i.e. the amount of material required to make the rope. The 13th Battalion have been making thin nets out of 'Tea Tree' bark for machine guns, as there is no stringy bark on the immediate coast & have found them most effective.[25]

While Hinder's idea of using stringy bark may have come from personal knowledge of the bush, it was also influenced by art. Art and nature are intertwined in the history of camouflage design. To comprehend the advantages of stringy-bark netting for camouflage, we can refer to the patterns of light and shadow falling through eucalypt forests, but we can also refer to 19th-century landscape paintings, including naturalistic scenes by the revered Frederick McCubbin. One admirer of McCubbin's work was art critic James Stuart MacDonald.[26] And not coincidentally, MacDonald was a camouflage artist with the Australian Imperial Forces (AIF) in France in 1918.[27] Nature as well as naturalistic painting stood him in good stead to undertake concealment and deception on the battlefields, and conversely, his intimate experience with camouflage in WWI provided a predisposition in

---

Development of Camouflage Organization in Australia, Sydney, NAA, Series C1707, Item 55, p. 5.

23 Charles Darwin's 1839 Journal quoted in Nicholas & Nicholas 2008, p. 39.

24 For Dakin's report on net-making see 'Voluntary camouflage net making in Australia' in W.J. Dakin, Camouflage report 1939–1945 (with appendices A–D), Canberra, AWM, Series 81 [77 Part 1], 1947, pp. 19–20.

25 Alec Hammond to Frank Hinder, 6 August 1941, Frank Hinder Papers, Sydney, Archive of the AGNSW, MS 1995.1, Box 4, Folder 10.

26 McCubbin, 'Some remarks on Australian art' in Macdonald 1986, p. 84.

27 Serle 1986, p. 251.

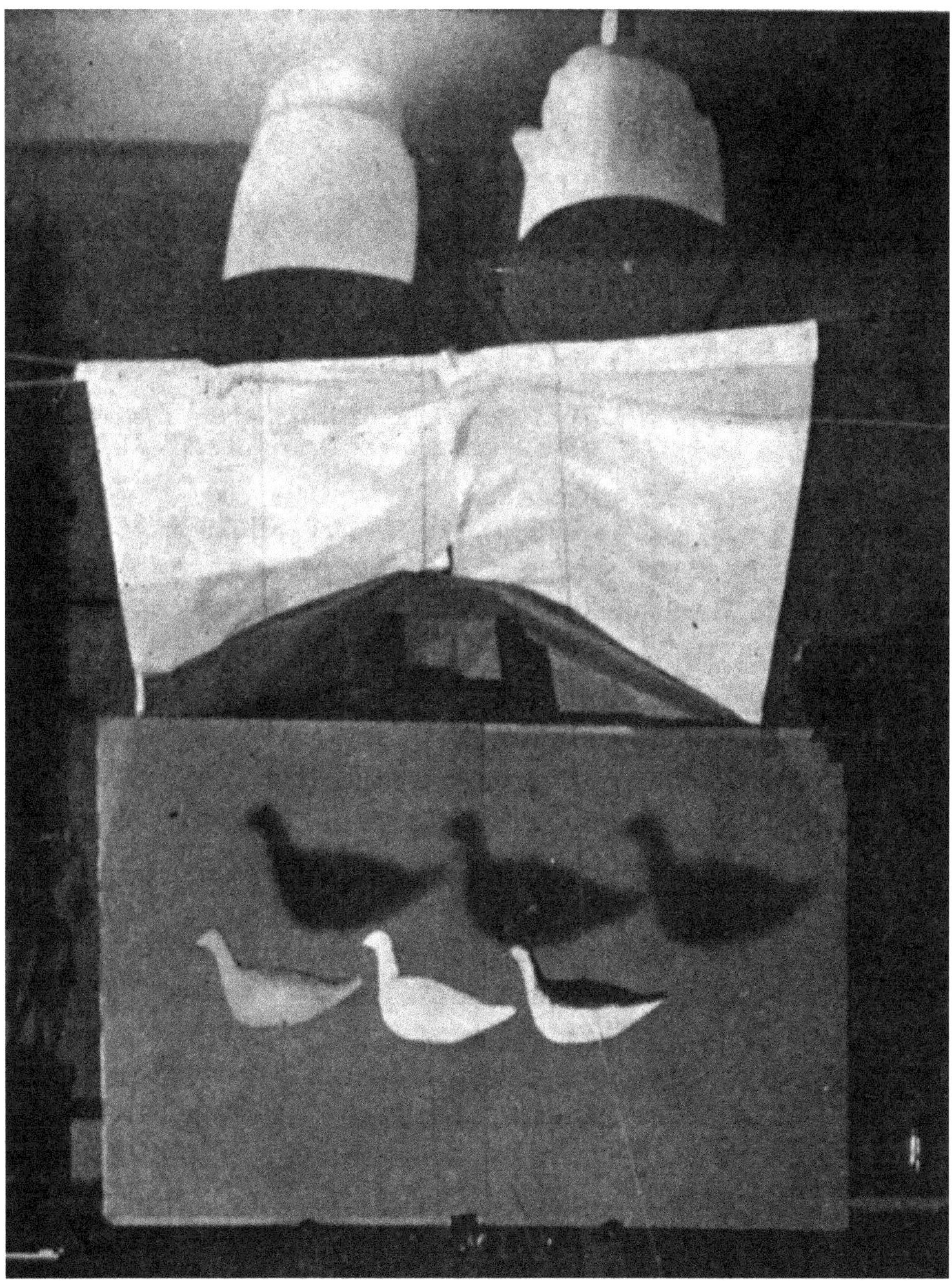

Fig. 3.11. Countershading demonstrated on three clay pigeons. Dakin 1942, p. 10. Photograph, W.J. Dakin.

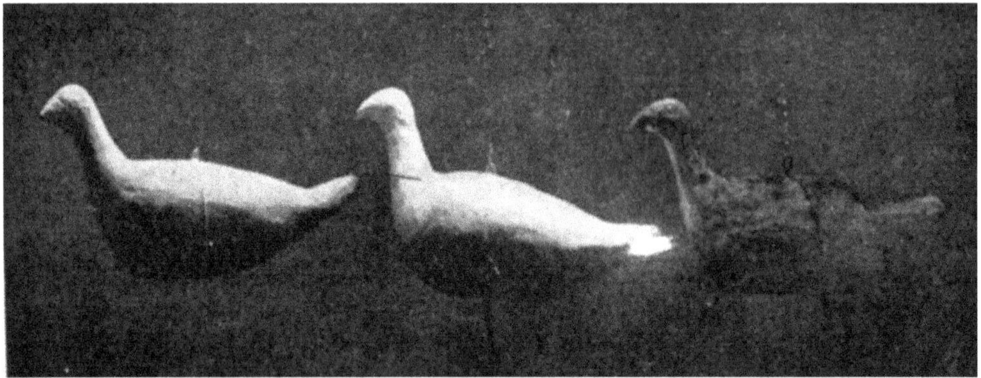

Fig. 3.12. Countershading demonstrated on three clay pigeons. Dakin 1942, p. 10. Photograph, W.J. Dakin.

taste for naturalistic paintings of dappled light. The painter's studio, therefore, was as much a laboratory for understanding the principles of camouflage as was the science laboratory, although it was the science setting that dominated experimental work conducted by the DHS in WWII.

The Zoology Department at the University of Sydney was the third important location for camouflage experimentation in Sydney. Between 1939 and 1941 the Sydney Camouflage Group conducted research and experiments in and around the University of Sydney and Dakin's office doubled as a laboratory and dark room. This was where he compiled the first edition of *The art of camouflage,* a small book that was published in 1941 with acknowledgment of the contribution of members of the group. When, however, it was reprinted in 1942 for military use, slight changes had been made to the text and it was published under Dakin's sole authorship. Both editions contain diagrams and photographs produced by the group's members, and others taken from British and European sources. Its key purpose was to elucidate four main camouflage principles and illustrate them with animal as well as military examples: realistic painting and colour resemblance to background, countershading, disruptive colouration, and effacement of shadows and their avoidance.

Realistic painting is given almost no space on the pages of *The art of camouflage.* At some point the group decided that painting illusionistic scenes on static structures, to make them appear as something different to their purpose, was a relatively useless type of camouflage. No matter how elaborate and effective the illusion, the structure remained and continued to cast shadows, thereby drawing attention to its whereabouts. Countershading, on the other hand, which is also referred to in the book as 'obliterative shading' (perhaps in deference to Abbott Thayer who coined the term) is discussed in detail. To illustrate the principle of countershading, Dakin took photographs of model birds in his zoology laboratory, imitating an experiment that Abbott Thayer conducted in 1909 (fig. 3.11). By darkening the top of the opaque, three-dimensional model and lighting it strongly from above, the bird appears flat rather than round (fig. 3.12). The DHS applied the principle of countershading to shoulders and arms of military uniforms to counter the effect of

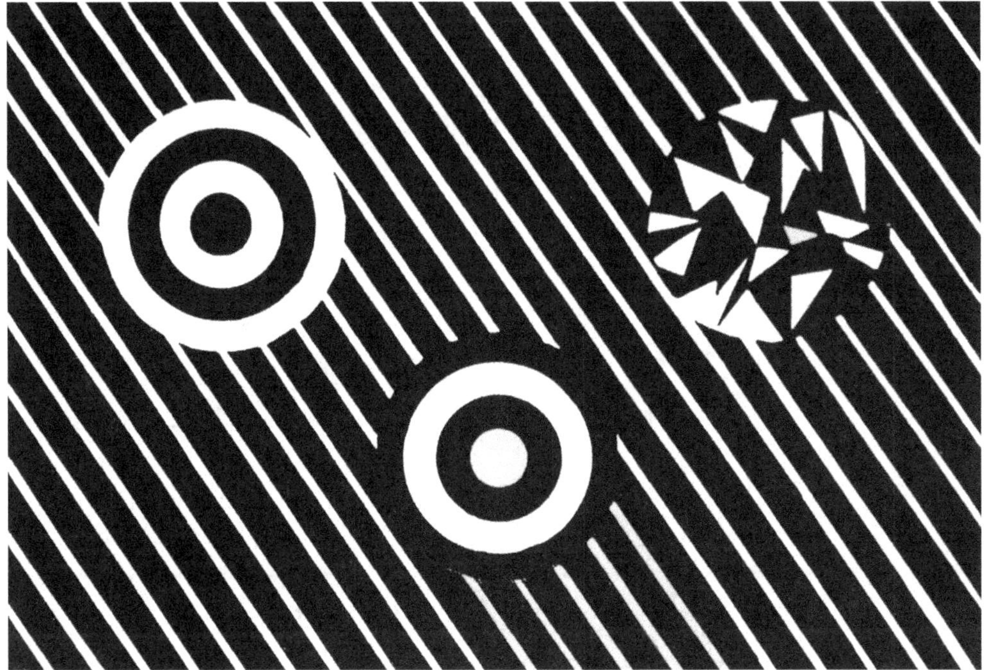

Fig. 3.13. Illustration of disruptive colouration from Dakin (1942), p. 12. Diagram by W.J. Dakin.

highlighting through illumination from the sky.[28] The idea was to optically flatten the human body and make it appear part of the general patterning of its surroundings.

To demonstrate the third principle, 'disruptive colouration', a series of photographs of Moorish idol fish and a black and white diagram illustrated what happens when the surface of an object is broken into patches of irregular shapes (fig. 3.13).[29] This was the theory that underpinned dazzle patterns on ships. When irregular patterns fuse visually with their background, the result is an object no longer legible in shape and outline. However, it was the effacement of shadows that occupied a dominant place in the work of the DHS (fig. 3.14). John Moore, for example, illustrated how to eradicate the effect of shadow formation around a building by extending its roofline with a screen, a principle that was applied to water and oil tanks where gradually curved nets removed the sharp clarity of the structure's cast shadow. And at Bankstown, Max Dupain applied the technique to plane concealment (fig. 3.15). Throughout the 1930s, Dupain's photography specialised in the doubling patterns of objects and their shadows, but on the airfields his challenge was to efface black reflections altogether, using nets that sloped to the ground.

Once Dakin became Technical Director of Camouflage for Australia he relocated his office and research team to the federal capital of Canberra. John Moore and Russell Roberts were left to run the NSW Directorate of Camouflage for the DHS from the University of

---

28 G. Howard Adams, 'Camoufleur's diary', 13 January 1945, in Camouflage General—Personal Diaries of Camouflage Office, Canberra, NAA, Series A453, Item 1944/17/2734.

29 Dakin 1942, p. 12.

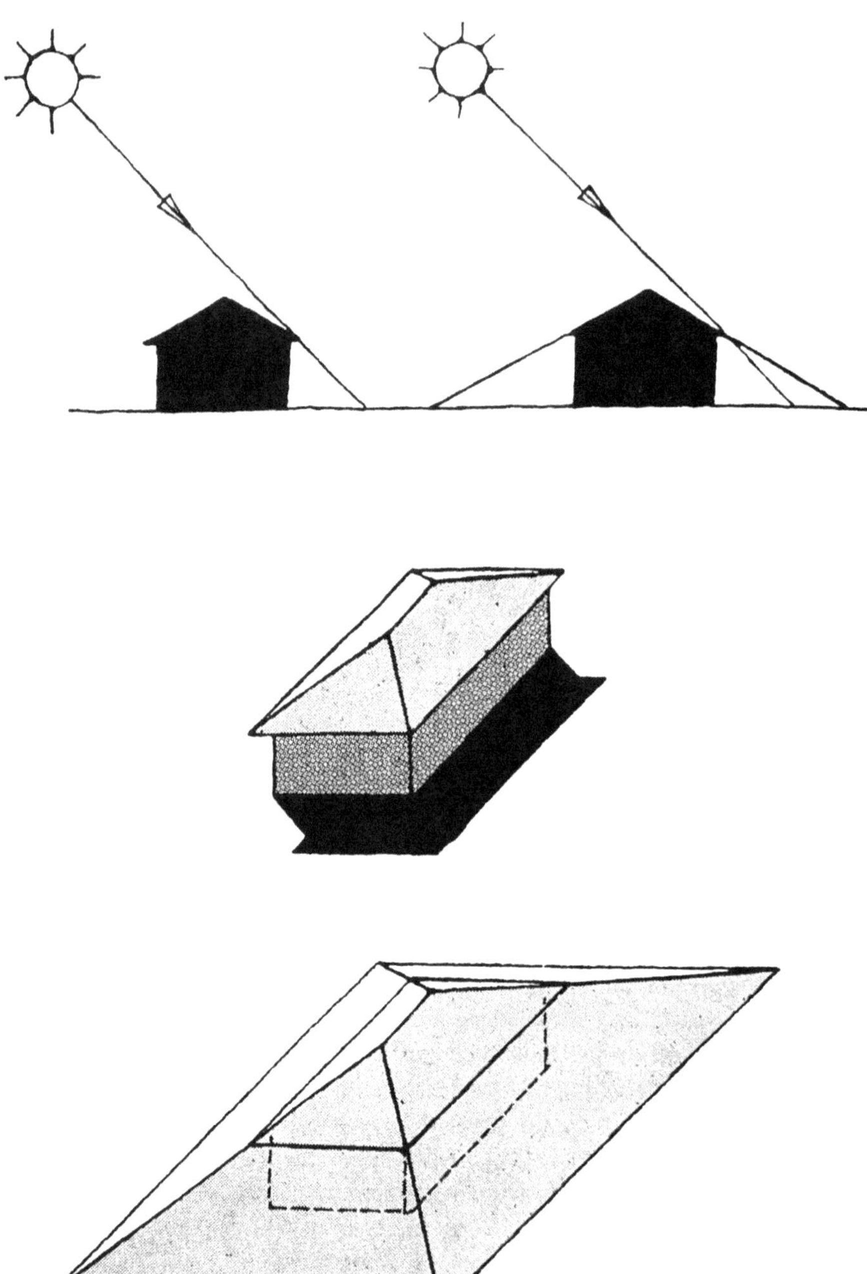

Fig 3.14. Shadow formation and eradication. Dakin 1942, p. 20. Illustration by J. Moore.

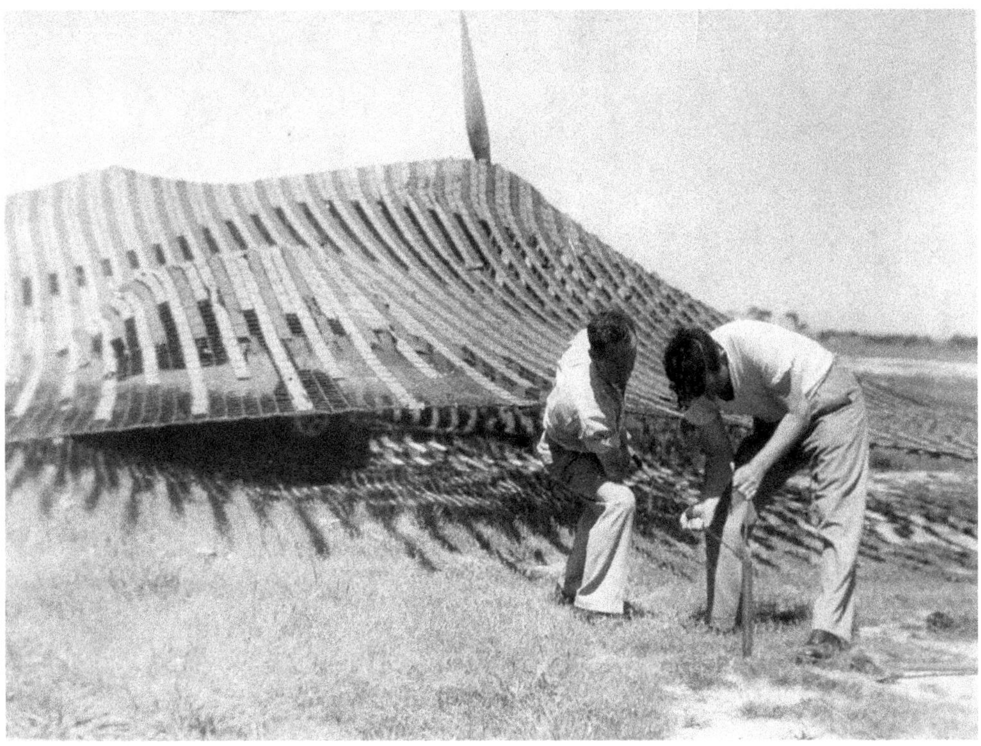

Fig. 3.15. Tying down a net for aircraft concealment at Bankstown, NSW, c. 1943. The collection of the NAA: C1905 T1, 1 [29]. Photograph, Max Dupain.

Sydney.[30] The University's science departments were already heavily involved with war-related research. The Physics Department, for example, ran an intensive course on defence radar.[31] The Botany Department investigated the use of flax crops in Tasmania and Victoria for canvas supply to Britain.[32] Similarly the Zoology Department specialised in camouflage.

From the Sydney office Moore and Roberts helped with the campaign in Papua and New Guinea by researching the vexed question of effective camouflage colours and patterns for the tropics. They worked in local bushland, using it as background for cut-out figures which they silhouetted against tropical plants to resolve how to balance tone with design 'against various types of jungle growth' (fig. 3.16).[33] Abbott Thayer had used the same method to understand animal camouflage, and had published his findings in *Concealing:*

30 'Camouflage', 3 July 1941, in Establishment of camouflage organisation and procedures, Canberra, NAA, Series A5954, Item 396/2.

31 Fielder Gill et. al. 'The "Bailey Boys": The University of Sydney and the training of radar officers' in MacLeod 1999, p. 470.

32 'Varsity works in war effort: all faculties contribute', *Honi Soit,* Thursday 15 May, Vol. XVIII, No. 9, 1941, p. 1.

33 A. Morley, 'Camoufleur's Diary 23 December 1944', Camouflage General–personal diaries of Camouflage Office, Canberra, NAA, Series A453, Item 1944/17/2734.

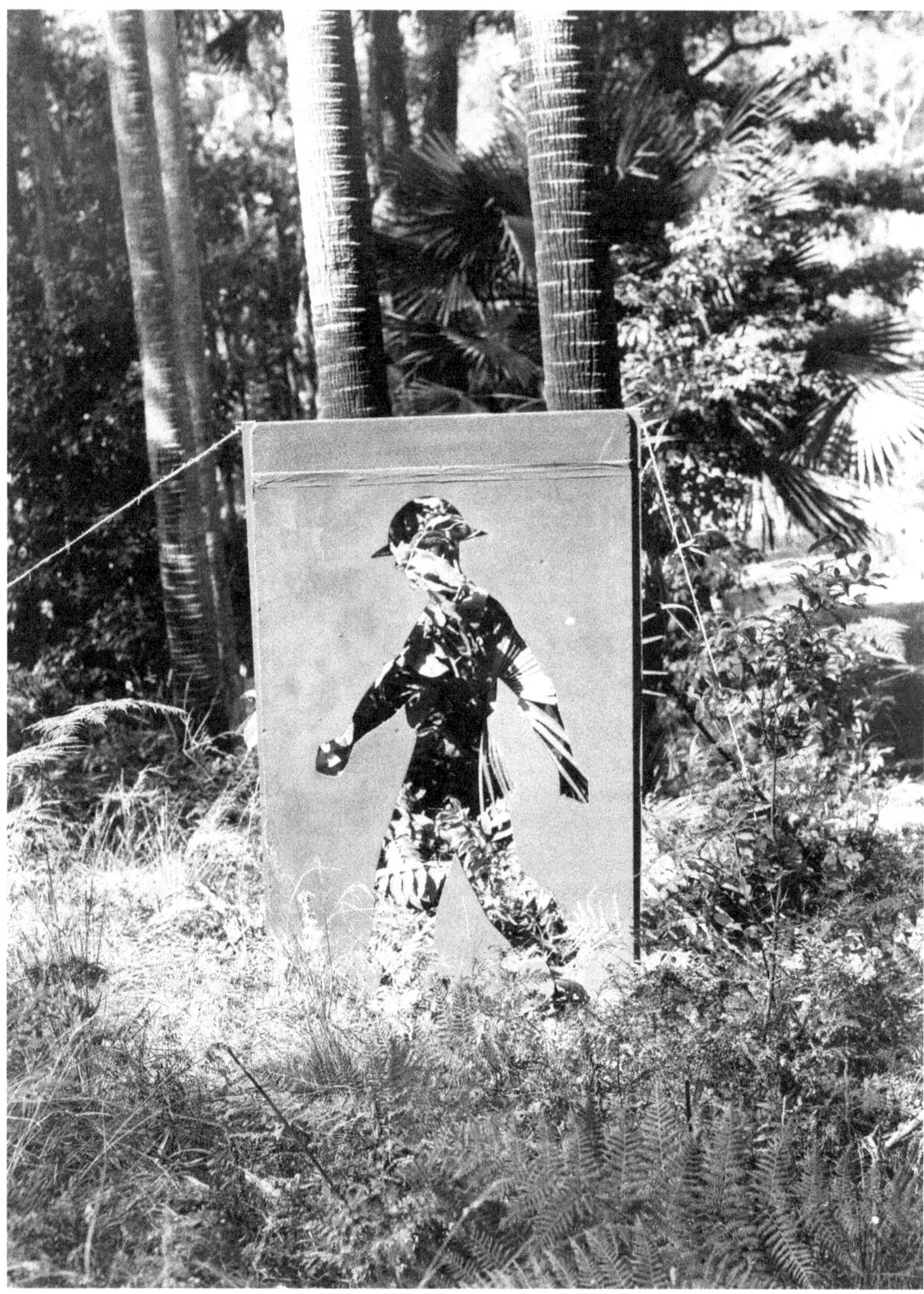

Fig. 3.16. Silhouette for designing camouflage uniform patterns and colours, c. 1943. The collection of the NAA: C1707, Item 01039.

coloration in the animal kingdom (1909). By placing cut-out silhouettes of zebras, striped and unstriped, against different backgrounds, Thayer had been able to judge the relative impact of skin patterns on concealment.[34]

Dakin's career in military camouflage resembles the trajectory of Thayer's, as it does the careers of other naturalists and zoologists including Hugh Cott. Following is a chapter devoted to William Dakin, and another that puts his work in the context of the history of ideas surrounding camouflage in the natural sciences. Dakin is a fascinating subject, but his life and work are not as well known as might be expected for a man who published widely in Australia not only on zoology and defence, but also on public education and health. His interests included the arts, particularly photography and landscape painting, and he was praised for making science popular through radio broadcasts designed for children. Alan Colefax emphasised these and other achievements in Dakin's obituary, one of only a few useful sources of biographical information.[35] But how were camouflage, zoology, education, health and war interconnected? And what relevance did theories of nature have for Dakin's role as the Technical Director of Camouflage for Australia in WWII? His appointment by the government certainly intrigued the Australian public. In 1941 a reporter from Adelaide marveled both at plans to apply empirical studies of animals to Australia's defence program, and at the deployment of William Dakin, a zoologist, to lead the work.[36]

---

34 Thayer 1909, p. 139.

35 Colefax also worked in the Zoology Department of the University of Sydney. See Colefax 1950, pp. 208–09.

36 'Animals will help us in camouflage: report from Adelaide', *Sun,* 2 August 1941, cited in Frank Hinder, 'Records of Lt F.C. Hinder, Camoufleur', Canberra, AWM, PR 88/133, File 895/4/182, Item 7 of 12.

PART 2

THE SCIENCE COMMUNITY

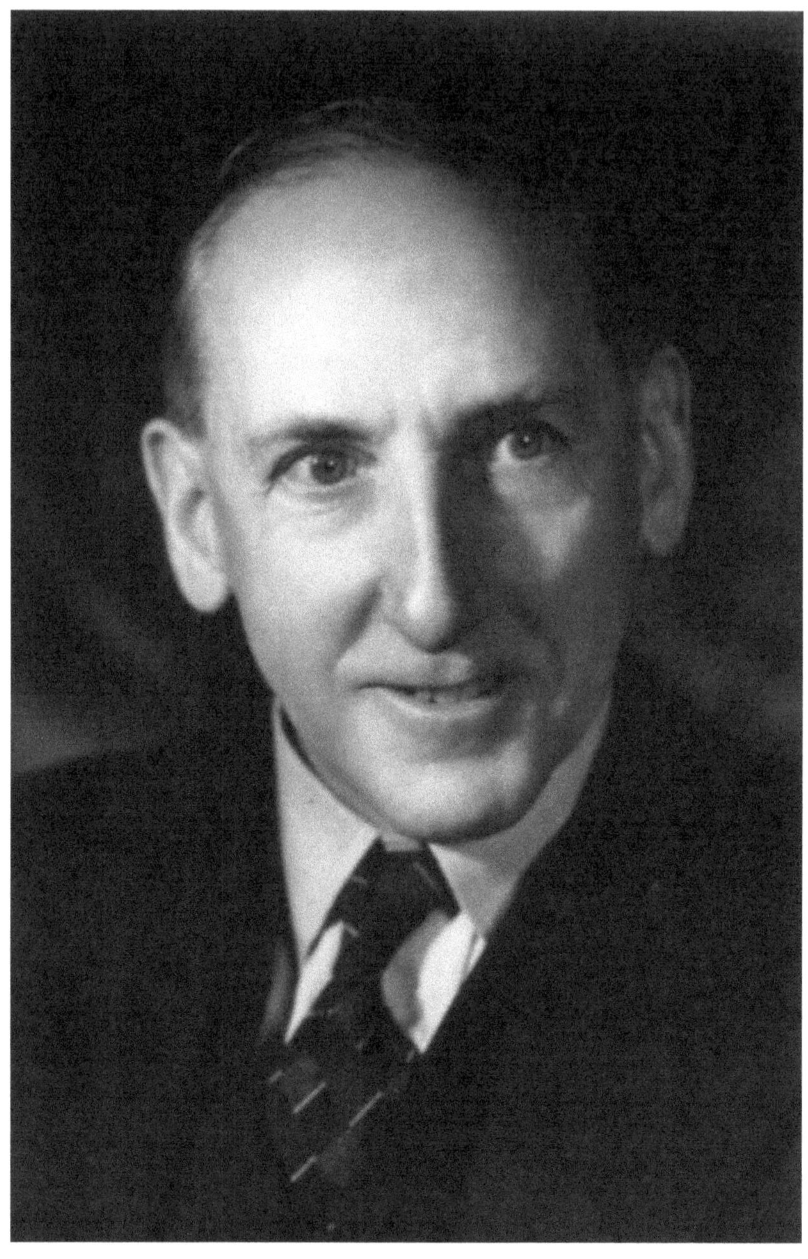

Fig. 4.1. Portrait of William Dakin, c. 1942–45. Photograph, Gervaise C. Purcell, courtesy of Leigh Purcell.

Chapter 4

William Dakin

One day, someone will write a biography of William John Dakin (1883–1950) that puts his writing on nature and warfare in the context of the history and politics of science. His work entailed a complicated mixture of Lamarckian evolution and the theory of heredity, eugenics, physiology and Darwinism, all of which he applied to the fields of science, education, public health and defence. In the meantime, two obituaries published in *Nature* and the *Australian Journal of Science,* a short entry in the *Australian dictionary of biography,* and Frank Hinder's memoirs, are the extent of commentary on his life. They paint a picture of a sometimes controversial man who was brilliant, had wide-ranging interests from the history of whaling to landscape painting, was a passionate photographer and a specialist in the Great Barrier Reef, who popularised science through radio broadcasts for children despite the disapproval of purist colleagues who believed academia should be free of popular application, and who never, in his lifetime, received proper recognition for his work as Technical Director of Camouflage for Australia in WWII.

Dakin waged a campaign in Australia in WWII to protect the nation and empire through advancements in camouflage defence, and he developed a theoretical basis for his work from the behaviours and colours of animals. Indeed, he brought a biological approach to a range of social problems, including health and warfare. Half his time as Technical Director was spent coordinating civil defence on the mainland, the other preparing troops and camouflage officers for service in the SW Pacific. But the problem of jungle camouflage preoccupied Dakin more than the other not least because, as the theatre of war shifted, the danger to the Australian mainland became less. As well as this, camouflage for tropical zones presented him with an opportunity to apply theories of biological camouflage to warfare in a geographical zone where making the artificial look like part of nature was imperative. The result, it will be seen, was Dakin's primitivistic vision for concealment and deception in the SW Pacific influenced by his desire for western soldiers to rediscover primal instincts. What is noteworthy about Dakin is that in texts on biology he wrote about war, while in texts on war he wrote about biology. In all his intellectual endeavours, nature, war and society were intertwined and this was true also for the earliest period of his career.

Dakin was born in Liverpool, England, on 23 April 1883 and began his academic life in 1907 at the University of Liverpool, specialising in the biochemistry of marine animals. But his research interests were both pure and applied and he was eager to work for industry, particularly agriculture and fisheries. It was the pearl fishing industry that brought him in 1913 to the University of Western Australia in Perth where he occupied the inaugural

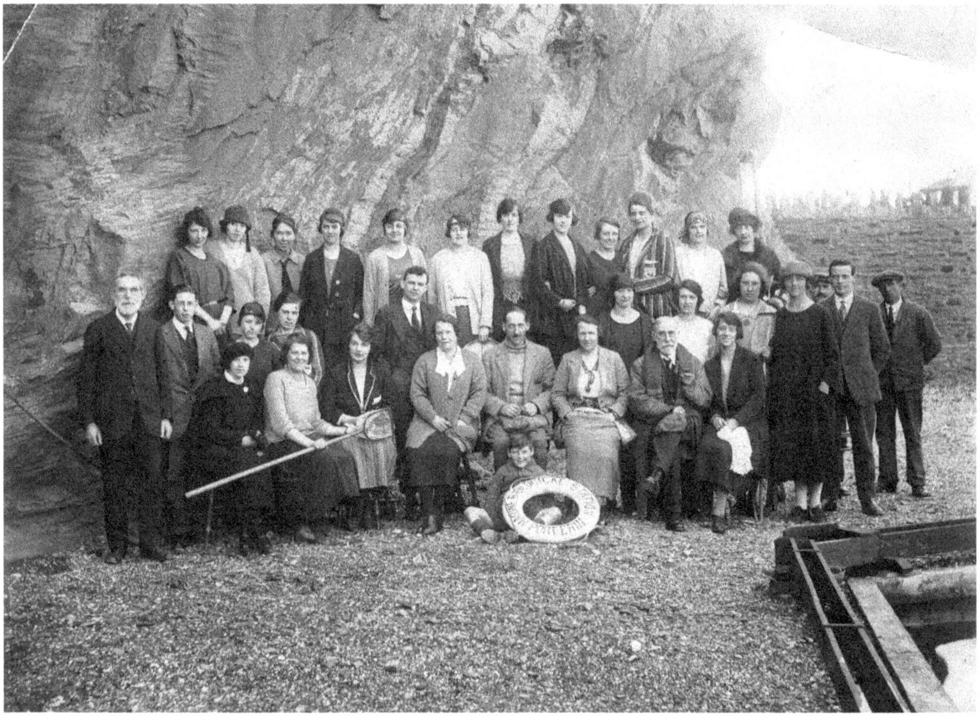

Fig. 4.2. Easter vacation class at Port Erin, UK, 1923, (Dakin seated in centre), Liverpool. Courtesy of University of Liverpool Library (University Archive: A301/2/32).

Chair of Biology. In 1920 he returned to the University of Liverpool as Derby Professor of Zoology and was appointed President of the Liverpool Biological Society. A photograph taken in 1923 shows Dakin seated in the centre of a gathering of scientists and students at the Marine Biological Station at Port Erin, Liverpool Bay (fig. 4.2).[1] Ever restless, Dakin migrated back to Australia in 1928 after accepting the Chair of the Zoology Department at the University of Sydney. In the 1930s he served as President of the Royal Zoological Society of New South Wales, President of the Linnean Society of New South Wales, and trustee of both the Australian Museum and the Taronga Park Zoological Gardens. At the outbreak of WWII he was seconded by the Australian government to live in Canberra and work for the DHS and when camouflage operations were wound back in 1945 he returned to the University of Sydney, retired in 1948 as Emeritus Professor, and died in 1950.[2]

Marine biology dominanted his research and his last book, *Australian seashores* (1952), published posthumously by his friends and fellow scientists, Isobel Bennett and

1 W.J. Dakin, 'The marine biological station at Port Erin', a booklet on marine biological stations and institutes written by Dakin and printed in Germany. Cited in the University of Liverpool's Special Collections and Archives, A 301/1/7.

2 All biographical information from three sources: Bygott & Cable 1981, pp. 190–91; Colefax 1950, pp. 208–09; Marshall 1950, pp. 751–52.

Elizabeth Pope, discussed connections between light and pigmentation in the evolution of camouflage mechanisms.[3] However, much earlier, during Dakin's inaugural lecture delivered at the University of Liverpool in 1921, he had addressed the audience on the role of zoologists in society and had argued that their function was to act as regulators of the natural laws of competition. Describing zoologists as society's 'special police', he characterised their job as enforcers of the laws of nature on humanity's terms, and their role to correct any imbalances between human and other animal competitors. Insects, he argued, presented the greatest threat of all animals to human livelihoods.[4] But the lecture was also an opportunity to inform his audience of the scope of his research, which for ten years had been in the field of the evolution of sight organs in scallops, a field he continued to specialise in throughout the 1920s.

In 1926 Dakin again published on the subject of visual organs in scallops, this time taking what he said was the 'unorthodox' position of claiming that the complex eyes of scallops are not the outcome of natural selection and chance based on external factors, as Darwin's theory of evolution would suggest, but are the result of inheritance and the passing on of physiological changes acquired during one scallop's lifetime to its progeny.[5] What stands out in the publication 'The eyes of pecten' (1928), and that has bearing on his future work in camouflage in WWII, is the importance of optics and the science of light, especially the impact of shadows, on his experiments. By passing shadows over scallops, ones shaped as starfish—to simulate the presence of the scallop's predator—Dakin was able to assess the animal's response and draw conclusions about evolutionary change.[6] The system of thinking surrounding Dakin's early scientific endeavours was focused on prey and predator, behavioural adaptation, light and shadow, visibility and invisibility. It also came to underpin his work in camouflage in WWII. Just as shadow was a vital consideration in his early zoological research, so shadow evolved into a subject in its own right in WWII, and one that dominated *The art of camouflage*. Shadow, he explained in the first and second editions, 'is often the greatest danger of all'.[7]

What also emerges from Dakin's 1926 publication on scallops is a view of himself as a contentious figure and discordant note within the mainstream of science. It was a self-image that played a significant part in how he conducted his campaign for camouflage defence in WWII. Indeed his life in Australia was characterised by conflict. The circumstances of Dakin's early years in Perth suggest he was never properly 'at home' in Australia. First his plan to establish a eugenics society in Western Australia failed.[8] Then, according to Bygott and Cable, he was refused permission to enlist in the AIF in WWI.[9] Before arriving in Australia he served with the Territorials as Second Lieutenant of the Lancashire and

3 Dakin 1952, pp. 78–79.

4 Dakin 1921, pp. 3–4.

5 Dakin 1928, p. 364.

6 Dakin 1928, pp. 361–62.

7 Dakin 1942, p. 18.

8 The failure was due to lack of public support according to Muriel Marion who reported this to the British Eugenics Education Society in 1933. See Wyndham 2003, p. 139.

9 Bygott & Cable 1981, p. 190.

Cheshire Royal Garrison Artillery in Britain.[10] As he was sufficiently trained and willing to enlist, it is unclear why Dakin was unable to join the AIF, but the majority of men who were refused were 'unfit' on the basis of age, height and health.[11] Instead Dakin was seconded during WWI by the Department of Public Health in Perth to improve the moral and physical wellbeing of Australians, using his extensive knowledge of physiology and scientific understanding of nutrition, respiration, excretion, exercise and reproduction.

Dakin's passion for physiology and zoology were combined in two books published in 1918, both of which were designed to educate Australian youth. The first was *Elements of animal biology* written for school children. It discussed the classification of animals from lower to higher orders and addressed the concept of survival of the fittest and good health in animals as a way of instructing its young readers on their own physiological wellbeing, cleanliness and correct social behaviour.[12] The other book, *Sex hygiene and sex education,* was co-authored with Everitt Atkinson, Commissioner of Public Health in Western Australia, and was written for the welfare of Australian adolescents to 'check immorality and sexual disease'.[13] Dakin and Atkinson were concerned that if venereal disease was left unchecked it would contribute, along with the low birthrate in the middle classes, to the eugenic concept of 'race suicide', by which they meant the demise of the Anglo-Saxon race. This matter is addressed in both books. The conclusion of *Elements of animal biology* appealed to the youth of Australia in 1918 to play their part in the survival of the British Empire which, it said, depended upon the 'imperative that we let no nation surpass us in physique, intelligence, and morality'.[14] Remember that Dakin had witnessed the drastic depletion of Australians in WWI and during that period many Australian doctors were concerned that the Australian race might 'wither away'.[15]

Biology was the common link to all fields of Dakin's work and the theory of evolution informs all his writing, particularly the idea that conflict and struggle are essential to the continuation of life. A passage from *Elements of animal biology* shows how seamlessly integrated were defence and weaponry into the natural order of the world:

> most animals have to search out their food and perhaps fight for it. They require acute senses, ever ready to perceive favourable or dangerous signs around them, and powerful weapons of defence to overcome their prey, or even defeat their own kind ... Let us keep

10 *The London Gazette*, 5 May 1911, p. 3443, www.london-gazette.co.uk/issues/28491/pages/3443, accessed 2 June 2010.

11 'Enlistment Standards First World War', *Encyclopedia*, Canberra, Australian War Memorial, www.awm.gov.au/encyclopedia/enlistment/index.asp, accessed 24 February 2009.

12 William Dakin's professional preoccupation with cleanliness in 1918 compares interestingly with Abbott H. Thayer (whose work on camouflage was influential on Dakin and is discussed in chapter 5). Both were anxious about physical and moral impurity. Elizabeth Lee examines Thayer's paintings of angels and argues that their purity and chastity cloaked Thayer's profound anxiety about disease and dirt which are revealed in 'persistent references in his letters and other writings to disease, prostitution, garbage excrement, and germs'. See Lee 2004, pp. 33–34.

13 Atkinson & Dakin 1918, p. 7.

14 Dakin 1948, p. 284.

15 J. Bostock & L.J. Nye quoted in Garton 1996, p. 160.

> before us for the present the picture of an intense struggle for existence which is being fought by all animals and plants.[16]

His career in publishing on camouflage also debuts with *The elements of animal biology* in a short section on 'Animal coloration: protective resemblance and mimicry'.[17] It makes sense he would publish on this subject during WWI when camouflage in nature first served as a model for newly formed camouflage units in the French, British, German and United States militaries. This was when the French term 'camouflage' made its first impression on the zoological community, including on William Pycraft in Britain (1868–1942) who later reflected how: 'During those dreadful years the word 'camouflage' was in constant use among us. And there were, surely, few who did not understand its meaning, few who did not realise that vast issues depended on it.'[18]

The question of the nature of animal versus the nature of human was raised urgently in WWI. Indeed the war showed a lack of distinction between animal and human that frightened and repulsed because it showed the savagery that evolution was meant to have overcome through natural selection and civilisation. Dakin, on the other hand, appears to have thought of war as evidence of human biological destiny; in war, by necessity, humans return to a primitive state of nature brought about by the struggle for survival. Instinct theory and the concept of the beast within were prevalent in all areas of intellectual endeavour at the time of WWI, including biology.[19] With this historical situation as background, when Dakin took up his position as Technical Director of Camouflage for Australia, he became focused on how to teach and demonstrate the logic of primitive masculinity for those seeking concealment in the tropical vegetation of northern Australia, Papua and New Guinea. The geography of those locations demanded, he argued, a physical state where 'being too dark is generally safer than being too light'.[20]

Always there is the suggestion in Dakin's advice to soldiers in combat zones in the SW Pacific that war demands a return to primitive origins. Yet his methods of approaching the subject were eccentric. Rather than present troops with camouflage in objective scientific terms, he chose literature, and Rudyard Kipling's story 'How the leopard got his spots', one of the *Just so stories* (1902), to instruct them. In 1942 Dakin was intrigued but concerned by the ingenuity of Japanese methods of concealment in New Guinea and Malaya: ' "I can smell giraffe, and I can hear giraffe, but I can't see giraffe." And that about represents the situation in a New Guinea or Malay rain forest if a yellow or light coloured individual is looking for a man of more sombre hue'.[21]

---

16 Dakin 1948, p. 282.

17 Dakin 1948, p. 128.

18 Pycraft 1925, p. 119.

19 Mitman 1997, p. 262.

20 W.J. Dakin and the Camouflage Directorate, 'Concealment, and camouflage of the individual in warfare', Canberra, DHS, 1944, in W.J. Dakin, Camouflage report 1939–1945 (appendix N), Canberra, AWM, Series 81 [77 Part 3], 1947, p. 9.

21 W.J. Dakin, 'Camouflage bulletin no. 7', 9 October 1942, in W.J. Dakin, Camouflage report 1939–1945 (appendix O), Canberra, AWM, Series 81 [77 Part 4], 1947, p. 7.

Kipling's story is about a zebra and giraffe who develop stripes and spots to hide among 'stripy shadows and blotched shadows in the forest', a transformation designed to escape detection by a leopard and an Ethiopian who in turn change their skins—the Ethiopian to blackish brown, the leopard to spots—to go undetected while they search for their elusive prey.[22] As the leopard repeatedly points out about the confusing disappearance of the zebra and giraffe, 'they haven't any form'.[23] Kipling was influenced by the camouflage thinking behind the adoption of khaki in the Boer War and, like his generation of artists and scientists, by the impact of Charles Darwin's theories of mimicry in *The origin of species* (1859). When Kipling wrote 'How the leopard got his spots', scientific theories on camouflage were being debated in the US, where Kipling was then living. Abbott H. Thayer, artist, naturalist and innovator of military camouflage in WWI who read and admired Kipling's story, was at the centre of the debate whether it is universally true that gradations of light and shade on animals lead to their becoming invisible to predators.[24]

Kipling's story about the sandy-coloured, veld-dwelling leopard and an Ethiopian man, who shift their hunting to the forest by adapting their colouration to the shadows and dappled light, gave Dakin an opportunity to teach soldiers about disguise and what it meant to behave instinctually. When the Ethiopian in Kipling's story became cunning and painted his skin black to hide in the forest, he regained his power as a hunter. This was one of the most important points of the story for Dakin. In a 1944 camouflage manual written for the military he spoke about 'The ten big sins of the hunter and the hunted'. Two of these are 'being a conspicuous colour or shape (or both)', and 'being a misfit in the background pattern'.[25] Like the polar bear who depends on the whiteness of its fur to catch prey in a white landscape of snow and ice, and like the tawny colour of the lion, the stripes of the tiger and the mottled colouring of moths, colour resemblance was for defensive and offensive purposes, he claimed. He urged soldiers to observe their surroundings by wearing grey, dark green and khaki, according to conditions because, as he cautioned, 'a sense of background must be developed. In other words (and the old story again) correct behaviour is the basis of all good concealment.'[26]

But for reasons of skin irritation and heat, Australian troops in the SW Pacific were reluctant to wear the DHS's new 'very dark brown-nature black-brown' cream—Skin Tone Commando Cream—to conceal the pinkness of their skin.[27] Initially too, they were also reluctant to wear green rather than khaki. In response, Dakin relayed reports from the jungle islands about the superior camouflage methods of Japanese soldiers who 'just

---

22 Kipling 1902, p. 53.

23 Kipling 1902, p. 52.

24 Thayer 1896; Thayer 1909, p. 135.

25 W.J. Dakin and the Camouflage Directorate, 'Concealment, and camouflage of the individual in warfare', Canberra, DHS, 1944, in W.J. Dakin, Camouflage report 1939–1945 (with appendices K–N), Canberra, AWM, Series 81 [77 Part 3], 1947, p. 3.

26 W.J. Dakin, 'Camouflage bulletin no. 7', 9 October 1942, in W.J. Dakin, Camouflage report 1939–1945 (appendix O), Canberra, AWM, Series 81 [77 Part 4], 1947, p. 7.

27 W.J. Dakin, Camouflage report 1939–1945 (with appendices A–D), Canberra, AWM, Series 81 [77 Part 1], 1947, p. 134.

melt into the trees and you can't pick them till you're right on them'.[28] Through bulletins and manuals prepared for the region, soldiers were told to think deceptively and were shown the importance of shifting the axis of the body to the horizontal position of lower evolutionary animals, using postures such as crouching and crawling. Dakin found soldiers too inhibited with their bodies and criticised them if they did not lie flat, telling them 'don't think it *infra dig* to crawl, and learn to crawl properly. You should be flat on the ground!'[29] He ordered them to creep, crawl, and slide on their stomachs like the creatures that filled his books on animal biology—the soldier-ants, snakes, frogs and lizards who live wary of enemies, stay close to the ground in a semi-subterranean 'hidden world', freeze still by instinct and hunt by night away from the glare of the sun.[30]

Interesting conceptual inversions took place when Dakin wrote about health and education in everyday society, and when he wrote about camouflage in war. A brutish side to Dakin emerged in his writing on war where he encouraged behaviours that were shadowy, primitive and instinctual, the opposite of those he encouraged in Australian youth in peacetime society. Whereas he taught youth that concealment and deception were negative in relation to civilised life, he taught soldiers that they were positive in relation to warfare. While he believed it imperative that in times of war the strong-willed are not afraid to lose their form by camouflaging themselves, and that it is permissible to act deceptively, in civilian life the idea of deception and lack of distinction among people he saw as detrimental to the wellbeing of the state. Likewise, in war it was vital for soldiers to liberate their instincts, but in civilian life Dakin considered it vital to suppress them. And while in war, silence and secrecy were paramount for preventing detection by the enemy, in his views on the raising of Australian youth Dakin regretted that the 'conspiracy of silence' surrounding sex hygiene and sex education was based on the incorrect assumption that 'silence makes for protection'.[31]

In civilian life Dakin admired and respected men who were conspicuous and stood out in relation to other men, who would step forward and achieve distinction, who were honest and stood tall. In war the converse was true; the man he admired was the one most likely to beat other men in war by strategically blending in and unselfconsciously disappearing, freezing, hiding in silence, becoming invisible, and moving sideways with stealth like the 'native' and the crab. In *Australian seashores* Dakin wrote:

> This is one of the first things to teach a soldier—especially the soldier brought up in a modern city. Many native tribes instinctively practise and recognize the value of the trick of remaining absolutely still when in full view of an enemy. This action is practised beautifully by many kinds of crabs …[32]

28 W.J. Dakin, 'Camouflage bulletin no. 8', 23 October 1942, in W.J. Dakin, Camouflage report 1939–1945 (appendix O), Canberra, AWM, Series 81 [77 Part 4], 1947, p. 6.

29 W.J. Dakin and the Camouflage Directorate, 'Concealment, and camouflage of the individual in warfare', Canberra, DHS, 1944, in W.J. Dakin, Camouflage report 1939–1945 (with appendices K–N), Canberra, AWM, Series 81 [77 Part 3], 1947, p. 32.

30 W.J. Dakin 1948, p. 272.

31 Atkinson & Dakin 1918, p. 42.

32 Dakin 1952, p. 74.

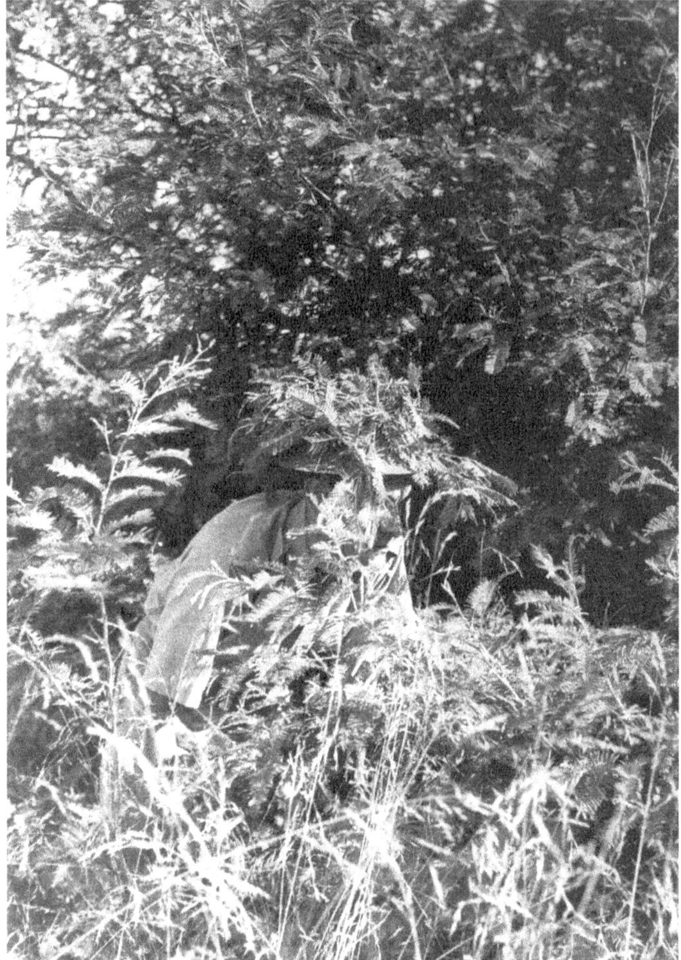

Fig. 4.3. DHS camouflage experiment with garnished hat, c. 1942. The collection of the NAA: C1905, 1[16]. Photograph, Gervaise C. Purcell.

A photograph of a crouching soldier dressed in jungle camouflage taken during an experiment by the DHS (in Sydney or Canberra), illustrates the types of correct behaviour in warfare that Dakin so often emphasised in camouflage manuals to Australian troops (fig. 4.3). He thought it commonsense to adorn a uniform with materials from the surrounding environment as crabs do when they try to avoid detection by arranging bits of seaweed on their backs. It illustrates the point that defence in human and animal realms were, to Dakin's way of thinking, one and the same nature:

> One could easily hide a soldier by attaching leafy branches to his uniform and hiding his round, smooth, tell-tale helmet with a small net supporting a few twigs and leaves, until the man resembled a small bush. But the man must play his part; one wouldn't expect to see a bush walking in the open.[33]

33 Dakin 1952, p. 82.

But he realised that all too often members of the military found it irrational, and certainly unnatural, to pin their hopes of survival on an ability to mimic the behaviour of animals. Their sense of self and even masculinity was enmeshed in the belief that humans are superior to animals. Nevertheless he did not let up, and consequently DHS camouflage manuals repeatedly point out how animals:

> choose dark corners as hiding places, during the day. Above all, they have learned through thousands of years of the struggle for existence that being seen is not merely a matter of colour but far more often a matter of injudicious movement and bad choice of resting place. The correctly coloured animals of the jungle have an instinct which automatically causes them to resort to the correct background and to remain immobile whilst they wish to be hidden. The soldier has to learn both these things.[34]

What drove Dakin? By all accounts he was irrepressible, irascible, stubborn but inspirational.[35] Colefax's obituary refers to a work ethic of 'tireless energy' and 'characteristic vigour'.[36] True enough, from 1939 to 1945 he was obsessed and fastidious about his role in the war effort, and about the contribution that camouflage would make to victory. Just as he was concerned that humankind, rather than insects, should possess the earth, so he looked at WWII through the same lens, and hence saw Germany and Japan as being in direct competition with the British Empire to possess the world—which made war part of the struggle for race and species domination. His worst fear in WWII was the defeat and demise of the Anglo-Saxon race and so the message he sent to Australian soldiers fighting in the SW Pacific was to 'wreak moral and physical ruin upon the invader'.[37] Frank Hinder observed that camouflage was like a religion to Dakin and likened him to Moses receiving the Ten Commandments and delivering people from a wilderness of disbelief at camouflage's power.[38] Camouflage had come to symbolise the road to victory through superior intelligence and, while the principles of concealment and deception were as old as life itself, Dakin was proud to think it was zoologists and naturalists who made camouflage a science fit for the modern age.

Without question, Dakin was motivated in his service during WWII by strong ties of kinship with Britain. Unlike many Australians who felt either indifferent or even antagonistic towards the British Empire at the start of WWII, Dakin was different.[39] But

---

34 W.J. Dakin, 'Camouflage bulletin no. 7', 9 October 1942, in W.J. Dakin, Camouflage report 1939–1945 (appendix O), Canberra, AWM, Series 81 [77 Part 4], 1947, p. 7.

35 Dakin's personality is commented on frequently in Frank Hinder, 'Records of Lt F.C. Hinder Camoufleur', Canberra, AWM, PR 88/133, File 895/4/182. Marine scientist Isobel Bennett—Dakin's secretary in Canberra during the war as well as a student of zoology at the University of Sydney—provided biographical information during an informal meeting in 2002.

36 Colefax 1950, p. 208.

37 W.J. Dakin, 'Camouflage in forward areas', in W.J. Dakin, Camouflage report 1939–1945 (with appendices P–R), Canberra, AWM, Series 81 [77 Part 5], 1947, p. 1.

38 F. Hinder, 'The ten demandments for a camouflage profit', Frank Hinder Papers, Sydney, Archive of the AGNSW, MS 1995.1, Box 4, Folder 10.

39 Dear & Foot 2001, p. 61.

he was not blindly loyal and wrote to Hubert Lazzarini, Minister of Home Security, that 'one great defect in the British Empire in the past has been the failure to make easy contact with scientific experts and other civilian specialists'.[40] Germany and Japan, he argued, understood completely the role of science in war but Britain and Australia had failed to wise up, to their detriment. The letter to Lazzarini is dated July 1942. It is important to put it in context. Dakin had just learnt of the Australian Army's decision to go its own way with camouflage; the services of the DHS were no longer needed by the army, which left Dakin without jurisdiction. He was shocked by the outcome and from July 1942 until the end of the war never regained composure, and never failed to take the opportunity to criticise the army for sidelining the knowledge and advice of civilians.

The war exhausted Dakin but it also took its toll on his research. In a cruel and double irony a large proportion of his work on the marine biology of plankton was lost at sea during WWII, in enemy action, as it made its way by ship from Australia to Britain for publication.[41] But what Dakin left which is of great interest to anyone with a fascination for camouflage (and is discussed in the next chapter) is evidence of a global network of scholars who were amazed and intrigued by the subject of camouflage in nature, art and war. William Dakin never cited Abbott Thayer or Hugh Cott in his books on zoology and camouflage, yet both men are manifestly present in the material he published and lectured with. But the omission of Cott and Thayer from Dakin's work raises an important subject: the highly competitive field of camouflage for zoologists and naturalists in the 20th century, and Dakin's place in it.

Before he died in April 1950, William Dakin was able to look back on two world wars and assess the importance of camouflage. In *Australian seashores* he concluded that 'the basic tricks of even military camouflage were already invented and in use by animals without backbones (Invertebrata), which lived on earth probably more than 300 million years ago'.[42] The epigraph to chapter 8 on 'Camouflage and living colour on the seashore' is a quote from Rudyard Kipling: 'I am the Most Wise Baviaan, saying in most wise tones, Let us melt into the landscape—just us two by our lones'. The source is, once again, 'How the leopard got his spots'. Dakin was dying of cancer which makes the idea of melting into the background, becoming invisible, and disappearing from the world, part magical and part melancholic.

40 W.J. Dakin to H.P. Lazzarini, c. 1942, in Camouflage—question of responsibility: army camouflage Canberra, NAA, Series A453, Item 1942/21/2632.

41 Marshall 1950, p. 752.

42 Dakin 1952, p. 73.

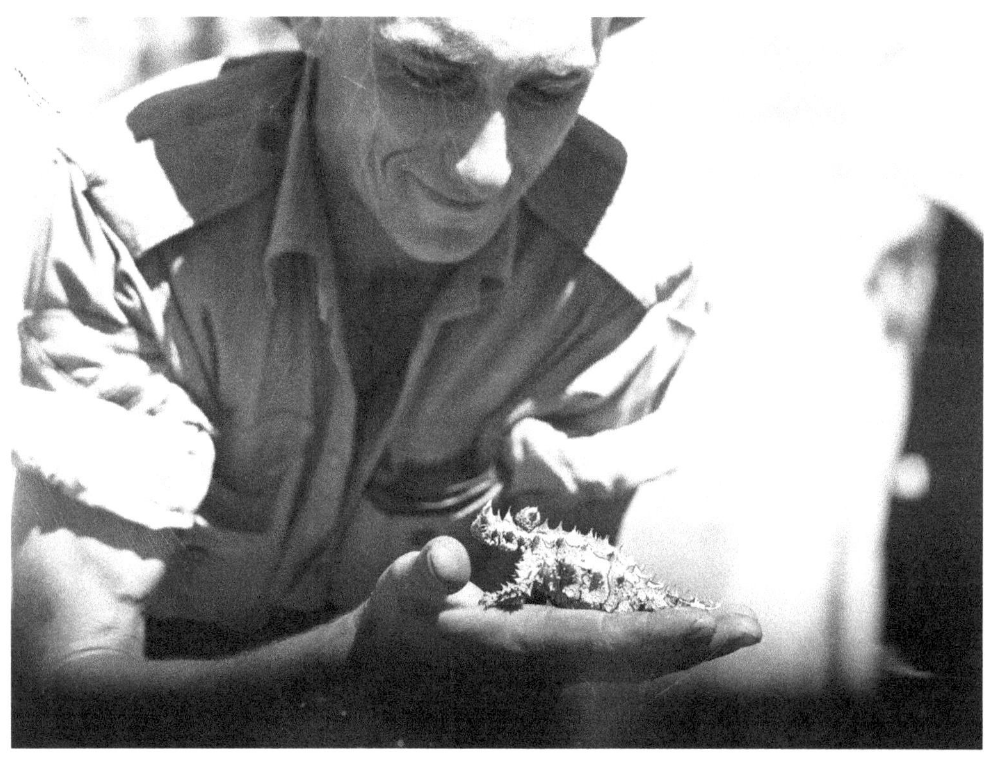

Fig. 5.1. ‘Mountain Devil’ military pet, Western Australia, 17 November 1942. Canberra, AWM, AWM 028810.

Chapter 5

Animal camouflage

It was in WWI when France established the first camouflage division in military history, that the full extent of camouflage's intertwinement of art, nature and war became evident. The French camouflage division used the chameleon as its mascot, and employed artists as its workforce.[1] Taking their lead from the French in WWI Australian artists working as camoufleurs for the DHS in WWII also chose an animal as emblem for their organisation, but faced with the widespread misunderstanding that camouflage is a strategy for the weak, chose the bolder motif of a tiger.[2] Animals took on a variety of symbolisms in both world wars, reflecting back on their human subjects whatever image was desired, from survival in adversity, to triumph through cunning. But symbolism aside, it was the wise soldier who sought solutions to the perennial problems of concealment and deception through direct observation of the outer skins and behaviours of animals, or so it was argued by two generations of naturalists and zoologists with an interest in things military. When the idea was put to the Australian government in 1941 and they appointed William Dakin to the role of director of camouflage operations for Australia and its territories, he joined a distinguished circle, including the late American naturalist Abbott H. Thayer and British zoologist Hugh B. Cott. Just why they thought animals had anything to teach people about warfare is best explained by Cott, who saw in both animals and people the same state of nature that Charles Darwin had written about, namely, 'that struggle for existence which is the lot of men not less than of animals'.[3]

As the historical event of Dakin's appointment as Technical Director came to public notice, troops in the deserts of Western Australia adopted a distant relative of the chameleon, the Australian mountain devil, also known as the thorny devil (*Moloch horridus*), as their pet (fig. 5.1). Nothing about this reptile is dangerous, yet it tries to suggest otherwise with a deceptively thorny skin and a habit of lowering its neck and pointing what looks like a second head to intimidate enemies before making its escape.[4] Other Australian animals use similar tactics, including the frilled lizard (*Chlamydosaurus kingii*) and according to Dakin these adaptations, which are designed to deceive and confuse, are the result of their struggle

1 Behrens 2009a, p. 86.

2 Notes on Australian camoufleur uniform, Frank Hinder Papers, Sydney, Archive of the AGNSW, MS 1995.1, Box 4, Folder 10.

3 Cott 1957, p. xv

4 Cronin 2008, p. 95.

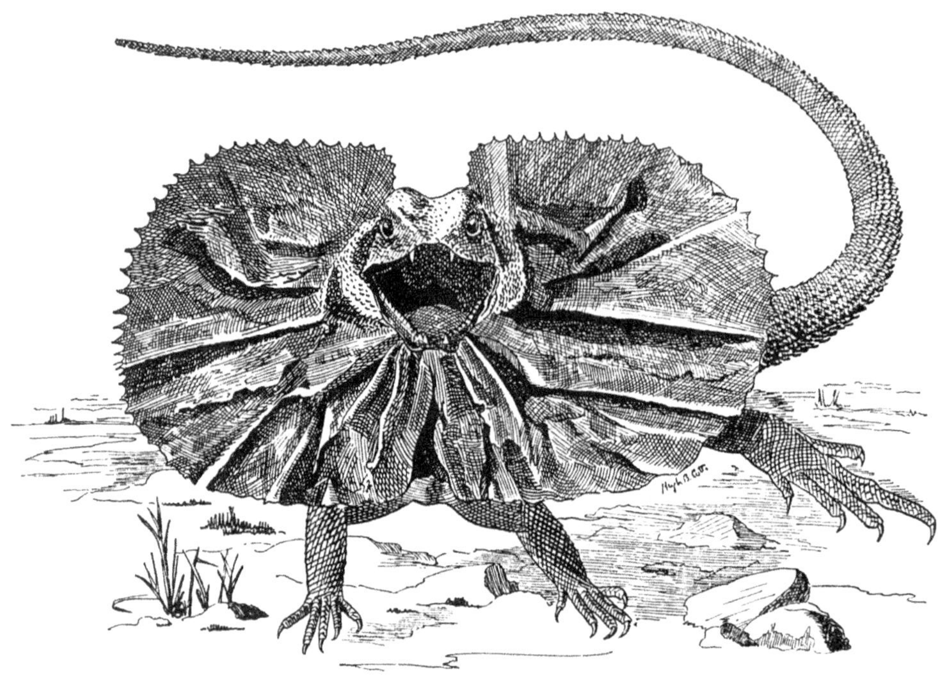

Fig. 5.2. Hugh B. Cott, Warning display of the frilled lizard (*Chlamydosaurus kingii*). Cott 1957 [1940], p. 218.

for survival on an ancient continent where life has seen 'ages of evolutionary modification'.[5]

Australian fauna has long been popular with western scientists of camouflage, including some who theorised military camouflage. Hugh Cott is a case in point and *Adaptive coloration in animals*—a standard textbook for camouflage experts in Britain in WWII[6]—includes an illustration of the warning display of the frilled lizard. Cott was intrigued with the way this antipodean lizard suddenly erects a fan of skin to ward off predators, making itself look twice the size. For his illustration he chose the moment when the animal adopts a state of menace, just before turning to run (fig. 5.2).[7] As previously mentioned, this method of animal deception is known as 'bluff' and is the same term applied to the great camouflage deception schemes of WWII. They too, relied on extrovert displays to conceal underlying weaknesses and none more so than Operation Hackney when a 'ghost force' held back Japanese forces on Goodenough Island in the SW Pacific with a non-existent brigade consisting of elaborate simulations of personnel and equipment made from hessian, coconut logs and tin.

Embedded then, in the histories of both world wars, certainly in the United States, Britain and Australia, is the story of the contribution of naturalists and zoologists to the

5 Dakin 1942, p. 8.

6 Goodden 2007, p. 14.

7 Cott 1957, p. 217.

advancement of camouflage defence. It is not accidental that two of military science's main camouflage principles—'countershading' and 'disruptive colouration'—were first theorised in relation to animals. As we have seen, both are explained in William Dakin's book *The art of camouflage*: disruptive patterns break up an animal's outline and make its form discontinuous and fragmented, while concealing colouration relies on optical illusions caused by light making the animal appear flat, rather than round, and therefore more easily overlooked.

As news spread during the early years of WWII that animals were once again to influence developments in weaponry the public was quick to question what use animal camouflage had to modern industrial war and how nature could possibly offer any guidance to progressive technological methods of dealing with political conflict. From Sydney to New York the press was fascinated by what seemed like the sudden but ironic importance of animals to defence. How incongruous, *The New York Times* argued, that despite modern progress and the evolution of civilisation, in times of war humans seek safety in nature, instinct, intuition and primal behaviours but at the same time claim that camouflage was invented by the military. In truth, it pointed out, camouflage is:

> older than the hills. The beetles and butterflies shaped like leaves, the spiders with the form and color of buds and lichens, the chameleon, the twiglike praying mantis, the tiger in his jungle of yellowish grass and black shadows—all started practicing camouflage long before army and navy men got around to it. Thus, with all his pryings into the mechanical, man still seeks his safety in simple, abiding nature.[8]

People from all walks of life were caught up in the current fever surrounding animal camouflage and human warfare, including Salvador Dalí, who was living in New York, and who, like many people, claimed he had the answer to modern camouflage design and the secrets of invisibility. In a short article published in *Esquire* magazine titled 'Total camouflage for total war' (1942), he wrote about the porous boundaries between art, war and nature in relation to camouflage, showing himself as something of an amateur naturalist when he described the wondrous mimicry of walking-leaf insects, and a surrealist when he claimed that 'just as the camouflage of 1914 was cubist and Picassan, so the camouflage of 1942 should be surrealist and Dalistic'.[9] He was struck to think how well animal mimicry compared with the surrealist strategy of visual doubling and deception.

By August 1941 the Australian press was fascinated to learn that the government had put a zoologist in charge of civil camouflage defence. *The Sun*, published in Sydney, reported that all over the country artists and scientists had turned to nature to solve problems of camouflaging Australia and that 'in these days of bomb and shell' it was kangaroos and emus—animals that had roamed the land for millions of years—and the multiplicity of insects in Australia (so used to fighting for their existence) that humans would now 'give fortunes to imitate'.[10] How strange, the article suggested, that lives could be saved and the

---

8 'Camouflage mysteries', *The New York Times Magazine*, 8 October 1939, p. 9, cited in Frank Hinder, 'Records of Lt F.C. Hinder Camoufleur', Canberra, AWM, PR 88/133, File 895/4/182, Item 4 of 12.

9 Dalí 1998 [1942], p. 340.

10 'Nature's aid in army camouflage', *The Sun*, 1 September 1941, p. 5.

war won by studying the habits of Australian fauna. In another example, a reporter wrote with some amusement how:

> Fish, birds and animals are to play their part in Australia's home defence programme. Their natural habits, by which they cleverly use camouflage as a protection against enemies, will provide the basis for the work of camouflage committees throughout Australia.[11]

But when the University of Sydney student newspaper *Honi Soit* tried to interview William Dakin on his new appointment to learn more about the relevance of animals to defence, its reporters were told the work was top secret and that loose talk about its details could jeopardise the DHS's operations. Still, Dakin took the opportunity to stress that it was not the military in WWI but 'a Professor of Zoology who first called the notice of the English government to the scientific aspects of camouflage'.[12] Who he had in mind is not made clear and the answer may even seem irrelevant, if it weren't for the competitive nature of camouflage's history and contentious claims about its origins. For there is no agreement that the source was a professor at all, or living in England as Dakin suggests. The contenders are British and American, and among British professors who can legitimately claim the distinction, two stand out: Edward B. Poulton (1856–1943), one time professor of zoology at Oxford University, and John Graham Kerr (1869–1957), a Scottish university professor.

Edward Poulton was the first scientist to argue for the significance of the principle of 'countershading' to animal invisibility, even before countershading's most famous US proponent, Abbott Thayer—a chronology that Thayer himself agreed with.[13] It is also agreed that Poulton's work on the science of natural camouflage impacted on the science of military camouflage in WWI. But John Kerr, who was a naturalist and teacher of Hugh Cott, claimed to have invented a disruptive camouflage design for ships in WWI, which he proposed to the British Admiralty.[14] In 1941 Kerr urged members of the House of Commons to recognise biologists as the most useful professionals for the design and research of camouflage for military purposes, and spoke against the deployment of artists and physicists.[15] In all probability Dakin was referring to John Kerr when he praised the Professor of Zoology who made biological camouflage widely known as a model for military defence in England. Dakin's friend, Professor A.D. Ross, a physicist who worked for the DHS, once wrote that Kerr was 'largely instrumental in directing attention to camouflage methods for warfare' which suggests this was the prevailing view among Australian scientists.[16]

11 'Animals will help us in camouflage: report from Adelaide' cited in Frank Hinder, 'Records of Lt F.C. Hinder, Camoufleur', Canberra, AWM, PR 88/133, File 895/4/182, Item 7 of 12. The source of the article is identified in a handwritten note as 'Sun' (*The Sun*), and the date of 2 August 1941 is noted. However, while the typeface of the article matches *The Sun*, the article was not published on this date.

12 'Varsity Works in War Effort: All Faculties Contribute', *Honi Soit*, Thursday May 15, Vol. XVIII, No. 9, 1941, p. 1.

13 Behrens 2009a, pp. 290–91.

14 Behrens 1999, p. 57.

15 Goodden 2007, p. 14.

16 A.D. Ross, 'Introductory', Camouflage—general camouflage school WA, Canberra, NAA, Series

But in the US the 'Father of Camouflage' is Abbott Thayer. Today scientists and camouflage historians, including Roy R. Behrens, widely agree that Thayer is the most important source of the principles of countershading and disruptive colouration and that his ideas had a decisive impact on camouflage for WWI and WWII. Countershading, or 'obliterative colouration', is also known as 'Thayer's Law' because in 1896 Thayer claimed it universally true that animals go unnoticed in their surroundings due to gradations of light and shade on their bodies leading to what he described as the effacement of themselves as solid objects. The spectator, he argued, will therefore 'see right through the space really occupied by an opaque animal'.[17] Thayer's work is set out in *Concealing: coloration in the animal kingdom* (1909), a book published by his son Gerald Thayer.[18] But his ideas were controversial, especially his claim that all animals are coloured for invisibility. Theodore Roosevelt engaged in a public debate in the US with Thayer and was incredulous that Thayer would claim that such conspicuous species as flamingos are coloured for concealment; he put it down to Thayer's subjective, artistic temperament rather than objective, scientific mind.[19] Many years later, Dakin also argued that 'the colour of an animal quite often seems to have no particular significance. It is an attribute of the possessor and that is all we can say about it.'[20]

Hugh Cott, who served in the British army in WWI, is credited with developing the principle of disruptive colouration in WWII, and making it what it is today—the most familiar of all military camouflage aesthetics, one that is highly visible in all parts of the globe through the vast array of patterns in contrasting tones, some striped some blotchy, on military uniforms.[21] In *Adaptive coloration in animals* Cott discussed coincident patterns on animal skins, including on the Australian mountain devil—*Moloch horridus* (the Australian reptile pictured in fig. 5.1)—referring to them as optical devices that make the animal hardly noticeable, especially when its eyes are closed and its mottled body is flattened against a mottle-coloured background. The optical effect of this type of camouflage, he argued, was suitable for large military objects and especially for ships, runways and aeroplane hangars, to fool the aerial gaze of the enemy.[22]

The basic premise of camouflage, though, came from Charles Darwin, Alfred Russel Wallace and Henry Walter Bates. The ideas they put forward in the 19th century, after observing nature in the wild, were not only scientifically groundbreaking but liberating for literary and visual artists, who became excited about the imaginative possibilities of invisibility and transformation. Little was understood about the colours and marking of animals before them. Darwin's revolutionary claim was that all life, plant and animal, is mimetic, not just human life as Aristotle had claimed.[23] In the first edition of *The origin*

A453, Item 1942/17/1957, p. 1.

17 Thayer 1896, p. 126.

18 Thayer 1909.

19 Thayer and Roosevelt engaged in a public argument. See Roosevelt 1910, pp. 501–21.

20 Dakin 1952, p. 74.

21 For a useful summary of camouflage's scientific history see Stevens & Merilaita 2009, p. 481.

22 Cott 1957, pp. 90–91.

23 Norris 1980, p. 1232.

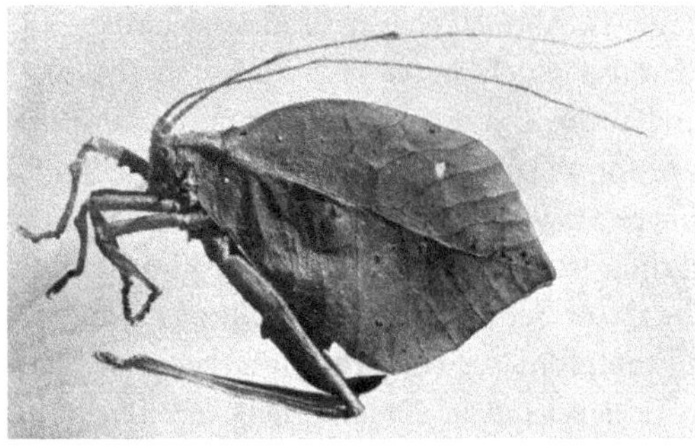

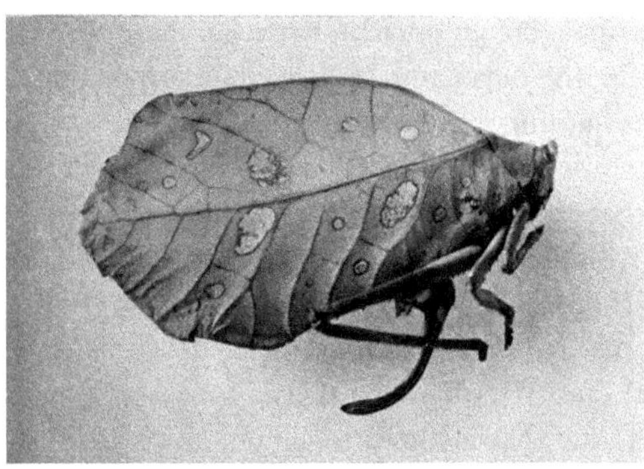

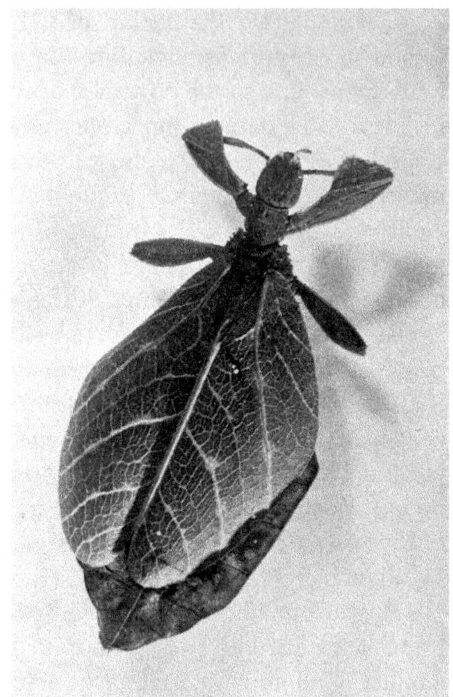

Fig. 5.3, Hugh B. Cott, Leaf insects: top left, *Cycloptera sp.*; below left, *Cycloptera excellens*; right, *Chitoniscus feedjeanus*.Cott 1957 [1940], plate 38.

of species (1859) he wrote about the 'tints' of animals, not as useless and trivial decorative details, but as essential devices that enable the animal to go about its life undetected, and provide it with a protective adaptation that is an outcome of natural selection.[24] Wallace's influence on Darwin is apparent in the fourth edition of *The origin of species*.[25] But in 1872, in the sixth edition, Darwin wrote his fullest exposition on the subject of concealment and deception, concentrating on insect mimicry and the resemblances of insects to a surprising array of forms:

> Insects often resemble for the sake of protection various objects, such as green or decayed leaves, dead twigs, bits of lichen, flowers, spines, excrement of birds, and living insects; but to this latter point I shall hereafter recur. The resemblance is often wonderfully close, and is not confined to colour, but extends to form, and even to the manner in which the insects hold themselves. The caterpillars which project motionless like dead twigs from the bushes on which they feed, offer an excellent instance of a resemblance of this kind.[26]

One of the most rewarding things about working in camouflage, said Alfred Wallace, was sensing that his discoveries about animals and 'their strange disguises as vegetable or

24 Darwin 1859, p. 84.

25 Darwin 1866, p. 506.

26 Darwin 1872, p. 181.

mineral substances' were destined to become an inexhaustible field for zoologists of the future.[27] If Wallace could have foreseen Hugh Cott's photographs of walking-leaf insects, he would have also seen his prediction come true for the field of aesthetics, since these photographs, detailing the false nibbled edges and trompe l'oeil water drops on the animal's wings, are among the most sensitive portraits of insects in natural history illustration (fig. 5.3). But he could not have foreseen that those characteristics of the animal world which intrigued both Darwin and himself, namely visual surprise, uncanny duplication, and metamorphosis, would also influence avant-garde artists, especially in the European surrealist movement, and would reverberate with generations of other artists around the globe including Frank Hinder, Adrian Feint and Max Dupain in Australia.

Mimicry is partly what attracted William Dakin to become an expert on the fauna of the Great Barrier Reef of Australia. He wrote of the paradox of the coral organism which was once mistakenly classified as stone, that branches and buds like a plant, and while properly classified as animal is immobile—a state, he conceded, that is 'very strange indeed to most people who think of animals as creatures which run, walk, fly, swim or crawl'.[28] Because he was trained as an intense observer of nature's mimicry Dakin felt suitably equipped to advise the Australian government on military camouflage in WWII and to claim as Kerr had in Britain, that naturalists and biological scientists had proven their value to military matters in WWI:

> The Biologist has played a very important part in Camouflage. In the first place, the Naturalist trained in outdoor observation has learned the tricks of concealment, the result of millions of years of animal evolution. The only works on Camouflage (although not then known by that name) before the last war were by Zoologists![29]

Yet while the modern history of the development of military camouflage is intertwined with biological science, it is also the case that scientists eager to contribute their advice to military organisations too often found their enthusiasm unreciprocated. This point has an important bearing on the history of William Dakin and the artists who worked in camouflage for the DHS. For example, when Abbott Thayer put his theories on concealment to the American and British militaries in WWI and argued he had a solution to protecting personnel and equipment, his ideas met profound scepticism, and while they eventually found acceptance, the protracted experience of arguing and demonstrating, persuading and politicking, left him in a state of nervous exhaustion from which he never recovered.[30] In 1925 a British zoologist, William Pycraft, wrote in Thayer's defence that he hoped, in the event of another war, that 'the Great War [had] taught the opponents of the coloration theory the need for reconsidering their position'.[31] But the professional militaries left camouflage to languish in interwar years and, with the onset of WWII, Hugh Cott described the state

27 Wallace 1988, p. 121.

28 Dakin 1955, p. 15.

29 Dakin 1942, p. 4.

30 Behrens 2009a, p. 346.

31 Pycraft 1925, p. 119.

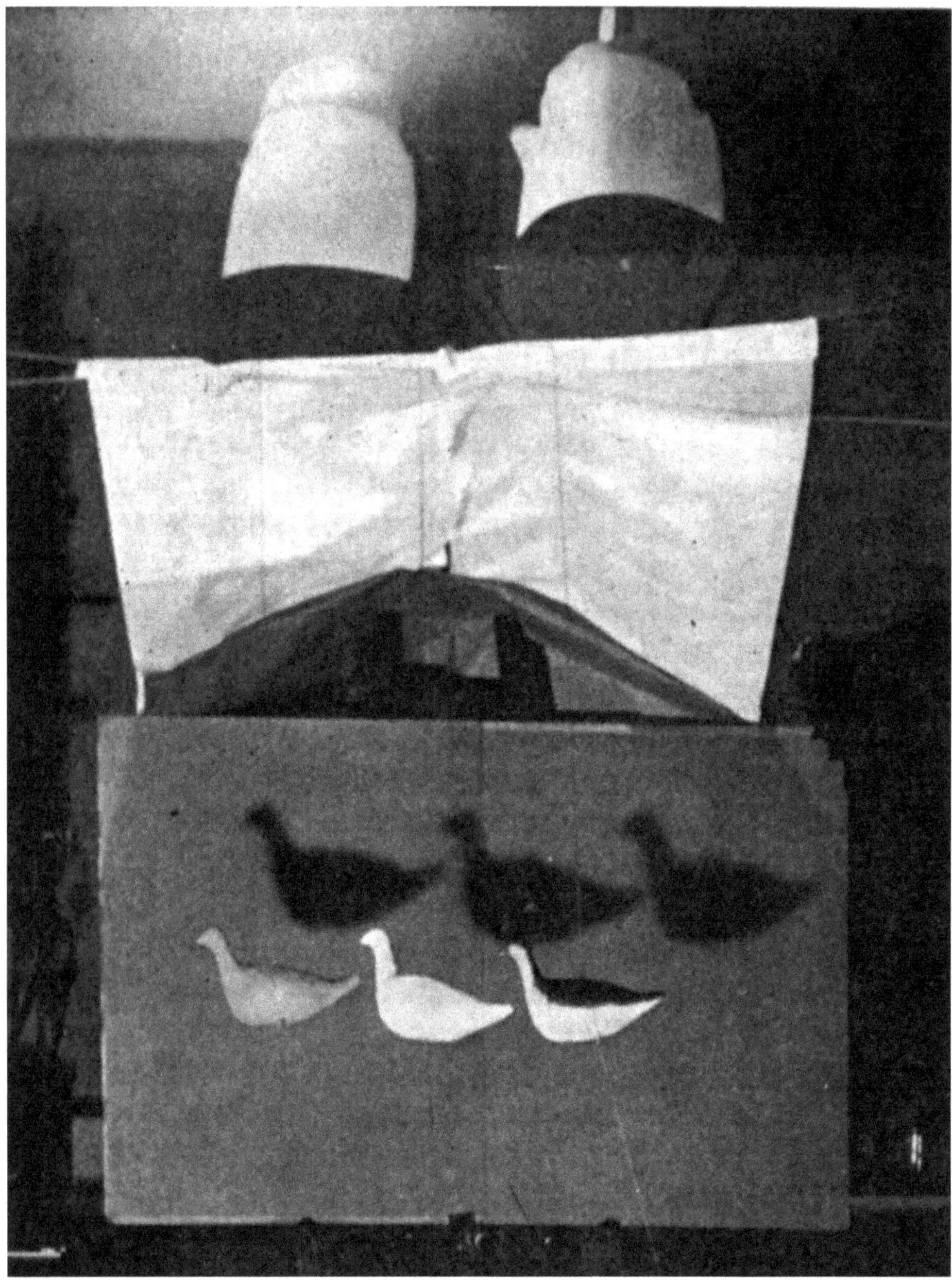

Fig. 5.4. Countershading demonstrated on three clay pigeons.Dakin 1942, p. 10. Photograph, W.J. Dakin.

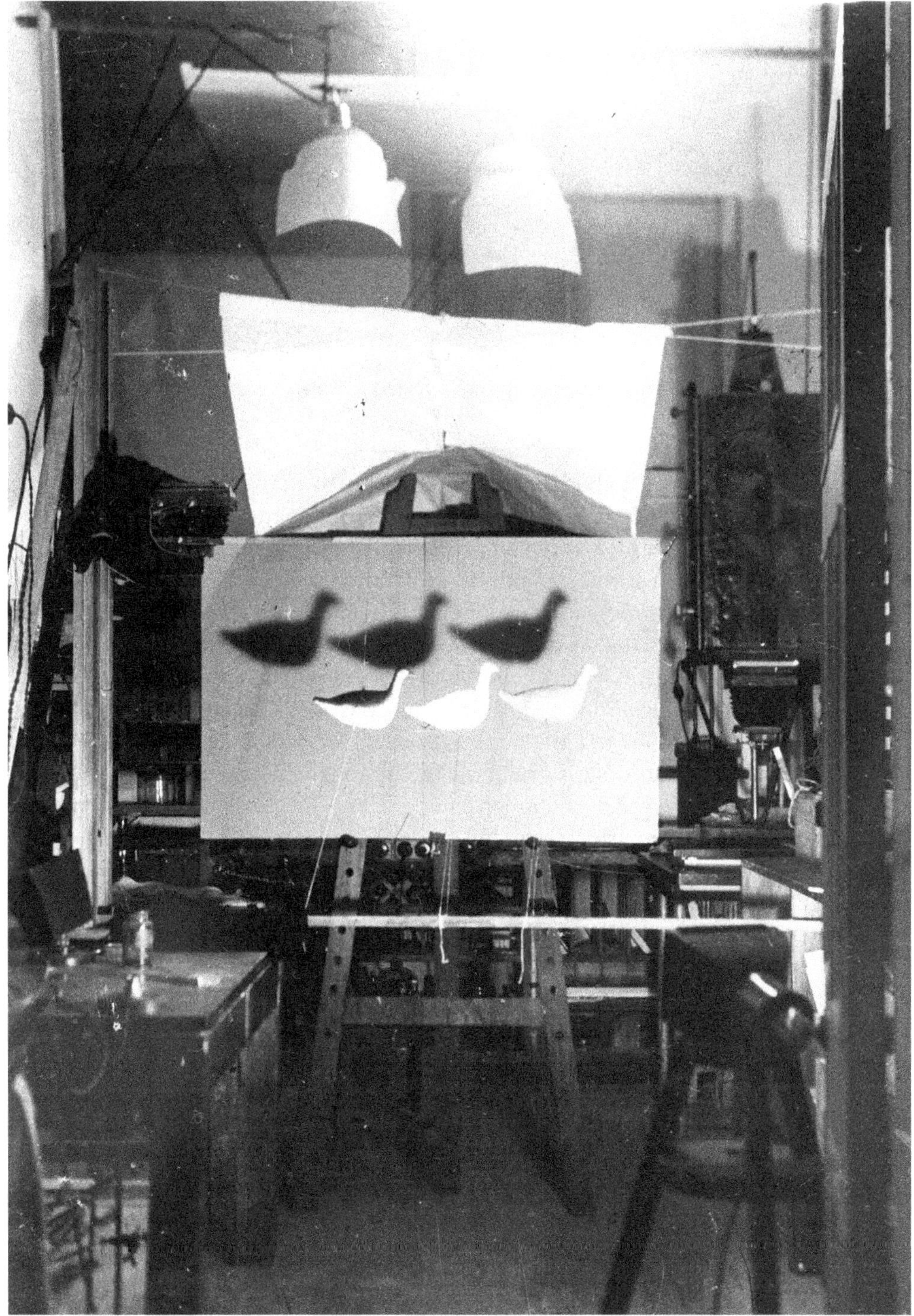

Fig. 5.5. Installation view of Dakin's countershading experiment at the University of Sydney, c. 1941. The collection of the NAA: C1907, 00977.

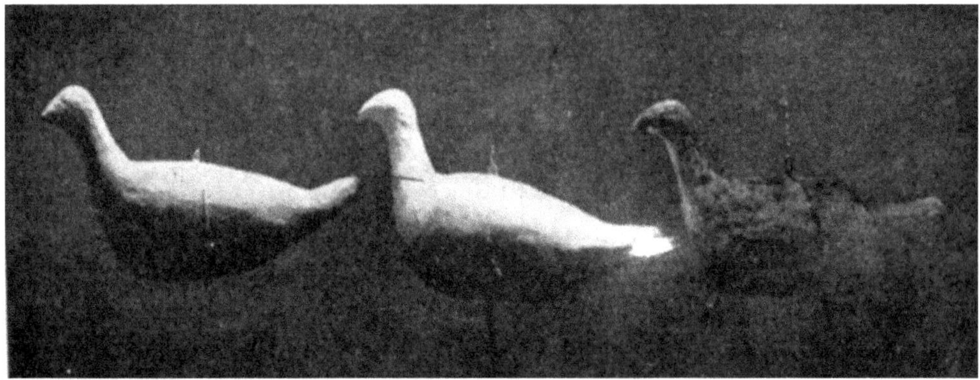

Fig. 5.6. Countershading demonstrated on three clay pigeons. Dakin 1942, p. 10. Photograph, W.J. Dakin.

of military camouflage as 'a child suffering from arrested development'.[32] Not only had military camouflage lain fallow since WWI, but as Dakin found out during the course of WWII, the same arguments for its military value had to be made all over again, and it was difficult to make the civilian voice, and the voice of the animal specialist, heard.

Dakin put as much information on animal camouflage, as was practically possible for a small book, into *The art of camouflage.* When first published in 1941 it served as a general guide to the useful application of camouflage's principles for military purpose. When it was published again in 1942, distribution was restricted to the Department of Defence and, throughout the war, this book remained a secret document. As stated earlier, the principle of countershading is explained through model birds, and the principle of disruptive colouration is explained through photographs of fish. Dakin set out an intriguing rationale for why birds and fish make useful models for military camouflage: it is, he said, because they 'suffer the same dangerous types of observation' from above and below as do aeroplanes and ships.[33] As he explained, if military planes and ships are coloured dark on their upper surfaces and lighter on their under surfaces, as birds and fish are, when observed from below against a light background they will become invisible, and when observed from above against a dark background they will also disappear.

When Dakin set up his experiments to test countershading on model birds in the Zoology Department at the University of Sydney, he carefully reconstructed the very experiments that Thayer had undertaken many years before (fig. 5.4–5.8). That Thayer's illustrations were well known in the zoology community and had made a considerable impact is evident in the fact that British zoologist William Pycraft reproduced them in their entirety in 1925 on the pages of his book *Camouflage in nature.* But while Pycraft's captions acknowledge Thayer as the source, Dakin's make no reference to Thayer, even though Dakin virtually copied the original experiments.

Yet Dakin was sure to know Thayer's work in detail. The American had installed demonstration models of countershading in museums in Oxford and Cambridge

32 Cott 1957, p. 438.

33 Dakin 1942, p. 61.

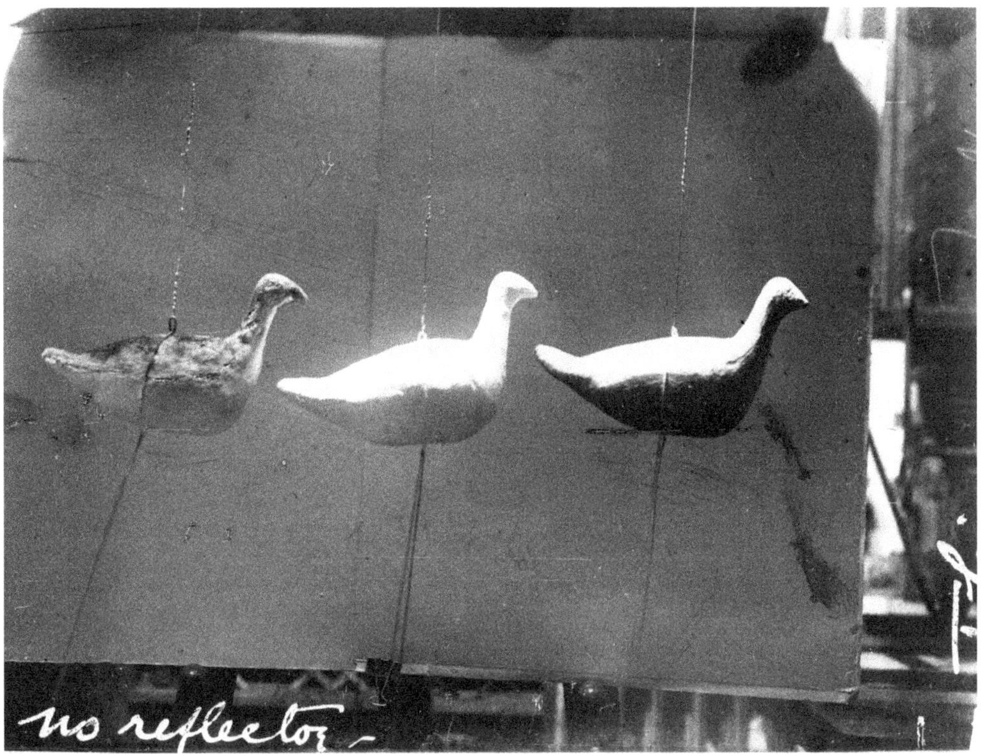

Fig. 5.7. Installation view of Dakin's countershading experiment at the University of Sydney, c. 1941. The collection of the NAA: C1907, 00978.

and in 1915 had visited not only Edinburgh but also Liverpool, Dakin's home, to give 'demonstrations of his theories to college professors and naturalists'.[34] With him he carried a suitcase containing a portable exhibition of concealing coloration applied to camouflage.[35] Although Dakin was living in Australia when Thayer visited Britain in 1915, he stayed in close professional contact with zoologists at the university and, when he returned there to take up his position as Professor of Zoology in 1920, Thayer's visit was still fresh in local memory. At the outbreak of WWII, Dakin also had the idea of assembling a portable demonstration kit—he called it a 'museum'—to raise awareness among civilians and troops of the significance of camouflage for protection and ultimately for a victorious outcome to the war.[36]

Not only did Dakin copy Thayer, he also copied Hugh Cott. There are uncanny visual similarities between Dakin's illustrations of black and white fish in *The art of camouflage* and Cott's illustrations of fish in *Adaptive coloration in animals*. Both sets of imagery illustrate the principle of disruptive colouration that Cott had developed and for which he

34 White 1951, p. 159.

35 Behrens 2002, pp. 55–60.

36 The 'museum' is mentioned in Frank Hinder, 'Records of Lt F.C. Hinder Camoufleur', Canberra, AWM, PR 88/133, File 895/4/182.

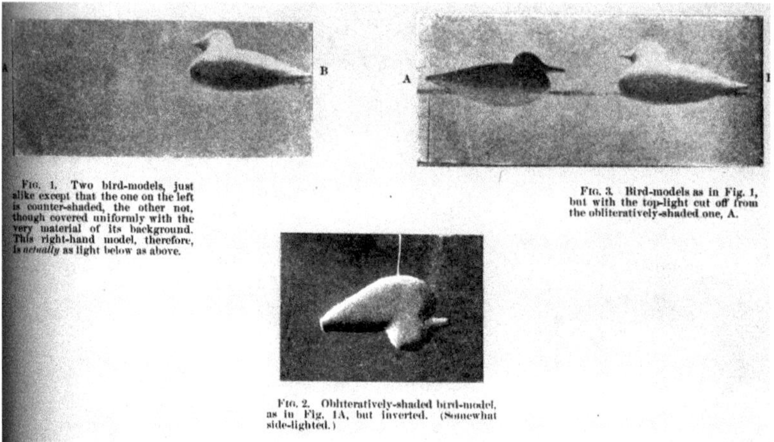

Fig. 5.8. Abbott H. Thayer, 'illustrations of obliterative coloration' (countershading). Thayer 1909, pp. 24–25.

is most often associated: the optical destruction and optical construction of shapes in the visual field (figs. 5.9 and 5.10).

As it transpires, Dakin owned a copy of Cott's book and what's more, he photographed plates from the book and made a collection of glass plate negatives for the DHS, to use as teaching aids (fig. 5.10).[37] Further, Dakin and Cott were in fact known to each other in the international zoology community, and Dakin's omission of a citation to Cott becomes all the more conspicuous by the presence of a citation to Dakin's research on evolutionary biology in Cott's bibliography.[38]

All of this demonstrates the point that camouflage as an applied science for warfare was a highly competitive field, reflecting the competitive nature of war. But one thing that stands out is that common to all was the importance of aesthetics and visual representation. Just as Charles Darwin's unique blend of logic and imagination is now recognised as having been influential on wider visual culture,[39] so books on camouflage by Abbott Thayer, Hugh Cott and William Dakin stand out as much for their pictorial qualities as for their science; as much for their patterns and designs, their textures and tones, and their sensibilities as for their information about animal behaviour. Abbott Thayer, after all, was primarily an artist of portraits and still lifes, not a naturalist, when he first published about camouflage in the scientific journal *Auk*. He said this equipped him with insight into the visual effects of shading and lighting on objects.[40] Cott was not only a zoologist but also an artist and photographer; today his book is admired as much for its illustrations as its text.

---

37 The negatives are labelled as the property of the office of the Technical Director of the Defence Central Camouflage Committee. See Glass plate negatives, Sydney, NAA, C 1907.

38 The citation reads: 'Dakin, W.J. 1921. Some Visual Organs and Their Bearing Upon Evolutionary Biology (An Inaugural Lecture). Liverpool, University Press, pp. 3–20', in Cott 1957, p. 444.

39 See Smith 2006.

40 Behrens 2009b, p. 501.

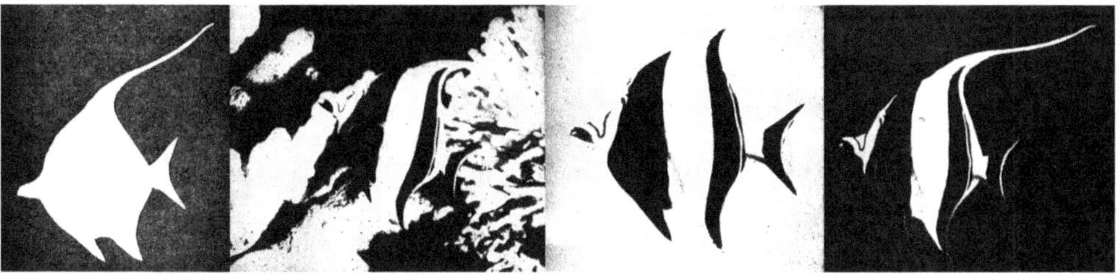

Fig. 5.9. Disruptive colouration demonstrated by Moorish idol, fish. Dakin 1941, pp. 13–16. Photographs by W.J. Dakin.

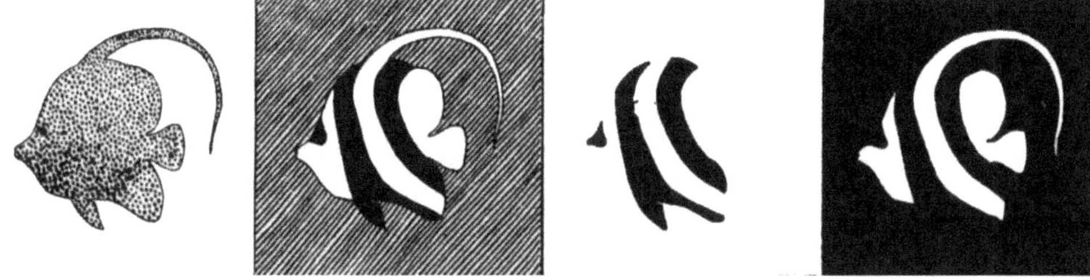

Fig. 5.10. Hugh B. Cott, 'Diagrams illustrating the principle of differential blending'. Cott 1957 [1940], p. 50.

Artists have been drawn to camouflage in nature because of the poetic suggestiveness of invisibility. Take, for example, an essay published in the US by John B.L. Goodwin in 1942 when the application of animal camouflage to warfare was making headlines, and where he proposed that 'if caterpillars can be unperceived against the grass, who knows what monsters may be indiscernible against the mimicked background of sky, or wall, or waterfall'.[41] It was the allure of the immaterial and the abstract that attracted Abbott Thayer, who Hillel Schwartz refers to as 'a highly visible crusader of the invisible'.[42] And the same seduction explains why Frank Hinder found natural history a satisfying complement to the theory of dynamic symmetry in modern abstract painting: because in both he sought invisible underlying structures and symmetries (fig. 5.11). Dakin, too, was enchanted by the invisible and wrote in one of his last books, *Great Barrier Reef, and some mention of other Australian coral reefs*, about those aspects of nature—he used the example of the clear water of a coral lagoon—that give 'the impression of not being there at all'.[43]

Without doubt, William Dakin believed camouflage studies had evolved from the biological sciences, or as he put it: 'there are fundamental principles in Camouflage which are definite and depend upon facts of Science'.[44] Yet Dakin too was talented across the arts as well as the sciences. When Alan Colefax published Dakin's obituary, he referred to

41 Goodwin 1941–1942, p. 9.

42 Schwartz 1996, p. 176.

43 Dakin 1955, p. 28.

44 Dakin 1942, p. 4.

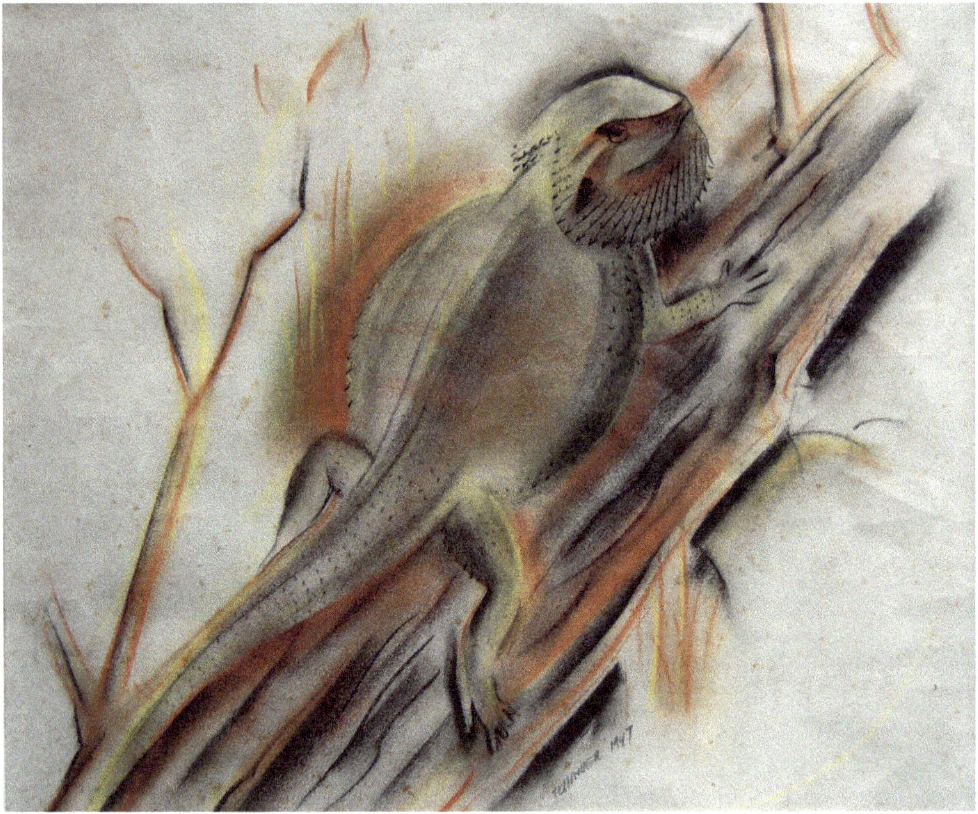

Fig. 5.11. Frank Hinder, *Bearded dragon*, 1947. Pastel on paper, 32 x 38.5 cm, private collection.

a man who was an expert in photography as well as 'a tireless campaigner for the wider teaching of Biology in schools, a fine pianist, and a landscape artist of considerable merit'.[45] Colefax celebrated Dakin's energetic life and described at length his position as Technical Director of Camouflage overseeing a committee of artists, engineers, and architects known as the Sydney Camouflage Group. But he also made reference to Dakin's difficulty at getting recognition for his wartime contribution. Even in death, Dakin's embattled relationships with the armed services remained an issue. But considering the sheer time and effort he expended on political wrangling over camouflage and who should control it, Dakin would surely have approved of Colefax's last word on the matter. When he was alive Dakin referred to his fractious relationship with the military as the 'clashing of authorities' and said it began early on, in 1940.[46] Although he claimed no responsibility for the animosity,

45 A.N. Colefax, 'Obituary; Professor William John Dakin, D.Sc., F.Z.S., *The Australian Journal of Science,* Vol. 12, No. 6, June 1950, p. 209. William Dakin is described as a keen photographer in Frank Hinder, 'Lt F.C. Hinder Personal Records', Canberra, AWM PR 88/133, Item 7 of 12.

46 W.J. Dakin, 'Difficulties in co-operation between the Services and the Department of Home Security' in W.J. Dakin, Camouflage report 1939–1945 (with appendices A–D), Canberra, AWM, Series 81 [77 Part 1], 1947, p. 140.

even painting a picture of himself as a victim, and although he tried to keep focused on the bigger picture—an Allied victory—his thoughts were constantly interrupted by a sense of injustice that scientists and civilians with a great deal to offer were treated poorly in times of war. But WWI history had forewarned him, and so in the tone of his wartime correspondence there is always the sense that he saw himself as a crusader.

PART 3

THE MILITARY CONTEXT

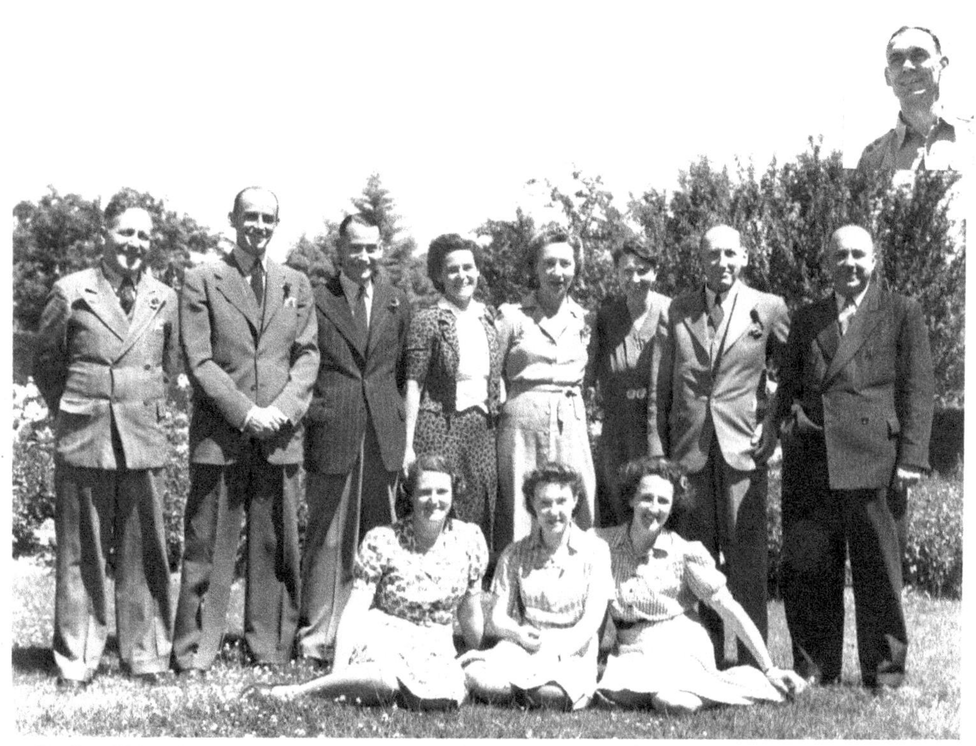

Fig. 6.1. William Dakin (2nd from right), Frank Hinder (superimposed top right), Margel Hinder (4th from left) and staff of the DHS Staff in Canberra. AWM, Frank Hinder Personal Records PR 88/133, 9 of 12.

Chapter 6

Policy and status

In 1939 William Dakin was buoyed by his successes in mobilising the Sydney Camouflage Group and in bringing together an unlikely combination of representatives from Sydney's modern art and design fraternity—among them Max Dupain, Douglas Annand, Frank Hinder, Sydney Ure Smith and Adrian Feint—with military officers including Group Captain De. La Rue of the RAAF, Captain B.S. Hussey and Lieutenant Colonel D.A. Whitehead of the army, and Lieutenant D. Barker of the Department of Defence. But in 1940 he assessed national preparations for camouflage defence and, feeling they were in disarray, put his response on record: 'this is not the way to run a war'.[1] Habitually blind to the impact of his often tactless communications on those around him, in 1941, in his new position as Technical Director of Camouflage, he prepared a report on the progress of military preparations that alleged that 'up to the beginning of the present year camouflage had played little or no part officially in relation to the Army in Australia'.[2]

The question was not whether he was right or wrong but whether a different approach was called for. The Australian army was a powerful organisation and its WWI history as the Australian and New Zealand Army Corps (ANZAC) at Gallipoli in 1915 is a defining part of Australian identity. Dakin often spoke spontaneously when he should have been circumspect, and he lacked sufficient diplomacy to mediate calmly between the DHS, the prime minister (who was also minister of defence), the armed forces and his own team of workers. Camouflage was a political football and the right to control it created rifts and arguments; diplomacy and negotiation were important to make the job work.

Nonetheless, Dakin's open and public disapproval of the poor organisation of camouflage by the army in particular, but also the airforce and navy, did not abate for the duration of the war, and even continued beyond 1945. In explaining the disputes more fully, what also needs to be addressed is both the sequence of events that put a civilian scientist in charge of camouflage defence, and the army's version of the history of camouflage's organisation in WWII, which differed to Dakin's. But how matters developed into a political wrangle can be put in perspective very quickly by referring first of all to the final report that Dakin submitted in 1947, signing off on his role as Technical Director.

When Dakin compiled the final report, outlining the history of camouflage in Australia in WWII, he hoped it would not only serve as an accurate record but also become the

1 William Dakin, 'Notes of Conference held at Premier's Department', 9 July 1940, in Notes of Camouflage, Sydney, NAA, Series SP 1048/7 Item s10/1/329, p. 8.

2 William Dakin, 24 October 1941, Camouflage and the Army, Sydney, NAA, Series C1707, File 62.

definitive reference on how to organise camouflage in the event of another world war. One chapter, titled 'Difficulties in cooperation between the Services and the Department of Home Security', warned that the self-ruling nature of the military was the chief problem encountered by the DHS, as was the military's unwillingness to break with convention. But it was also, he argued, 'that curious atmosphere which surrounds a man in uniform and which has been built up by long tradition during a time when total war was never dreamt of', an air of superiority over civilians.[3] The report stressed the army's lack of cooperation, failures in policy, absence of military attention to camouflage techniques between the two world wars, and lack of interest in camouflage by senior officers whom he believed demonstrated 'serious disbelief, or dangerous toleration'.[4] But most of all Dakin regretted how, for reasons of status, the armed forces ignored the advice of civilians, even though it was plain that 'the awakening to the importance of modern camouflage technique in Australia came entirely from the civil side'.[5]

There is no question that the five years Dakin spent as Technical Director wore him out. Even in 1947, with the war over, feelings about the injustice of the circumstances continued to plague him. Perhaps it led to his protracted illness (cancer) and relatively premature death three years later. One thing is for certain, once Dakin and his team of camouflage officers were put to work in daily military life, once they set aside their professions, studio practices and commercial businesses to be part of the war effort, their lives were ruled as much by the clash between army and civilian authorities as by responsibilities with camouflage. True, the circumstances of their lives throughout the war could have been much worse; at least they were spared combat. Instead they fought a war inside a war, and with little or no training in military organisation and administration, or in military etiquette, they were also social outsiders. As Dakin's report makes clear, officers in his department were looked upon as inferiors.[6]

To answer the question of how this team of scientists and artists got itself in the position of working so closely with military organisations and of channeling their ideas into the protection of civilians, but also of fighting forces, an expansion of chapter 2 is needed. For there is much more to say about the confusing sequence of events that led to the government establishing the DCCC and to Dakin's appointment as Technical Director of Camouflage. The place to begin is with the first public announcement, in Sydney, of the government's plan to develop an urgent camouflage strategy for all states and territories, which was published in July 1941.[7]

It was in 1940 that the work of the unofficial Sydney Camouflage Group was drawn to the attention of Prime Minister Robert Menzies, who invited Dakin and representatives

---

3 W.J. Dakin, Camouflage report 1939–1945 (with appendices A–D), Canberra, AWM, Series 81 [77 Part 1], 1947, p. 147.

4 W.J. Dakin, 'Camouflage in Australia', 31 March 1942, in Accreditation and attachment of camouflage officers', Canberra, NAA, Series A5799, Item 27/1943.

5 W.J. Dakin, Camouflage report 1939–1945 (with appendices A–D), Canberra, AWM, Series 81 [77 Part 1], 1947, p. 1.

6 W.J. Dakin, Camouflage report 1939–1945 (with appendices A–D), Canberra, AWM, Series 81 [77 Part 1], 1947, p. 40.

7 'Camouflage plan in all states', *SMH*, 16 July 1941, p. 8.

of other civilian-based camouflage committees in the states of NSW and Victoria to a meeting. The delegation included Victor Tadgell from NSW and Daryl Lindsay from Victoria. The following were outcomes: the appointment of Tadgell, an original member of the Sydney Camouflage Group, to fulltime camouflage duty with the army, attached to the RA; the establishment of the federal DCCC in July 1941 within the Department of Defence; the establishment, in August 1941, of State Defence Camouflage Committees; the establishment of a Camouflage Research Station at Georges Heights, Sydney; and the appointment of Dakin to the Central Committee in Canberra to liaise with the armed services and other committees, to execute policy, and advise the army, airforce and navy on all camouflage matters. The DCCC was a joint committee of the Commonwealth government that included army, airforce and navy personnel as well as civilian representatives. Later, following the re-organisation of camouflage administration in Britain, the committee was removed from the Department of Defence and placed under the DHS.[8]

When the government appointed Dakin as Technical Director it described him as 'undoubtedly the outstanding authority on Camouflage in Australia at the present time'.[9] Each state and territory was assigned a deputy director of camouflage to work under Dakin. Two army representatives were appointed chairs of camouflage committees in Darwin and Port Moresby, and each deputy director (selected from both public office and academic life) was a specialist in either an art-related or science-related field. The list included Daryl Lindsay (Keeper of Prints, National Gallery in Melbourne) in Victoria, Victor E. Greenhalgh (artist) in Victoria, R.A. McInnis (city planner) in Queensland, Louis McCubbin (Director National Gallery in Adelaide) in South Australia, Alexander D. Ross (Professor of Physics) in Western Australia, Dr Joseph Pearson (Director of the Tasmanian Museum and Art Gallery) in Tasmania, and John D. Moore (architect) in NSW.[10]

From 1941 to early 1942 when the Department of Defence established the DCCC with subcommittees in each state, morale was relatively high and it was an exciting time for innovations in camouflage training, research and fieldwork. Everyone followed policy. But in February 1942 the army insisted on separating from the DHS, preferring instead to control its own camouflage research and operations. In the separation the DHS lost its share of the Camouflage Research Station at Georges Heights, a facility that had been installed with the most modern equipment to support the science and art of camouflage. Following its severance from the army, in April 1942, the Department of Home Security moved headquarters to the Canberra Hotel in the federal capital.[11] In May 1942 Dakin and his main officers, including Frank and Margel Hinder, Douglas Annand, Gervaise Purcell, Eric Thompson, Frank H. Moloney, and Dakin's secretary and scientist Isobel Bennett, relocated

8 Information in this paragraph is taken from W.J. Dakin, Camouflage report 1939–1945 (with appendices A–D), Canberra, AWM, Series 81 [77 Part 1], 1947, pp. 1–55, and also from Assistant Secretary (Civil Defence), 'Camouflage', 3 July 1941, Establishment of camouflage organisation and procedures, Canberra, NAA, Series A5954, Item 396/2.

9 Assistant Secretary (Civil Defence), 'Camouflage', 3 July 1941, Establishment of camouflage organisation and procedures, Canberra, NAA, Series A5954, Item 396/2, p. 10.

10 Mellor 1958, p. 535, footnote no. 8.

11 W.J. Dakin, Camouflage report 1939–1945 (with appendices A–D), Canberra, AWM, Series 81 [77 Part 1], 1947, p. 55.

to the capital.[12] But they were a mobile bunch and members were sometimes stationed in Canberra and at other times in towns and villages throughout Australia and its territories. A group photograph shows a well-dressed team of ordinary civilians, an unlikely lot to work in secret on methods and techniques to achieve the concealment of fighting troops in Darwin, Townsville, Papua and New Guinea (fig. 6.1).

By March 1943, 80 camouflage artists with the DHS were in the field, mostly attached to the RAAF. But despite the extensive assistance given to the airforce by the camouflage section, historian D.P. Mellor noted that 'officers on R.A.A.F. aerodromes were not always willing to cooperate with civilians, who in any event were only there as advisers'.[13] In November 1943 the RAAF ceased camouflage treatment of its establishments south of the Tropic of Capricorn (with the exception of radar stations) and no longer required assistance from camouflage artists with the DHS. Dakin diverted those who were now redundant on airfields to research.[14] In 1944 the navy followed the example of the army and decided that 'the services of the Camouflage Section, Department of Home Security, were no longer necessary for research and advice'.[15] As a result, camouflage operations were wound back, and by December 1944, of the 15 camouflage officers still with The Department of Home Security, six worked in the Research Section at the University of Sydney, two were stationed at an Instructional Pool at Townsville, four were in Darwin and three in New Guinea.[16] The work of the DCCC was considered complete in early 1945, before war ended.

The biggest insult, Dakin later reflected, was how the army had acted 'as though a civilian control had been forced on the Services' when, from the earliest days of the Sydney Camouflage Group, he had always encouraged cooperation.[17] Army officers were instrumental in drawing up the original policy on camouflage, at which time all agreed that the DHS would occupy a consulting role for the armed services and share the facilities at Georges Heights. In 1947, Dakin's final report argued:

> this keenness of civilians should have created no antagonism. In all countries national defence was developing civil awareness of the dangers of modern war and organisations were being created which would have to undertake some duties identical with those of the Services. It certainly became clear early in 1940 that this was a Total War.[18]

---

12 For a brief history of this time as it affected Isobel Bennett see Allen 1992, pp. 551–62.

13 Mellor 1958, p. 542.

14 A.W. Welch to Department of Air, 8 February 1944, Camouflage personnel NSW Camoufleurs staff appointments, Canberra, NAA, Series A453, Item 1942/28/2298.

15 H.P. Lazzarini, 'War Cabinet Agenda', 14 August 1944, Establishment of Camouflage organisation, Canberra, NAA, Series A1308, Item 710/1/27, p. 2.

16 Departmental history; Camouflage organisation—administration, Canberra, NAA, Series CP 951/1/1, Item vol. 1, p. 5.

17 W.J. Dakin, Camouflage report 1939–1945 (with appendices A–D), Canberra, AWM, Series 81 [77 Part 1], 1947, p. 2

18 W.J. Dakin, Camouflage report 1939–1945 (with appendices A–D), Canberra, AWM, Series 81 [77 Part 1], 1947, pp. 1–2.

He was bitterly disappointed in the way the military used rank, discipline, status, image and history to diminish his own morale and the enthusiasm of his team of experts. Hadn't he been willing, for the sake of Australia, to be taken away from his own scientific research and hadn't he too been 'prepared to put all he knew and all his future activities at the disposal of the Services if it were so desired'?[19] It is worth backtracking briefly and drawing attention to a letter to the chief of general staff dated June 1942. It expresses Dakin's mounting frustration about the army's proposed changes to the newly formed camouflage policy:

> I have been surprised to see a General Routine Order issued by General Sir Thomas Blamey, dated June 5th, which, in regard to Camouflage, seems to indicate that in future all Army Camouflage is to be a matter for Army alone. This seems to cut right across the Camouflage Regulations and Organisation set up by the Commonwealth Government ... I shall, of course have to bring up the whole matter to the Prime Minister and Cabinet, and should be glad of some more information before doing so.[20]

Three days after this letter was written, Dakin's supervisor, Minister Hubert Lazzarini, wrote his own letter to Prime Minister John Curtin explaining the position that had developed between the army and the DHS, and supporting Dakin's argument that camouflage was 'now a matter for science and scientific experts together with architects and artists'.[21] The prime minister responded on 8 September 1942 by forwarding his separate communications with the Minister for the Army.[22] It was a request for amendments to the National Security (Camouflage) Regulations, giving more authority to officers of the defence forces.[23] This was the point at which the greater part of the DHS activities became connected with the RAAF, the army having gone its separate way. A deflated Dakin wrote his own letter to the prime minister, regretting the change in policy, which he took as an unfair punishment for a situation where, as he saw it, 'our one sin is that we are a civilian body'.[24]

But the more he thought about it, the more the term 'civilian' seemed misleading and perhaps the root of the problem. Several members of his national team including Louis McCubbin from Adelaide, and original members of the Sydney Camouflage Group, including Adrian Feint, had had previous army and camouflage experience in WWI, yet were technically 'civilians' in WWII. He decided that the term 'civilian' was often meaningless,

---

19 W.J. Dakin, Camouflage report 1939–1945 (with appendices A–D), Canberra, AWM, Series 81 [77 Part 1], 1947, p. 11.

20 W.J. Dakin to General V.A.H. Sturdee, 13 June 1942, Camouflage—question of responsibility: army camouflage, Canberra, NAA, Series A453, Item 1942/21/2632.

21 H.P. Lazzarini to John Curtin, 'Australian Camouflage', 16 June 1942, in Camouflage—question of responsibility: army camouflage Canberra, NAA, Series A453, Item 1942/21/2632.

22 John Curtin to H.P Lazzarini, 8 September 1942, in Camouflage—question of responsibility: army camouflage, Canberra, NAA, Series A453, Item 1942/21/2632.

23 Minister for the Army F.M. Forde to Prime Minister John Curtin, undated c. 1 September 1942, Camouflage—question of responsibility: army camouflage, Canberra, NAA, Series A453, Item 1942/21/2632.

24 W.J. Dakin to John Curtin, 30 September 1942, in Camouflage association with army & War Cabinet addendum, Sydney, NAA, Series C1707 Item 19, p. 2.

that it was too often used as a term of insult, and that in reality it was only 'of consequence in that it indicates status (or lack of it)'.[25] Dakin disagreed with the argument that military culture requires autonomy and its own set of ethics, attitudes and expectations; rather, he thought that, in the catastrophe of total war, it should allow itself to integrate members of the wider society.

Yet for most people today looking back at history, the army's position would seem more than reasonable. Simply put, that position was that the camouflaging of military installations was a fundamental army responsibility.[26] Likewise the training of its own members was considered the responsibility of the army's School of Military Engineering, and the camouflage training units run by the RAE. But throughout the war the military had tensions of its own and Dakin's campaign to accelerate the use of camouflage weaponry was possibly low on the list of priorities.[27] The navy, airforce and army were poorly equipped with modern ships and planes, and personnel were under-trained.[28] In the army, a two-part structure of professionals as well as militia looked upon each other as rivals, and other stresses were caused by the transition of recruited and conscripted members from the civilian community.[29] When camoufleurs from the DHS were added to this mix—a group who were neither fighting forces nor properly support forces for fighting forces—it is little wonder that members of the regular forces were confused. As non-combatants working in a field under the direction of a zoologist, camoufleurs represented a completely unfamiliar sphere of military activity in army, navy and airforce contexts. Nevertheless, military authorities denied that they were uncooperative or that they had ignored the matter of camouflage in preparations for war.

According to the Department of Defence Co-ordination, the early history of the organisation of camouflage in Australia began in 1938, before the Sydney Camouflage Group came into existence. In the Defence Department's version of history, the Defence Committee requested a report from the Service Boards and Munitions Supply Board in 1938 on secret measures to camouflage industrial plants, and at this time the Department of the Army sought advice from the British army regarding its development of camouflage. The Defence Committee in Australia again raised the subject in January 1939 in relation to protection of naval and military establishments. And the army is said to have put a policy in place on camouflage in 1939 that covered field units, military installations, coastal defences, and large civilian installations. The Defence Committee claimed that, once Dakin's chairing

25 W.J. Dakin, Camouflage report 1939–1945 (with appendices A–D), Canberra, AWM, Series 81 [77 Part 1], 1947, p. 9.

26 H.P. Lazzarini, 'War Cabinet Agendum: Camouflage', 14 October 1942, in Establishment of camouflage organisation and procedures, Canberra, NAA, Series A5954, Item 396/2, p. 3.

27 Tensions within the military are discussed in Grey 2008, pp. 142–47.

28 Penglase & Horner 1992, p. 1.

29 At the outbreak of WWII the Australian Military Forces had a small regular force known as the PMF (Permanent Military Forces), a large militia of part-time citizen soldiers for the defence of Australian territories known as the CMF (Citizen Military Forces), and a recruited volunteer overseas fighting force known as the 2nd AIF (Australian Imperial Force). There were artists who worked for camouflage in the CMF, amongst them James Cant, who are not included in this study. See Radford 2006, p. 188.

of the Sydney Camouflage Group came to attention in 1939, the Department of the Army offered to assist Dakin, not the other way around.[30]

When the Minister for the Army, G.A. Street, learnt of the group, he sent Dakin a letter inviting him to consult with camouflage experts at Victoria Barracks, but cautioned that 'the art of camouflage is a part of military science which has been developed only after considerable research and experiment'.[31] Nevertheless, a recommendation was made in 1940 to utilise civilian expertise, including that of artist Frank Hinder—then deployed by the Department of the Interior in Sydney—to expedite preparations for camouflaging places of national and military importance, such as petrol tanks, gas works, water supply buildings, and any highly visible structure in close proximity to defence works. But from the military's point of view the army, airforce and navy were responsible for the development of a camouflage committee to estimate costs and make decisions on civic priorities.

Yet while the services were confident that they were in control of camouflage organisation from an early date, D.P. Mellor's assessment of its development for civic defence differs. His viewpoint supports what thousands of pages of faded war reports written by Dakin claimed: the military organisation of camouflage in Australia was ad hoc and chaotic.[32] Indeed it took a review of camouflage organisation in the region by US General Douglas McArthur, Supreme Commander of Allied Forces in the Pacific (advising Prime Minister John Curtin in 1942) before the problem caused by both civilian and military involvement in camouflage was fully addressed. Whereas McArthur once thought it would be advantageous to combine the camouflage activities of the Australian government with the US and Allied airforce 'into one activity', reports from the Air Corps that 'the camouflage activitites of the Australian government are rather complicated' convinced him otherwise.[33] But he issued a camouflage Directive to the Allied air, naval and land force in the SW Pacific spelling out the responsibilities of commanding officers. They were instructed to 'regard camouflage as an instrument of war' and he reminded them of the role of the DHS's DCCC 'from whom advice and assistance in reference to the technical design, accomplishment and supervision of camouflage works may be received on request'.[34] Dakin was pleased that his organisation was properly recognised.

But if camouflage organisation represented a difficult nuisance for the armed forces, it is also true that there were times when it represented a crisis of identity for camoufleurs in the DHS. Members of combat or combat support corps, aircrew or ground staff, and

30 All information and terminology in this paragraph from Assistant Secretary (Civil Defence), 'Camouflage', 3 July 1941, Establishment of camouflage organisation and procedures, Canberra, NAA, Series A5954, Item 396/2.

31 B.A. Street, Minister for the Army to W.J. Dakin, 22 November 1939, Camouflage organisation, Sydney, NAA, Series SP 1008/1, Item 469/2/28.

32 Mellor 1958, pp. 534–36.

33 Lt Col. Ray T. Elsmore (Allied Air Forces, Southwest Pacific Area), 'Status of Camouflage Activities', 24 June 1942, Allied Air Forces South West Pacific Area, Canberra, National Archives of Australia, Series A705/2, Item 62/4/88, p. 3.

34 General MacArthur, 'Camouflage Directive', 23 July 1942, in W.J. Dakin, Camouflage report 1939–1945 (with appendices K–N), Canberra, AWM, Series 81 [77 Part 3], 1947, appendix L.

other branches of military organisations—regardless of whether they were volunteer, conscript or regular, professional or amateur, experienced or inexperienced—at least had a common bond through the purpose of combat. But camouflage officers with the DHS were not part of this community. They were simply an adjunct group of civilians serving the camouflage needs of the armed forces, if and when required. The situation became critical in 1943 when camoufleurs with the DHS were first stationed in the fighting area of New Guinea. Their reports gave Dakin 'fullest evidence that our men are stigmatised as civilians notwithstanding their excellent work in fighting areas'.[35]

For a considerable time, officers with the DHS working on the Australian mainland, but also Papua and New Guinea, had no accreditation and no uniform, experienced unsafe and inequitable conditions, and lacked clarification on repatriation rights and on compensation for dependants in the all too likely event of flying accidents while on survey and reconnaissance missions.[36] Death by plane accident was the fate that met camoufleur Victor Bragg, killed in June 1942 while on service for the DHS in NSW.[37] And Frank Hinder was lucky to escape incineration in a bomber crash at Rabaul in 1941.[38] From a safety point of view Dakin was right to argue that the situation for his officers was untenable, but also that the status of his men was at stake:

> At the present moment, to take one example, our Camouflage Officers in the Darwin area are under Air or Army discipline, they are conducting essential work for the R.A.A.F, at the request of the R.A.A.F, yet these men (already subjected to enemy bombing) have no status for compensation. They have most serious difficulties in carrying on their work because no one outside the R.A.A.F officers immediately concerned can decide their rights, privileges, or standing. They may even be arrested as having no right to be in the area. They have no legal uniform. Local officers will probably state that they cannot be permitted to go to other forward areas, such as the islands—yet war correspondents and war artists etc. are able to go. Unless able to go forward they will be of little use where they are, and no other men trained so highly in aerodrome camouflage are available in Australia.[39]

By March 1943 after submissions to chiefs of naval, general and air staff, and the prime

---

35 W.J. Dakin, 'Relations of camouflage section to the services', 3 November 1943, Establishment of camouflage organisation and procedures, Canberra, NAA, Series A5954, Item 396/2.

36 In August 1941 the War Cabinet considered a submission to review compensation in the event of death caused by air accident during camouflage assignments. See Department of Defence compensation for Dependants of Civilians, Canberra, National Archives of Australia, Series A 663, Item 052/1/48.

37 See entry for V. Bragg, 'Staff Record: Camouflage Section', Canberra, Department of Home Security, NAA, Series A691, Item 1, accumulation dates 1 January 1942–31 December 1945.

38 W.J. Dakin, 'New Guinea and Island Area', Professor Dakin's camouflage report, Sydney, NAA, C 1908, Item 5.

39 'Minute by Defence Committee at Meeting held on Friday 12 March, 1943', Accreditation and attachment of Camouflage Officers of Department of Home Security, Canberra, NAA, Series A5954, Item 396/9, p. 1.

minister, it was agreed that camoufleurs be given accreditation, and be assigned as officers attached to the RAAF and to field units, but remain personnel of the DHS.[40]

Perhaps Dakin's most contentious claim was that the fighting services, especially the army, had given camouflage no attention whatsoever between WWI and WWII. Yet in the course of his duty he encountered plenty of evidence to support the claim. One colleague at the University of Sydney, who came to Dakin's defence, was Professor Eric Ashby, a botanist who was also busy with war-related research but found time to complain to the press that 'camouflage is still regarded in some quarters as a hobby rather than as an instrument of war'.[41] The main conceptual obstacles to widespread application of camouflage principles, Dakin argued, were the misconceptions that it was an optional addition to everyday military activity, and a sign of weakness. He found himself constantly making the argument that camouflage was essential to war and 'not a decoration to be stuck on something afterwards'.[42] To be effective it had to be incorporated into design. And by no accounts, he argued to government ministers, should it be looked upon as 'a sort of joke'.[43] He was not alone in criticising the ignorant who looked upon camouflage as unnecessary and comical. In Britain, artist Roland Penrose, who was seconded to teach the Home Guard the main principles of concealment and deception, also emphasised that 'camouflage is no mystery and no joke. It is a matter of life and death—of victory or defeat.'[44]

Dakin complained about how the fighting forces had failed to comprehend that camouflage is not just for protection to help the weak; when used well it conceals intention and strength until such time as both are brought into action.[45] Freezing still, becoming invisible, losing distinction, faking and deceiving, were, he argued, signs of martial power not martial timidity. But, from WWI throughout WWII, the easy acceptance of camouflage in military culture was impeded by its association with makeup and deception. The persistent inference that, by aiming for invisibility camouflage was effeminate, and further, that it belonged to the less serious realms of hobbies, decorations and jokes, was exacerbated in the highly masculinised context of military culture. In that context, the idea of distinction made more sense socially, intellectually and visually, than did invisibility. These attitudes form the background to the camoufleurs' image as outsiders and to their battle to secure both accreditation and an identifying uniform. It is a subject that deserves its own chapter.

---

40 Arthur S. Drakeford (Minister of Air) to John Curtin (Minister of Defence), 31 March 1943, Accreditation and attachment of Camouflage Officers of Department of Home Security, Canberra, NAA, Series A5954, Item 396/9.

41 Ashby 1941, p. 4.

42 W.J. Dakin, 27 October 42, Preplanning camouflage, Sydney, NAA, Series 1707, Item 57, p. 1.

43 William Dakin, 'Notes of Conference held at Premier's Department', 9 July 1940, in Notes of Camouflage, Sydney, NAA, Series SP 1048/7 Item s10/1/329, p. 5.

44 Penrose 1941, p. 102.

45 W.J. Dakin, Camouflage report 1939–1945 (with appendices A–D), Canberra, AWM, Series 81 [77 Part 1], 1947, p. 6.

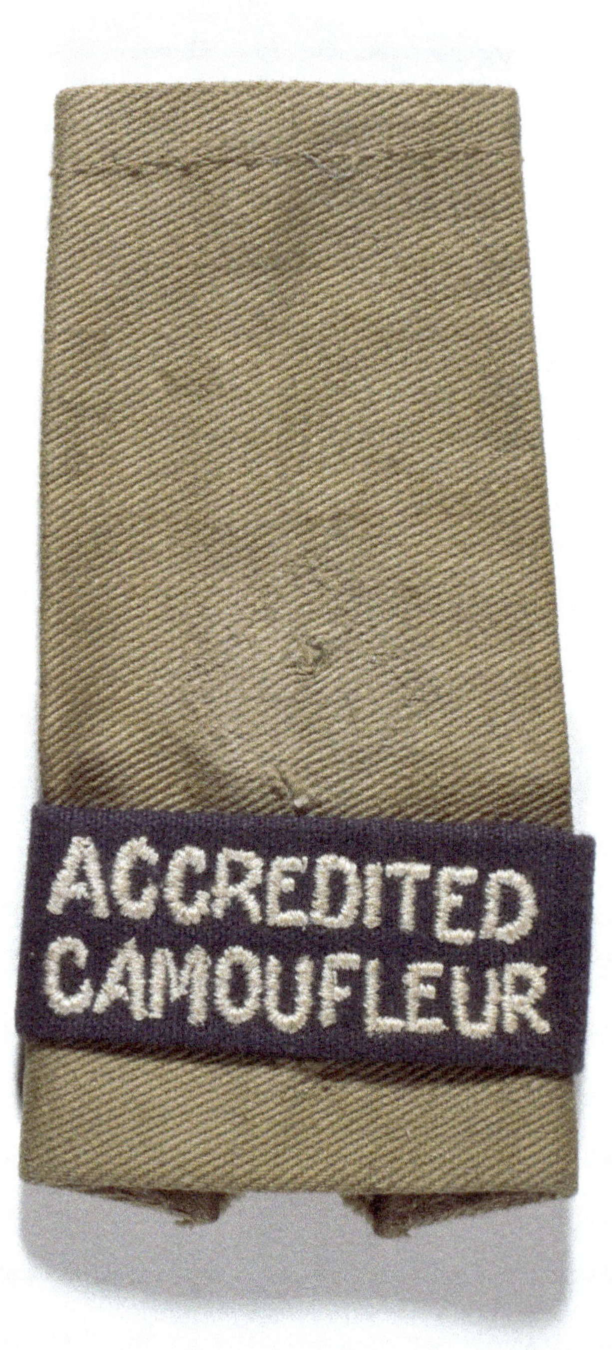

Fig. 7.1. 'Camoufleur', shoulder flash, 1943. Frank Hinder Papers. Collection: Archive of the AGNSW.

Chapter 7

Image

From an image point of view, the camouflage team in the DHS faced two problems. One was the absence of a uniform and therefore the unambiguous identity that a uniform projects, as well as the inward sense of being part of a community that a uniform makes possible. The other problem was the negative impact on identity of what Dakin referred to as the demeaning scepticism surrounding camouflage itself.[1] Dakin and his team tried their best to support and protect the armed forces, but there was little gratitude shown, until 1944. One year before the end of the war, he wrote to Frank Hinder that 'the Dep of Air is officially highly regarding us now & counting us as their camouflage section'.[2] What was surprising to him was that, despite camouflage's obvious advantages for animal survival, and despite significant military developments in the past 50 years—such as the introduction of khaki uniforms in the Boer War—modern militaries seemed not to be camouflage-minded. Whether this reflected the number of untrained recruits and conscripts going to war, or greater reliance on technological weaponry, the fact remained that feelings against camouflage ranged from apathy to suspicion.

A great deal more should have been learnt from WWI. When aerial reconnaissance technology became sophisticated, camouflage was no longer optional. It was vital to take notice of shadow, tones and shapes on the ground, to make effective use of camouflage nets for concealment, to deploy cunning use of decoys, and interpret aerial photographs. But from Dakin's point of view with the onset of WWII, when Australia finally followed other Allied countries and undertook camouflage measures, it was like starting from scratch. Education programs were needed to orientate military personnel to the main principles of concealment and deception and dispel misconceptions that camouflage was somehow effeminate and decorative or that it signified passivity, weakness, timidity, cowardice, even unfair combat. Personnel had to recognise that, by taking control of the colours and forms of their environment, they would have mastery over space, and therefore more control over the enemy.

Yet there was something deep-seated in western thinking that made it difficult for camouflage to shake its image of weakness, something linked philosophically and politically

1 W.J. Dakin, Camouflage report 1939–1945 (with appendices A–D), Canberra, AWM, Series 81 [77 Part 1], 1947, pp. 5–6.

2 William Dakin letter to Frank Hinder, 6 September1944, Frank Hinder Papers, Sydney, Archive of the AGNSW, MS 1995.1, Box 4, Folder 10.

with a dislike of mimesis, imitation, copying, mimicry and other conditions and states that impede—what the philosopher Plato established as one of life's ultimate goals—our apprehension of reality.[3] Among the great thinkers in modern history who stand out in this regard is Friedrich Nietzsche who Margot Norris claims possessed a 'mimeophobia' and who disagreed with Charles Darwin that the evolution of mimicry in animals was a positive force; rather it represented a loss of will to power and favoured the mediocre and the average, like the butterflies of the Amazon that survive by blending, deceiving, masking and disguising individuality.[4] Moreover, in the philosophy of war itself, there is no consistency on the intellectual strengths of camouflage. Of considerable influence on western military theory is Carl von Clausewitz's treatise *On war* written between 1816 and 1830. The author questioned the value of deception in warfare, referring to stratagems such as the sending of false reports to the enemy as feeble in comparison to straightforward military dealings like persuasion, intimidation and force. Camouflage tricks were a sign of weaker forces, he said, who are driven by desperation and looking for 'an infinitesimal glimmering of hope'.[5] On the other hand, the teachings of ancient Chinese military writer Sun Tzu, author of *The art of war*, whose series of short essays written in the 6th century BC first came to attention in the west in the late 18th century, and informed the Imperial Japanese Army in WWII, advised soldiers that the basis of all warfare is deception: they must adapt to their terrain, turn the lay of the land including the vegetation to their advantage and use simulation and dissimulation to 'hold out baits to entice the enemy. Feign disorder, and crush him.'[6]

Early in the war William Dakin heard troops say that camouflage was 'stuff for children' and 'kindergarten stuff'.[7] Insofar as anything pertaining to children was taken as a sign of the infantile and immature, this inferred that camouflage was analogous to some primitive infancy stage of warfare before civilisation's advances with technologies and machines. Were camoufleurs therefore the equivalent of children and naïves, and was he himself seen as merely playing at war (fig. 7.1)? Clear to Dakin was that in the gendered, macho world of military life, camouflage was at the feminine end of the social spectrum:

> The title 'Home Security' may be regarded as a somewhat misleading expression. It tends, perhaps, to emphasise a passive attitude. Whatever the name may indicate, those in control of Camouflage realise very clearly that its closest association is, and must be, with the perfection of a fighting machine, as it always has been.[8]

Home, security, passivity; the chain of signifiers led anyone sceptical about camouflage's purpose and power to only one conclusion, one reached in the US as well in WWII and

3 See Golden 1975, pp. 118–32.

4 Norris 1980, p. 1233.

5 von Clausewitz 1918, p. 207.

6 Sun Tzu 2008, p. 10.

7 W.J. Dakin and the Camouflage Directorate, 'Concealment, and camouflage of the individual in warfare', Canberra, DHS, 1944, in W.J. Dakin, Camouflage report 1939–1945 (with appendices K–N), Canberra, AWM, Series 81 [77 Part 3], 1947, p. 48.

8 W.J. Dakin, 'Camouflage bulletin no. 5', 11 September 1942, in Camouflage report 1939–1945 (appendix O), Canberra, AWM, Series 81 [77 Part 4], 1947, p. 3.

described by military historian Seymour Reit as the misconception that it was a method for achieving 'purely *protective* concealment'.[9] It was a false impression that was prevalent in WWI as well, and even topical in the years preceding WWI. The dispute that arose in 1910 between Theodore Roosevelt and Abbott Thayer over whether every animal is camouflaged for concealment, including large mammals and predators, was, in the view of Alexander Nemerov, precipitated by Roosevelt's conviction that camouflage is 'a form of effeminate cowardice, a mere defensive strategy [that] all but announced an unmanly desire to hide instead of fight'.[10]

One point to take from the dispute between Thayer and Roosevelt is the extent to which masculinity and heroism were bound up in the ethos of visibility. It affected many different representations of war, from textual to visual and it reverberated in Australia too in WWI in relation to a war painting by the distinguished Arthur Streeton (1867–1943). As an Official War Artist, Streeton's brief was to visit the battlefields of France and commemorate Australian participation in the war. But when he unveiled *Bellicourt tunnel* (also known as *The struggle for Bellicourt, 29 September 1918*) with strangely few soldiers visible in the landscape, Streeton found himself involved in the moral politics of their invisibility (fig. 7.2). He explained his aesthetic decision to Walter Baldwin Spencer in terms of camouflage itself, and argued that 'true pictures of Battlefields are very *quiet* looking things. Theres [sic] nothing much to be seen—everybody & thing is hidden & camouflaged.'[11] Certainly the area where he painted, Bellicourt, was a maze of slit and zigzag trenches where men hiding below ground were often undetectable to the eye from '40 yards away'.[12] A generous critic might have thought *The struggle for Bellicourt, 29 September 1918* an authentic representation of a war landscape in which the military world was largely hidden below ground, but instead the invisibility of soldiers, and the absence of overt heroic action and commentary on the human cost of war, made the painting seem superficial. Even in the late 20th century it was accused of being 'picturesque without being profound'.[13] Without visible human figures the painting had no gravitas.

It was common in WWI to think it manlier for soldiers to stand out rather than conceal their whereabouts, whereas those who campaigned for camouflage in WWII argued that bravery was useless if it meant becoming a target.[14] The trouble was that when camouflage civil defence was taught in Britain and Australia, the audiences included survivors of WWI who had to be convinced of the necessity for concealment and deception. Roland Penrose, who prepared a camouflage manual for the Home Guard in Britain, addressed his older readers, people with traditional views on warfare, by acknowledging that 'the idea of hiding

9 Reit 1978, p. 65.

10 Nemerov 1997, p. 79.

11 Galbally & Gray 1989, p. 152.

12 British-born golf course designer Alister Mackenzie discussing military entrenchments in *Golf illustrated*, quoted in Behrens 2009a, p. 266.

13 Reid 1977, p. 11.

14 For example, American artist and camoufleur John G. Browning judged that any man 'boasting to show—or try to show—how "army" he is makes him much less a man'. John Browning cited in Behrens 2009a, p. 72.

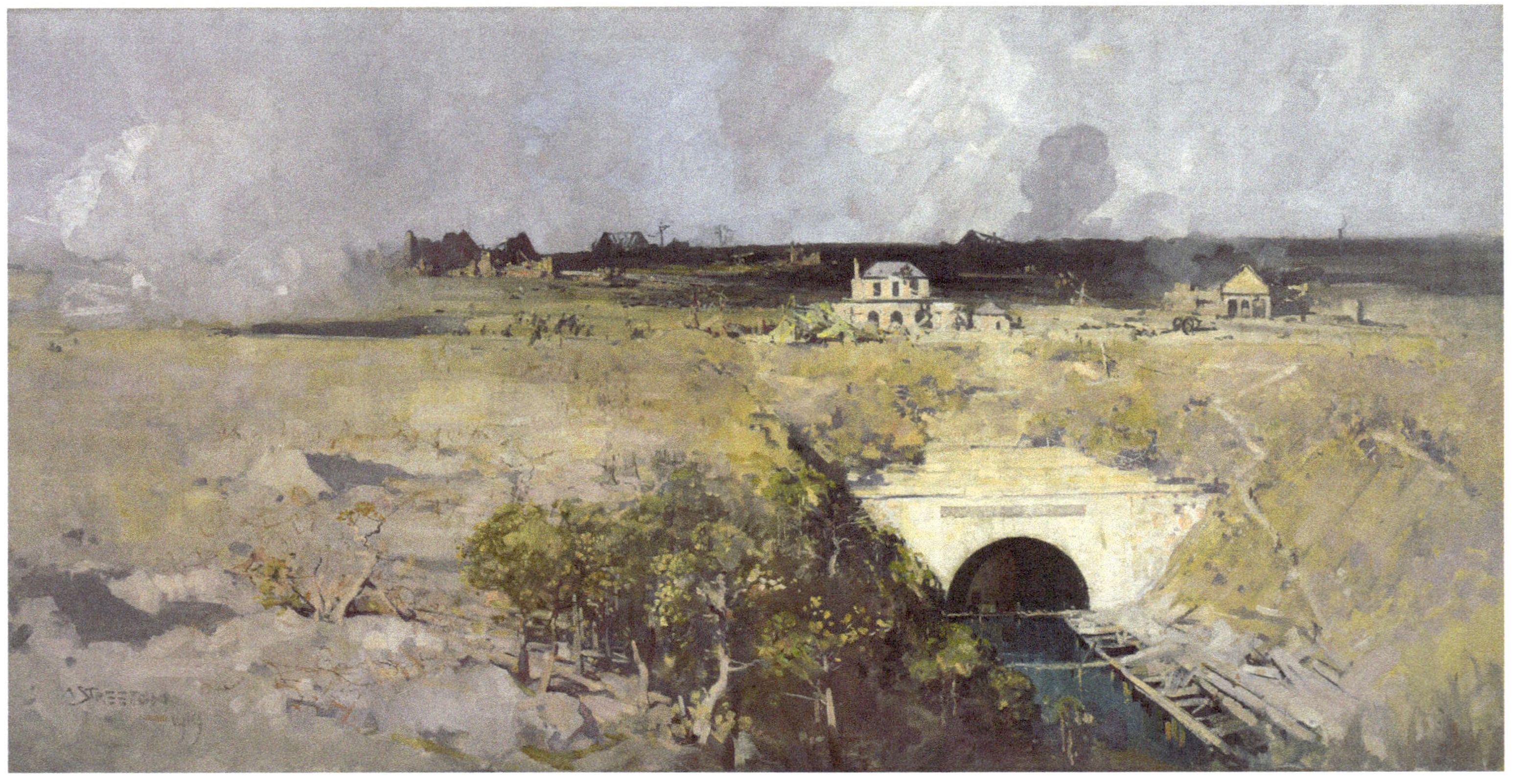

Fig. 7.2. Arthur Streeton, *Bellicourt tunnel (The struggle for Bellicourt*, 29 September 1918). Oil on canvas, 122.5 x 229.5 cm, 1919, Canberra, AWM, ART12437.

from your enemy and the use of deception may possibly be repulsive', but went on to explain that the enemy was gaining the upper hand by using concealment and deception, leaving no option but for everyone to embrace camouflage strategies 'however revolutionary these may appear to those who believe only in ancient traditions'.[15]

But how did camouflage earn a reputation as something so abject that it came to be seen, in Penrose's works, as 'repulsive'? Elizabeth Kahn, writing about WWI, traces the origins of the word 'camouflage' and finds it once had a devious meaning in French popular literature. Apparently, Guirand de Scevola, the artist and founder of the first French camouflage division, liked to remind people that before the war 'camouflage' was a term for criminals hiding from the police.[16] Furtive, shady, cunning—camouflage was all of these but in addition it meant 'faking'. These dark associations only increased with time, and as WWII dragged into its sixth year, camouflage's image evolved into a sign of humanity's increasing stealth and deception. The amazing camouflage deceptions of WWII where whole armies and towns were concealed from view were good reason, writes Hillel Schwartz, for people to ask 'where, then are our own skills at disguise, decoy and deception leading us?'[17] But returning to a more innocent time, one less reliant on deception, where warfare was more honest and simple and focused on intimidation and force, was also impossible. Winston Churchill came to the defence of camouflage in WWII when he said 'it is perfectly justified to deceive the enemy'.[18] The British, then, had the 'Magic Gang' that included the stage magician Jasper Maskelyne, and a team of engineers and artists who made the Suez Canal 'disappear', 'moved' the port of Alexandria, and created a 'phantom army' before the battle at El Alamein.[19] On 11 November 1942 Churchill addressed the House of Commons about Alamein, a battle that changed the tide of WWII, and marveled at the camouflage success of the 10th Armored Corps who made dummy duplications of its tanks and equipment and then left them behind while it 'moved silently elsewhere in the night'.[20]

War historian Guy Hartcup, attributes all advances with concealment and deception in WWII to artists, and is adamant that 'the artist, with his understanding of the subtleties of colour, tone and texture and his ability to draw on visual memory, has probably contributed the most to military camouflage in all its forms'.[21] But as artists in the DHS found, it was far from easy for their profession to be accepted and valued. Mistrust and cynicism about camouflage aside, an additional impediment to camoufleurs having a healthy self-image was the way artists, as a social group, were poked fun at for dedicating their lives to art, an aspect of culture that others saw as unnecessary. Even artists who signed up for the armed forces in WWII were ridiculed. Sali Herman, for example, joined the army and worked in camouflage, but as Frank Hinder noted more than once, 'slighting remarks about artists'

15 Penrose 1941, p. 4.

16 Kahn 1984, pp. 148–49.

17 Schwartz 1996.

18 Winston Churchill, 11 November 1942, quoted in Hartcup 1979, Chapter 6 epigraph.

19 See Breuer 2001, pp. 94–98.

20 Col. John Ohmer, 'US Report on Camouflage', 15 September 1943, Air intelligence reports, Sydney, NAA, Series C 1707, Item 36, p. 6.

21 Hartcup 1979, pp. 8–9.

plagued Herman.[22] Indeed the social disharmony of military life forced Donald Friend to adopt emotional and physical disguises, and he did not fail to notice how camouflage was a matter of his own survival. Friend witnessed horrific sights while in the army, particularly in Borneo, but the adjustment to military life was an ordeal in itself:

> like the camouflaged chameleon, I have taken on the ghastly tints of my environment and have been infected, as with so many plagues, with a monstrous repertoire of dull virtues, that have drained my soul of colour.[23]

Artists had a dubious reputation and in truth the high proportion of painters, designers, photographers and architects in Dakin's group was something that niggled at him also. On the one hand he was happy to second members of the arts community to the DHS, but on the other he conceded that many misunderstandings had arisen as a result of a trend worldwide for artists to sign up for camouflage work in WWII:

> Outside the Services there was a very definite idea prevailing that all Camoufleurs were artists. With the outbreak of war with Japan and the rush that followed to use manpower in the most needed ways, artist organizations began clamouring for the appointment of young male artists as Camouflage officers despite the fact that they had no training. This was particularly the case in Melbourne and Sydney.[24]

That rush to work in camouflage, coupled with preconceptions of the artist as someone who paints beautiful landscapes, explains why in Britain, the US and Australia, camouflage gained a reputation as a dilettante's attitude to war.[25] Indeed in all Allied countries in both world wars, when artists were involved, camouflage ran a greater risk of being seen as a form of unnecessary decoration. Further, the keenness of artists to work in camouflage didn't just lead to accusations of camouflage as useless decoration compared with the structural necessities of war, it also suggested hobbies and Sunday painters. Camouflage, argued the British Ministry of Air, 'cannot be learned in a day, nor does six months training at an Art School carry full qualifications for its practice'.[26] Similarly in 1942 the US magazine *ART-news* released a report on 'The first official War Department Release on Camouflage and the Artist', where the writer, Captain Spierer of the US Army, cautioned that 'when an artist enters a camouflage battalion he is not going to find an easel awaiting him to paint pretty pictures.'[27]

---

22 Frank Hinder, 'Records of Lt F.C. Hinder Camoufleur', Canberra, AWM, PR 88/133, File 895/4/182, Item 7 of 12, p. 7.

23 Donald Friend, 26 December 1942, in Gray (ed.) 2001, p. 160.

24 W.J. Dakin, Camouflage report 1939–1945 (with appendices A–D), Canberra, AWM, Series 81 [77 Part 1], 1947, p. 37.

25 An American army officer in WWI was horror-struck by the deployment of artists for military duties: 'Oh God, as if we didn't have enough trouble! They send us artists!' See Behrens 2002, frontpiece.

26 British Air Ministry, 'A few simple camouflage maxims', Camouflage information from England, Sydney, National Archives of Australia, Series C1707, File 58.

27 Spierer 1942, p. 9.

It is one of those quirky anomalies that while men rather than women specialised in camouflage, camouflage was perceived as a feminine type of warfare relying on the so-called womanly activities of masking, subterfuge and deception. As previously mentioned the French verb 'camoufler' translates as the act of putting on makeup for the stage. It was no coincidence that in the US during WWII camouflage was known as 'the glamor girl of civilian defense'.[28] And it is not by chance that when the DHS initiated the manufacture of Skin Tone Commando Cream to cover the pink skin of jungle-fighting troops in the New Guinea Area in dark brown, it consulted a cosmetic manufacturer, a physiologist and an actress. Yet when he discussed the idea of developing a camouflage cream with the armed forces, the army referred the idea to a house-painting specialist. Dakin found this laughable; the house-painting industry specialised in chemicals and solvents, not how to cover and protect human skin.[29] He settled for 'a make-up cream which girls had been using for colouring legs to resemble silk stockings'.[30] Bear in mind that only 30 years earlier, Australian men who wore wrist watches were deemed effeminate,[31] and it comes as less of a surprise to learn that Australian troops in the tropical jungles—both army and RAAF—preferred to let nature take its course rather than wear Skin Tone Commando Cream or the green dye distributed by the army for skin colouration, a situation that D.P. Mellor found worthy of note in his official history of WWII:

> Curiously enough, men in forward areas were disinclined to make use of this aid to concealment, one reason being that they felt that the bearded and dirty faces of those who had been trekking through the jungle for days offered just as good a means of escaping detection.[32]

So vital, however, did Dakin think the cream for individual concealment in Papua and New Guinea, that when Bob Curtis became Officer in Command for camouflage in the SW Pacific area he was given the special task of campaigning for its adoption and ensuring its distribution to troops.[33] But the reports always came back negative; men in forward areas were concerned that wearing cream on the skin would inhibit perspiration.[34] In short, while camouflage cream made an impression on the VDC (fig. 7.3) men in the field preferred more primal methods, rubbing dirt into their skin and using the grime of combat life to keep their faces in shadow.

---

28 Edward M. Farmer in Schwartz 1996, p. 201.

29 W.J. Dakin, 'Camouflage and Army', 2 October 1942, Camouflage—question of responsibility: army camouflage Canberra, NAA, Series A453, Item 1942/21/2632, p. 5.

30 W.J. Dakin, Camouflage report 1939–1945 (with appendices A–D), Canberra, AWM, Series 81 [77 Part 1], 1947, p. 134.

31 McQueen 1978, p. 90.

32 Mellor 1958, p. 541

33 R.E. Curtis letter to A.W. Welch, 26 December 1943, Camouflage—general—camouflage organisation in New Guinea, Canberra, National Archives Series A453, Item 1943/17/1510 Part 1.

34 W.J. Dakin, Camouflage report 1939–1945 (with appendices A–D), Canberra, AWM, Series 81 [77 Part 1], 1947, p. 134.

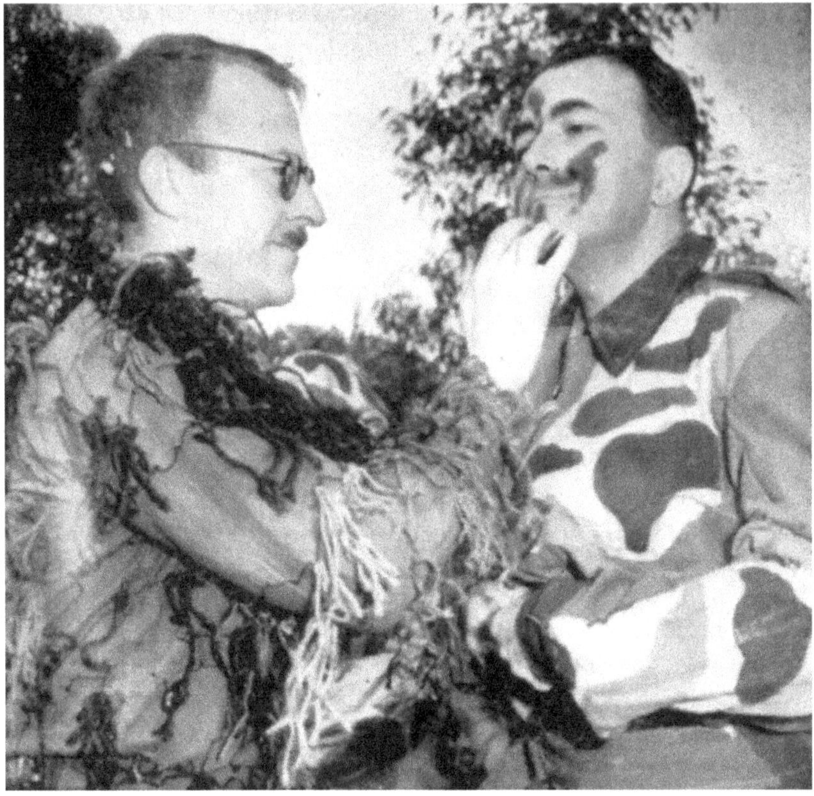

Fig. 7.3. Camouflage demonstration. On Guard with the Volunteer Defence Corps (1944), p. 77.

To return to an earlier point; the struggle to win the right to wear a uniform preoccupied camoufleurs in the DHS for the first three years of the war. Charles O'Harte, for example, felt like an unwelcome outsider when he returned to town after camouflage assignments in remote northern Australia. At 51, and having begun the war as an artist with the Labour Squad in Sydney, O'Harte was among the DHS's oldest volunteers for tropical service. In March 1943 he wrote to his supervisor in Townsville that, while he was glad to serve his country, he was disappointed that camoufleurs 'are nobody's business, our official standing is undefined, we have no place where we can contact anyone that can vouch for us'.[35] Hotels turned him away; it was hard to believe a man without a uniform who claimed he was attached to the airforce. O'Harte's letter was forwarded to Dakin in Canberra who added it to the growing pile of evidence to support his push to get accreditation with the airforce, and a uniform with insignia for identification.

To say that in the army, navy and airforce the uniform was a mark of distinction is an understatement. It symbolised a special class of warrior-hero whose status embodied ideals of masculinity, and this was true regardless of the services being populated by women

35 Charles O'Harte to D.H. Wilson, 25 February 1943, in Dispersal hangars, Sydney, National Archives of Australia, Series C1707, Item 4, p. 2.

as well as men. What camouflage specialists encountered when they tried to work inside the army, for example, was a dress code that denoted a self-contained world with its own history, one tied to the Anzac legend and the sacrifices of WWI, and one therefore linked to the foundation of modern Australia.[36] But as well, a military uniform communicated belonging and a shared set of values and ambitions. To Frank Hinder it meant the difference between being an outsider and being 'one of the boys'.[37] With a uniform he could get on with his work without looking conspicuous. The alternative was to stand out on airfields, military camps and training centres, environments where artists, like women, were out of place and where men who were noncombatants were often made to feel inferior. In WWI, as well, those behind the lines felt inferior and guilty.[38] With camouflage—along with its civilian practitioners—suffering from a poor image, a uniform offered the kind of social disguise within military circles necessary for self-protection and self-esteem, and without it, as Dakin pointed out more than once, camoufleurs were put 'in a curiously inferior position when on Service objectives'.[39]

Finally, however, uniforms were issued. The first consignment of khaki clothing arrived in May 1943, and at the same time the DHS's camoufleurs received accreditation with the RAAF. It meant they were permitted to wear the insignia 'Accredited Camoufleur' as a shoulder flash (fig. 7.1).[40] But no sooner had the labels been prepared when the DHS changed the title from 'Camoufleur' to 'Camouflage Officer'.[41] The new title was thought manlier, and more military. The word 'camoufleur' had always been the butt of jokes as the following comment from Hinder shows:

> transferred to the Dept. as a 'camofleur' or 'camopansy' as we called ourselves. (Not that we were the latter so far as I know!).[42]

Without doubt the name 'camouflage officer' had a stronger and more proactive ring, while 'camoufleur' compounded that passivity that Dakin spoke of in relation to the title 'Department of Home Security'.

It was no accident that when the DHS's new uniforms arrived, and the old floral title 'camoufleur' was replaced, its officers chose for their emblem the motif of a tiger. Since they were not permitted to wear airforce insignia, badges, armlets or buttons, Douglas

---

36 Garton 1998, p. 86.

37 Frank Hinder, 'Records of Lt F.C. Hinder Camoufleur', Canberra, AWM, PR 88/133, File 895/4/182, Item 7 of 12.

38 See Thomson 1995, p. 137.

39 W.J. Dakin, Camouflage report 1939–1945 (with appendices A–D), Canberra, AWM, Series 81 [77 Part 1], 1947, p. 40.

40 A.W. Welch to Department of Air, 5 May 1943, 'Accreditation of camouflage officers of Department of Home Security', Accreditation and Attachment of Camouflage Officers, Canberra, NAA, Series A453, Item 1943/28/566.

41 A.W. Welch to Officer-in-Charge (New Guinea), 15 September 1943, Camouflage—general—camouflage organisation in New Guinea, Canberra, National Archives Series A453, Item 1943/17/1510 Part 1.

42 Frank Hinder, 'Records of Lt F.C. Hinder Camoufleur', Canberra, AWM, PR 88/133, File 895/4/182.

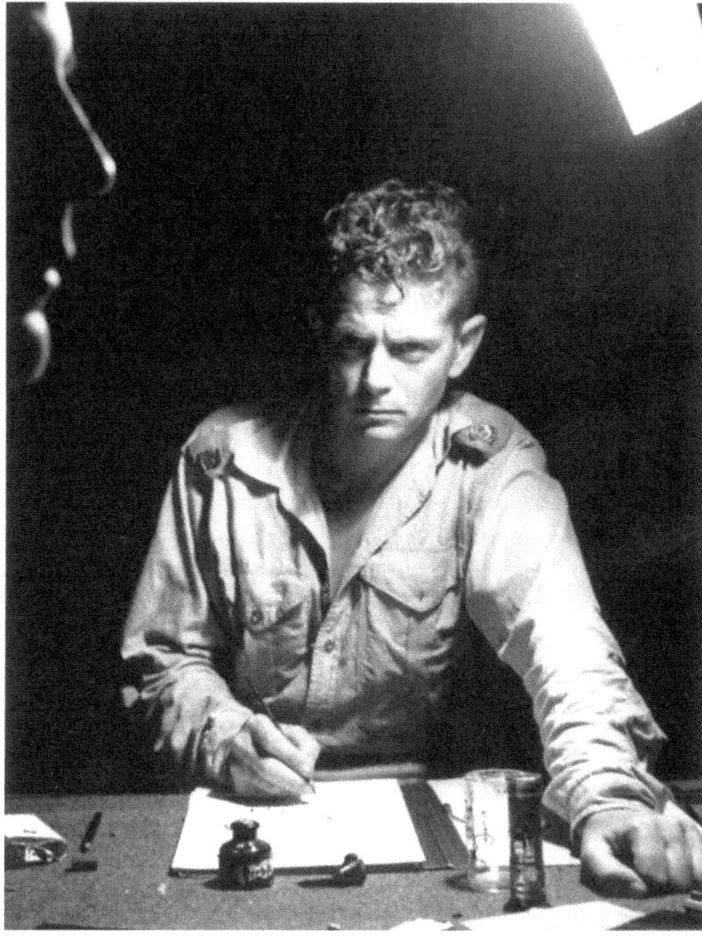

Fig. 7.4. Clement Seale sketching Captain R.E. Curtis wearing the camouflage officer shoulder badge, c. 1943. AWM P02186.004. Photograph, Max Dupain.

Annand emblazoned the tiger on shoulder flashes for identification.[43] It told the world that this team was vigorous and dynamic, that its work was a matter of life and death, and that camouflage, which is both defensive and offensive in war, was there to play a part in victory (fig. 7.4).

At every opportunity members of the DHS incorporated images of tigers into education and training material because, as Frank Hinder was keen to stress in training sessions, 'the tiger conceals to attack!'[44] (fig. 7.5). Posters were designed showing the animal suddenly emerging from its hiding place and propelling forward on its prey (fig. 7.6). And when

43 See note referring to ' "uniform" finally worked out—shoulder flash of tiger and palm leaf designed by Doug Annand' in Frank Hinder Papers, Sydney, Archive of the AGNSW, MS 1995.1, Box 4, Folder 10.

44 Frank Hinder, 'Teaching aids for individual concealment', Frank Hinder Papers, Sydney, Archive of the AGNSW, MS 1995.1, Box 4, Folder 10.

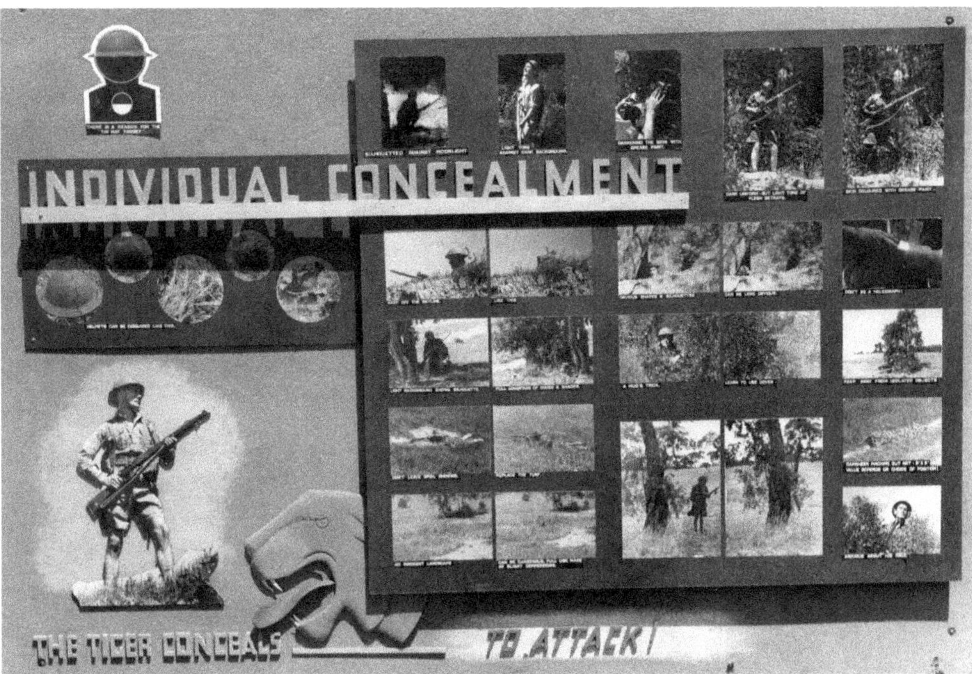

Fig. 7.5. *Individual concealment*, part of a portable camouflage instruction installation. Canberra, AWM, Hinder Personal Records PR 88/133, 9 of 12.

airforce personnel opened their pocket camouflage manuals prepared by the DHS they were reminded that:

> Camouflage in war is a duty and responsibility in which everyone in the Services may play a part. It is not a cloak solely developed for the protection of the weak. The tiger is protectively coloured. Offensive action needs concealment both for its preparation and consummation.[45]

In view of how difficult life without a uniform was for men in the DHS it is almost inconceivable how women got along in civilian clothes. As it turns out from a list of over 100 people, only one woman in the camouflage section of the department had duties in field and operational camouflage. This was Margaret Morison, an architect from Perth, who served as a camoufleur for the Western Australia area.[46] Five women were listed as officers with the NSW Directorate of Camouflage, but they were not given field duties and their contribution remains unknown. In the case of Mrs R.A. Curtis, however, the work was secretarial.[47] Other women worked in camouflage while in the army, such as Sapper

45 Dakin quoted in Mellor 1958, p. 543.

46 See entry for Margaret Morison, Department of Home Security Staff register—camouflage section, Canberra, NAA, Series A691, Item 1, accumulation dates 1 January 1942–31 December 1945.

47 Women listed for New South Wales are Mrs R.A. Curtis, Miss N. Dansey, Miss M.T. Scott, Mrs I. O'Brien, and Miss M.I Longworth. See John D. Moore to DHS, Canberra, 5 March 1943, Camouflage

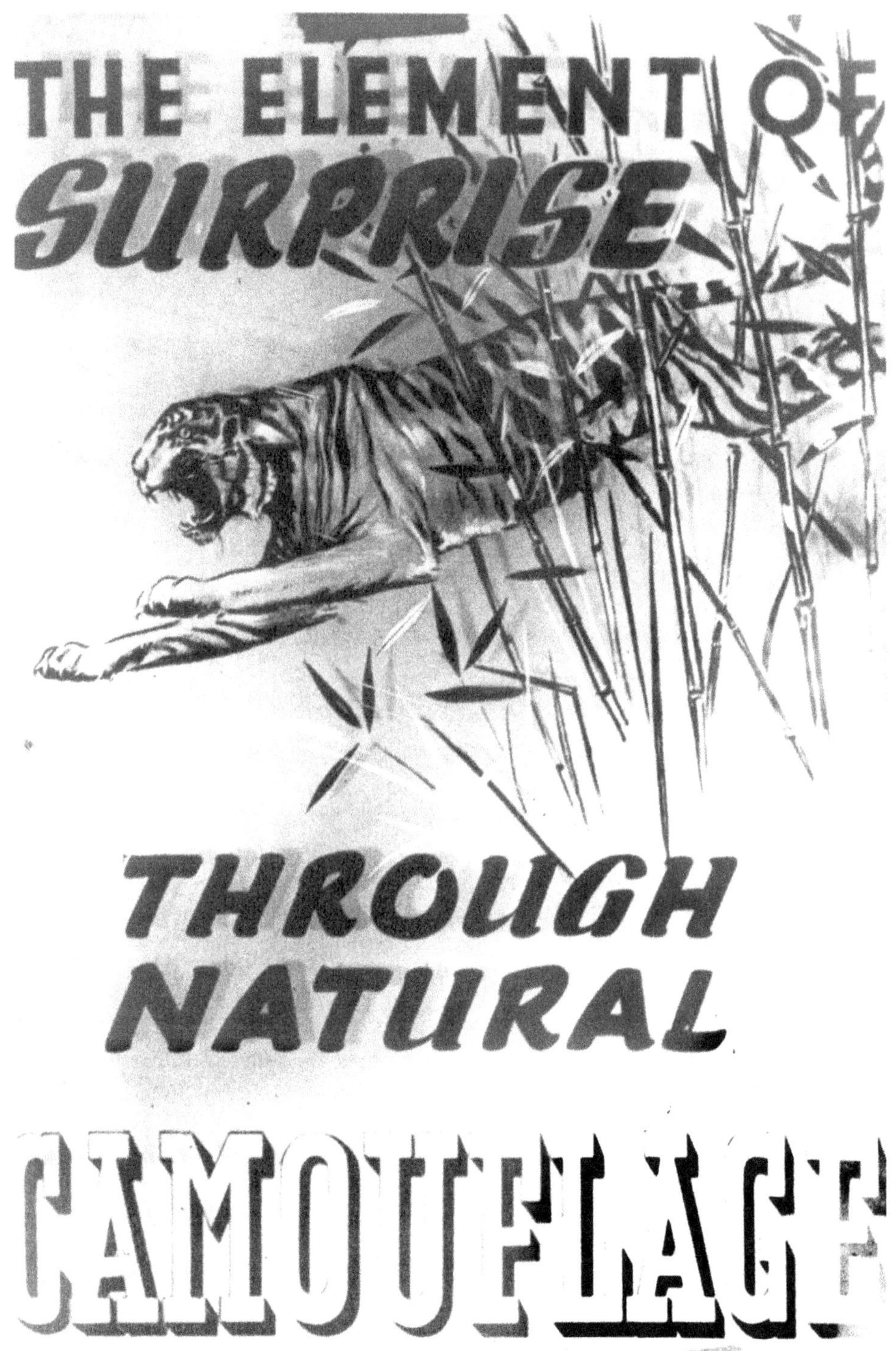

Fig. 7.6. *The element of surprise through natural camouflage*, DHS poster, c. 1942. The collection of the NAA: C1905 T1, 4 [1].

Anne Ross.[48] And one woman's name comes up in relation to the camouflage section of the Department of the Interior, Loma Latour (1902–1964) who was stationed at Bankstown and was indeed disadvantaged by lack of uniform. When she was 'desperately anxious to go to Queensland as a camoufleur', John Moore decided against it on the basis that she would be made unwelcome 'on the aerodromes in the position of camoufleur in women's civilian clothes'.[49] As in the art world that Latour was part of, women were kept on the fringes of camouflage operations and were singled out for special comment and different treatment. Perhaps it was Loma Latour or Margaret Morison who made Frank Hinder laugh when he remembered:

> the woman artist who was on the committee and who was responsible for advising that the best material for making ground patterns on the aerodrome runway was Indian Ink![50]

And the virtual absence of women from the Sydney Camouflage Group and the DHS is made all the more conspicuous by the fact that camouflage attracted many modernist designers—Douglas Annand, Frank Hinder and Ronald Rigg are three—yet those who dominated the field of modern design in Australia in the 1930s were women.[51]

Regardless of the gender imbalance, camouflage duties had a reputation as a soft option, and this served to confirm for its critics why artists were keen to sign up for it. But a wartime life in camouflage came about in unpredictable and circuitous ways for Australian artists, architects and designers. They belonged to professions that were otherwise galvanised against war, especially following the carnage of WWI. The artist community was active in the protest against fascism in the years surrounding WWII.[52] But few appealed for military exemption on conscientious grounds. The dual antagonisms they felt to 'militarism' on one hand, and to 'shirkers' on the other were, for most individuals, impossible to reconcile. There was pressure to be brave and patriotic, to voluntarily sign up for service, devote body and soul to the war, and be visible in doing so. The alternative was to take the 'un-Australian' path of conscientious objection and other acts of invisibility.

---

personnel NSW Camoufleurs staff appointments, Canberra, NAA, Series A453, Item 1942/28/2298.

48 For a photograph of Anne Ross see AWM Collection, Photograph 015581, cas.awm.gov.au/Item/015581, accessed 10/03/10.

49 John D. Moore to A.W. Welch, 13 May 1942, Camouflage—Personnel NSW Camoufleurs staff appointments, Canberra, NAA, Series A453, Item 1942/28/2298.

50 Frank Hinder, 'Records of Lt F.C. Hinder Camoufleur', Canberra, AWM, PR 88/133, File 895/4/182, Item 9 of 12.

51 Butler & Donaldson 2008/9, p. 133.

52 In 1942 the Contemporary Art Society in Melbourne held the *Anti-Fascist Exhibition* with a catalogue written by Albert Tucker. Tucker is arguably the greatest of Australia's commentators on WWII through his paintings and drawings, many of which were produced in response to the experience of being conscripted into the Militia, hospitalised for illness, and exposed to many physically and mentally mutilated soldiers at Heidelberg Military Hospital in the Army Medical Corps. See Mollison & Bonham1982, pp. 32–34.

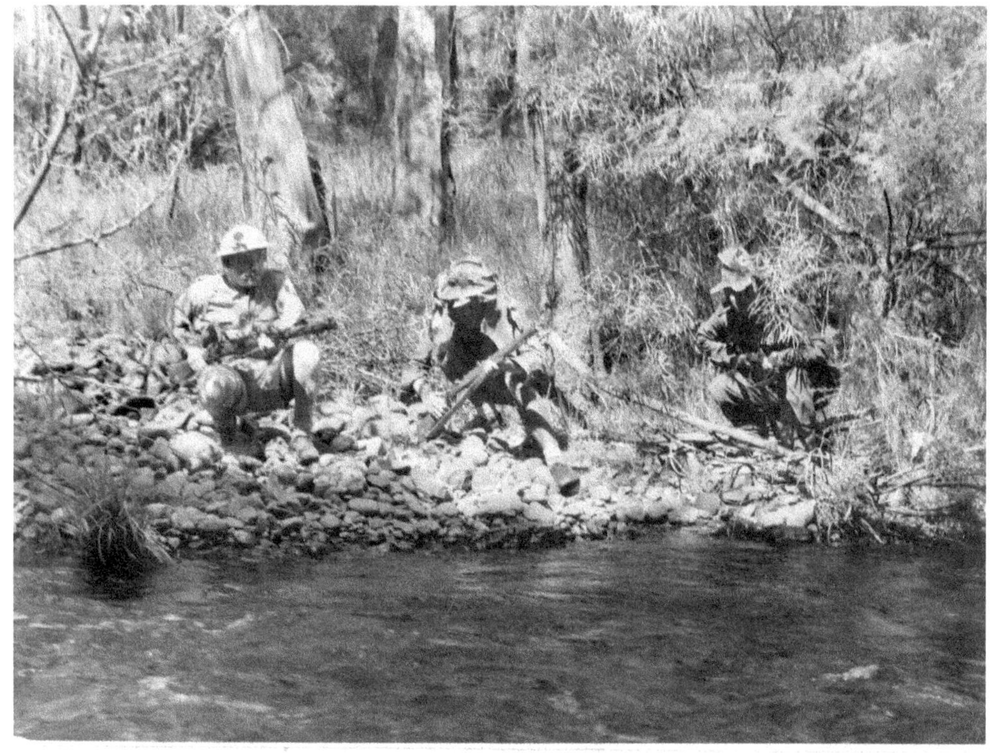

Fig. 8.1. Canberra experiment with (left to right) Eric Thompson, Frank Hinder and a Mr Hargrave. Canberra, AWM, Hinder Personal Records PR 88/133, 9 of 12.

Chapter 8

Conscience

> An epidemic of destruction sweeps the world today. The mind of civilized man is set to stop it. What the enemy would destroy, however, he must first see. To confuse and paralyse this vision is the role of camouflage. Here the artist and more particularly the modern artist can fulfil a vital function for opposed to this vision of destruction is the vision of creation (Arshile Gorky, c. 1943).[1]

In 1938, as Douglas Annand, Russell Roberts, Max Dupain and Adrian Feint worked on their concept of a modern design space for the Australian Pavilion at the 1939 New York World's Fair, the American abstract-expressionist painter Arshile Gorky began a set of murals titled *Aviation*, also for the World's Fair.[2] Within four years, all had applied to work in camouflage as their social contribution to the war effort.[3] Without doubt, the trend to seek deployment to this field, where the artist's visual intelligence seemed well suited to the utilitarian purpose of war, was international in scale. And as Gorky's manifesto (quoted above) for a civil defence course in New York shows, the alliance of modern artists and war could be justified by claiming camouflage to be the way to safeguard progressive society and culture, and by asserting its natural relationship to art. But the social principle that impelled artists to work in this military-related field also risked being seen as a lack of principle by anyone of a mind that such collusion signalled the militarisation of progressive art. Postwar writers especially have been both circumspect or—in the case of social philosopher Paul Virilio—hostile towards connections between the avantgarde and the violence of war.[4] And Harold Rosenberg, when discussing Gorky's 'schizophrenia' at assessing the social responsibility of working for the war enterprise, looked back at the artist's decision to work in camouflage as a sad reminder of 'the passion of American art twenty-five years ago to prove it had a place in man's progress, even in his increased capacity to kill himself'.[5]

Indeed the question of artistic conscience was an ever-widening debate in Australia in the years leading up to WWII, prompted first by the rise of fascism in the 1930s. It raised

1 Rosenberg 1962, p. 133.

2 For a discussion of the Australian Pavilion see Stephen 2006, pp. 29–40.

3 For an interesting range of quotes and comments relating to Gorky's camouflage venture see Behrens 2009a, pp. 165–66.

4 See 'A pitiless art' in Virilio 2003.

5 Rosenberg 1962, p. 92.

the problem of whether it was morally right for modern artists, in an age threatened by totalitarian dictatorship, to make art about self-referential investigations of expressionism and abstraction, at the expense of direct political comment.[6] By 1939 the vanguard artists of Australia had seen or were aware of Picasso's *Guernica* (1937), a painting provoked by the bombing of the Basque town of the same name by Germany during the Spanish Civil War; in 1937 it was exhibited in the Spanish Pavilion at the Paris International Exposition. They were aware of works like Otto Dix's *Der Krieg* (1925), an expressionist painting created in response to WWI and typical of a new realm of thought and a new pictorial language of emotion brought to the protest of political violence after that war. The matter of art's function was polarised into whether art should be for art's sake, or society's, and which, out of individualism and socialism, stood for the morally right path to achieve artistic freedom and political freedom.[7]

Also pressing as a matter of ethics throughout the 1930s in Australia was the need to curtail unwarranted negativity towards modern art and create a culture of respect. It was seen as a time to build support for an intelligent, critical engagement with art that challenged the old guard whose reactionary views were epitomised by the critic Lionel Lindsay, a man who referred to modernism in art as 'a freak'.[8] It was the conservatives, those who held a view of true art as opposed to pseudo art, that caused the William Dobell 'scandal' in 1944 when Dobell's award-winning painting of a fellow camouflage worker, *Portrait of Joshua Smith,* was denounced in court as nothing but a caricature.

With the onset of WWII, however, an additional question for artists was the morality of military service. For Albert Tucker, a painter who is celebrated today for depicting the human cost of the war era, the thought of another world war put him into a 'state of confusion about responsibility, social issues, attitudes that one should take, and the obligations of a painter'.[9] Like many of his associates, Tucker was called up for the army. But just how many Australian artists and designers went to war after 1939 is uncertain. Some were conscripted, as Tucker was, to serve in the geographical area of Australia while others were sent to the SW Pacific zone. Some volunteered for service abroad with the AIF. Many were seconded to specialist departments, such as the Military History Section, specifically for their skills in art and design, and some who enlisted, and others who did not enlist, were later commissioned as Official War Artists.[10]

Artistically and politically, those central in the early organisation of camouflage in Australia in WWII were a mixed bag of modernists, academic realists, socialists,

---

6 See the matter discussed in Adrian Lawlor, 'Should art be intelligible?', originally published in *Arquebus*, reprinted in Stephen, McNamara & Goad (eds.) 2006, pp. 142–48. Also see the communist position in Noel Counihan, 'How Albert Tucker misrepresents Marxism', originally published in *Angry Penguins* No. 5, 1943, reprinted in Stephen, McNamara & Goad 2006, pp. 436–44.

7 Eagle 1989, p. 195.

8 Lindsay 1942, p. 1.

9 Albert Tucker interviewed in Mollison & Bonham 1982, p. 35.

10 Among Official War Artists were Sali Herman, Nora Heysen, Donald Friend, Arthur Murch, and Robert E. Curtis. See 'Second World War Official War Artists', *Encyclopaedia*, Canberra, AWM, www.awm.gov.au/encyclopedia/war_artists/artists.asp, accessed 24/2/09.

democrats, nationalists and internationalists. On the surface at least, they appear to have had few reservations about participating in the war effort. Many, after all, were proactive in forming the Sydney Camouflage Group even before the announcement of war. And it was not as if they were young and naïve; Adrian Feint was forty-five at the outbreak of WWII, had achieved distinction for war service in France in WWI and was settling into middle-age and a life of taste and connoisseurship.[11] But with war imminent Feint became highly motivated, joined the Sydney Camouflage Group, and was among the artists whom John Moore described as exemplary for being 'willing to give up their present occupations and take full time work as camoufleurs'.[12] John Moore was an older member too, being fifty-three on entry to the DHS. In general, camoufleurs with the department belonged to an older age group than fighting soldiers, who were usually eighteen when conscripted. Staff records show that most members of the DHS were between twenty-eight and thirty-five years of age, with the exception of George Bell, an influential figure in Australian art and education, who put himself forward for service to the department at the age of sixty-three.[13]

When Douglas Annand, Frank Hinder, Max Dupain, John Moore, Arthur Murch, Russell Roberts, Adrian Feint, Robert Curtis and others met in Sydney Ure Smith's office to talk camouflage in 1939, their decision to contribute to war's function and put art and design at the service of defence and attack, to assist the daily, operational aspects of war activity, put them in a different position to that of Official War Artists. The latter were deployed to make personal representations of war for future commemoration and remembrance, and to make others reflect on the human cost of political conflict. The men and women who became war artists, some combatants others not, were part of a long tradition where states and rulers—the most famous is Napoleon, but Queen Victoria is another example—selectively commissioned artists to depict military victories and military careers. From their service they produced works of art that often transcended the base materiality of war by focusing on loyalty, bravery and heroism. But the camoufleur's production was never intended for public display, or to be looked at as an aesthetic object in its own right. Consequently their status as artists in the war was quite different and how they are remembered by history is different too.

When art historian, Elizabeth Kahn, looked into the history of European camoufleurs in WWI and found them all but invisible—a situation she attributed to disapproval of war itself by art historians working in the years postdating WWII—she wrote that 'art and war are often seen as antithetical: the first constructive and beautiful, the second destructive and hideous'.[14] It was an issue Frank Hinder also thought about, and in one of his many quirky poems addressed the conundrum of the camoufleur caught within the dichotomised politics of abject war, and beauty. With typical black humour Hinder portrayed the camoufleurs

11 See Martin 2009, p. 28.

12 John D. Moore to A.W. Welch, 6 March 1942, Camouflage—Personnel NSW Camoufleurs staff appointments, Canberra, NAA, Series A453, Item 1942/28/2298.

13 See entries for George Bell and John D. Moore in Department of Home Security Staff register—camouflage section, Canberra, NAA, Series A691, Item 1, accumulation dates 1 January 1942–31 December 1945.

14 Kahn 1984, p. 3.

of WWII as social outcasts, although just who made them that way is a little unclear in the poem, whether it was the arts community, the military community, the postwar generation, or all of the above:

> … Peal out great bell!
> Tis the death little-knell
> Of the camoufleurs.
> … That race of overpaid untouchables
> Sludge in the eye of beauty—
> Blue venom on the lips of life—
> Dirt in the concrete of The Architect's brain…[15]

While it is unlikely, given the patriotic fervour of the war years, that the arts community in the 1940s maligned Australia's camoufleurs for involving themselves in the application of art to military purpose, in later years two members of the DHS felt they needed to set the record straight on matters of war and peace. In his 1986 autobiography Max Dupain wrote that he was a 'pacifist' when WWII descended and further, with reference to war, that he 'scorned the heroics of it all and despised the negativism created by brute force'.[16] Camouflage for Dupain was a means of contributing to war in a less brutal way. Similarly, in *Great Barrier Reef, and some mention of other Australian coral reefs* a book written towards the end of his life, William Dakin wrote of his 'hatred of war'.[17]

Nevertheless it comes as a shock to see photographs of progressive thinkers and leaders in modern art and design in Australian history, including Frank Hinder and Eric Thompson, dressed in camouflage military garb in 1943, simulating war action (in the federal capital of Canberra thousands of miles from the war zone), and brandishing guns (figs. 8.1 & 8.2). Art history, a discipline that concentrates on the peacetime work of artists of the war generation, does not prepare us for these sorts of anomaly. Yet as a camoufleur, Hinder was given target-shooting practice and taught how to throw grenades; and he later wrote that he was grateful for the compulsory military training he had obtained as a school pupil at Newington College between 1916 and 1918. School had given him insight into military procedures and protocols.[18] Newington boasted the oldest cadet corps in Australia, designed to develop in boys an interest in the army and provide a foundation in military knowledge. But Hinder was also glad that he and his camouflage colleagues never had to fight in the war and 'face up to the real thing'.[19] When Dakin seconded him from the CMF

---

15 Frank Hinder, untitled poem, Frank Hinder Papers, Sydney, Archive of the AGNSW, MS 1995.1, Box 4, Folder 10.

16 Dupain 1986, p. 15.

17 Dakin did, however, qualify this by saying that war can lead to important scientific discoveries, arguing that the atomic bomb testing on Bikini Atoll had furthered science by revealing evidence of Darwin's theory of reef subsidence. See Dakin 1955 [1950], p. 130.

18 Frank Hinder, 'Records of Lt F.C. Hinder Camoufleur', Canberra, AWM, PR 88/133, File 895/4/182, Item 7 of 12.

19 Frank Hinder, 'Records of Lt F.C. Hinder Camoufleur', Canberra, AWM, PR 88/133, File 895/4/182, Item 9 of 12.

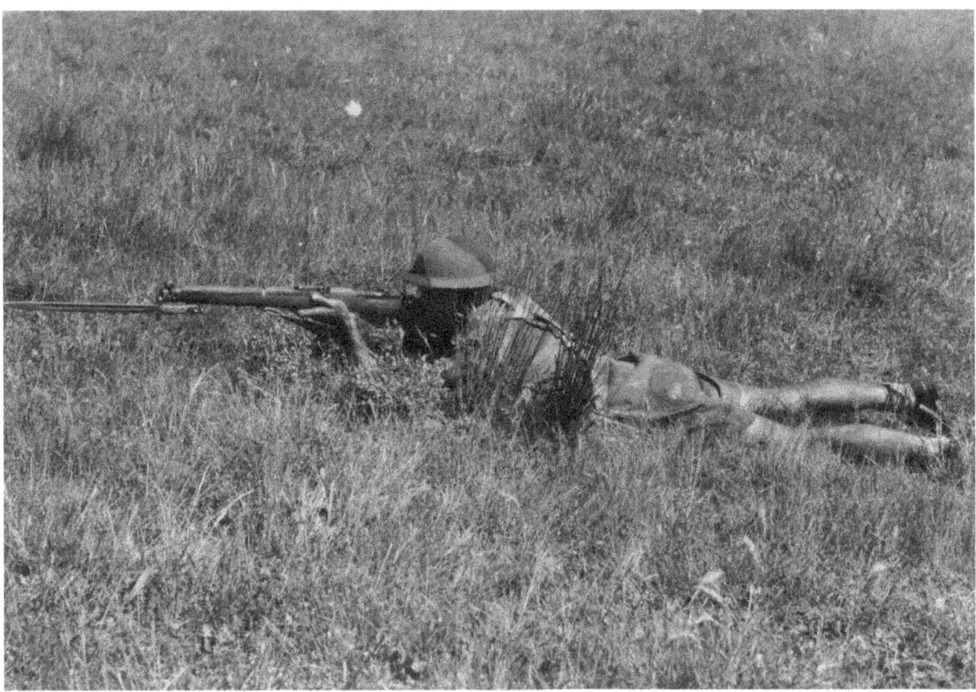

Fig. 8.2. Frank Hinder camouflage experiment. On verso: 'Skin tone', ACT, F.C.H '41'. Frank Hinder Papers. Collection: Archive of the AGNSW.

to the DHS, Hinder was relieved. Like his colleagues Sidney Nolan and Albert Tucker, he felt palpably out of place in the army.

Nevertheless the artists who worked in camouflage were committed to working for the national good. They shared a high sense of national duty, saw themselves as defending democracy, wanted to shield and safeguard citizens against external violence, and were committed to border protection against Japanese invasion. Camouflage was a way of bringing specialist skills to national protection. Their predecessors in WWI had done the same, and set an example. When artists in the Sydney Camouflage Group published their research in *The art of camouflage* in 1941, they defined camouflage in gentle terms as 'art (with a scientific basis)' applied to armies in the field.[20]

A poster by Daryl Lindsay titled *Which way?* embodies the spirit of those times (fig. 8.3). For his concept Lindsay took Prime Minister Robert Menzies' 1941 appeal for Australians to 'devote themselves body and soul' in support of the war, and turned it into visual propaganda.[21] The poster challenged Australian men in the years around 1940 to choose the right path in wartime by following the figure of the soldier in khaki uniform, not the lesser man in the white suit. The choice was between a purposeful, earthy hero striding to the future, or a shirker. From the time of the Anzacs the shirker was the 'non-

20 Dakin (ed.) 1941, p. 3.

21 Menzies' radio broadcast of 17 June 1941 quoted in Penglase & Horner 1992, p. 34.

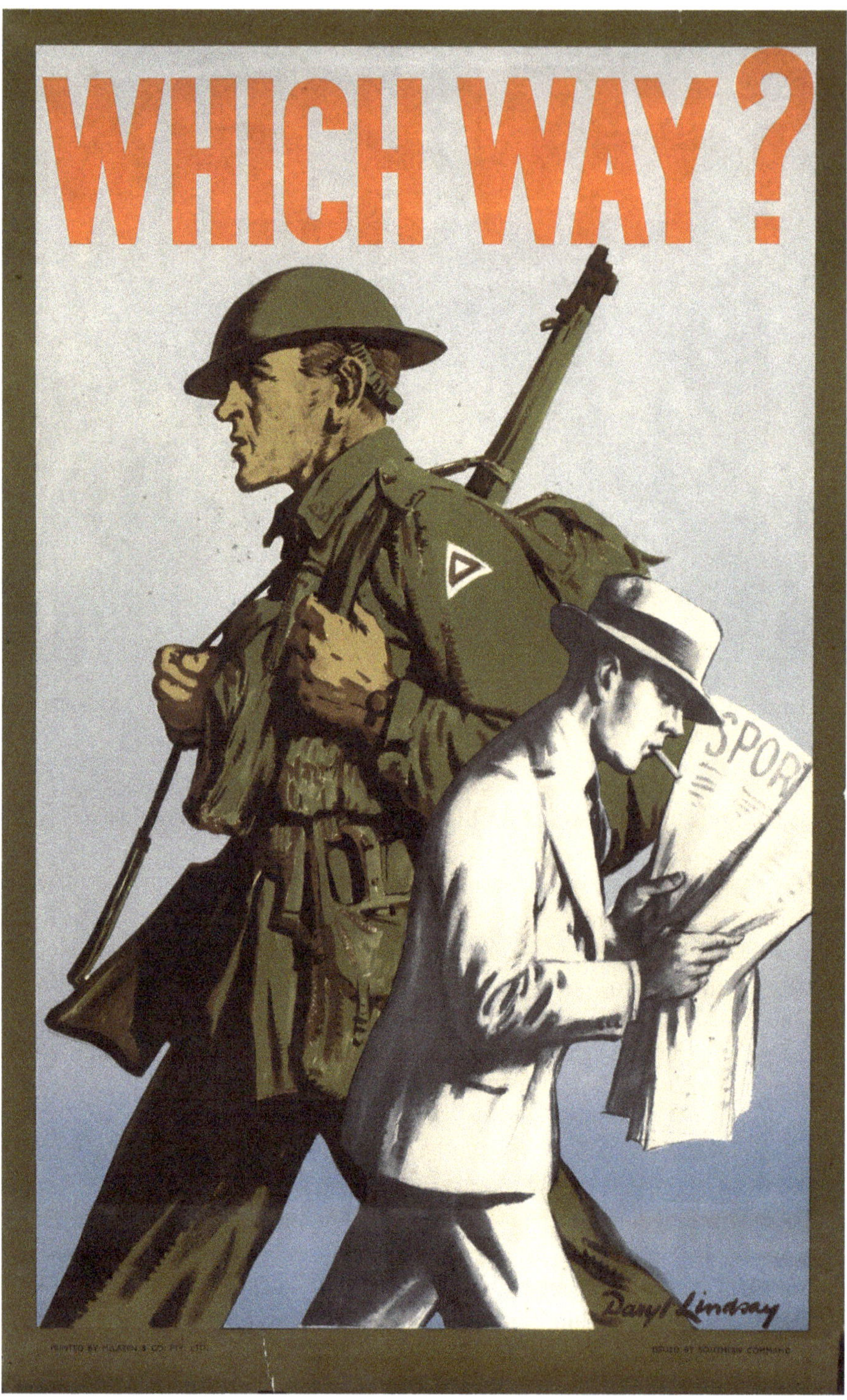

Fig. 8.3. Daryl Lindsay (for Australian Commonwealth Military Forces), *Which way?* Recruitment poster, offset lithograph on paper, 100.6 x 63.6 cm, c. 1939–1943, Canberra, AWM, ARTV06721.

man' despised for his supposed cowardly, selfish behaviour and diminished sense of duty.[22] Shirkers included pacifists and able-bodied men who looked like they should have been doing active service. Consequently when Japan entered the war in late 1941, the prevailing feeling was that all able-bodied individuals who did not contribute to the war effort were undesirable and unwanted.[23]

At least one camoufleur in the DHS, Ralph Shelley, was a conscientious objector who believed that the artist's duty is to redeem violence, not contribute to it. But in the prevailing climate of 'one in, all in', a slogan emblazoned on recruitment posters for the AIF, relatively few people applied for registration on the grounds of conscientious objection, because they ran the risk of prosecution, persecution and imprisonment. Shelley applied and failed to get exemption from military service and was then stationed at Richmond aerodrome in Sydney, working in camouflage. But once there he refused to take the oath pledging his service to the Military Forces of the Commonwealth of Australia and was evicted from the department by Dakin, who wrote on Shelley's file that it was 'undesirable to employ conscientious objectors on the staff', and added a character assessment, written by John Moore, who described Shelley as 'characterless & colourless & cannot be classed as useful'.[24] Shelley stood for everything undesirable about the man in white on Daryl Lindsay's poster. Not only had Moore and Dakin lived through WWI as adults, they were of a generation to remember, and be influenced by, the Universal Service Scheme in Australia which was in place between 1911 and 1929 to provide military training for all young men aged twelve to twenty-six.[25] For the middle-classes at that time, 'the ideal boy was defined by performance in cadets, loyalty to King, country and empire, and by physical fitness'.[26] But these were ideals that Shelley turned his back on.

Daryl Lindsay was also part of Dakin's inner camouflage circle. This well-respected artist of the old school had social connections with Ure Smith as well as with Prime Minister Robert Menzies. They shared a vision of the role of camouflage in civil defence. Daryl Lindsay served in WWI in France and by WWII occupied an honorary position as Keeper of the Prints at the National Gallery of Victoria. It was in 1940 that he joined Dakin's delegation to Prime Minister Robert Menzies to discuss the future of camouflage in the war.[27] Lindsay was then appointed Deputy Director of Camouflage for the State of Victoria and under Dakin's supervision was charged, along with other state deputy directors, to assist the DHS 'to do all it possibly can to further the war work'.[28]

---

22 Garton 1995, p. 191.

23 Smith 1989, pp. 16–18.

24 See entry for Ralph D. Shelley in Department of Home Security Staff register—camouflage section, Canberra, NAA, Series A691, Item 1, accumulation dates 1 January 1942–31 December 1945.

25 The Universal Service Scheme is outlined by the AWM; see 'Conscription', *Military Organisation and Structure*, www.awm.gov.au/atwar/structure/army.asp, accessed 20/10/09.

26 Crotty 2001, p. 81.

27 'Camouflage', 3 July 1941, in Establishment of camouflage organisation and procedures, Canberra, NAA, Series A5954, Item 396/2, p. 3a.

28 'Relations of Camouflage Section, Department of Home Security, to the Services', 3 November 1943, Establishment of camouflage organisation and procedures, Canberra, NAA, Series A5954, Item 396/2.

Artists may not have had much social standing in Australia in 1939, but the value of artists to war became very topical. The fear of Japanese invasion galvanised the nation to such an extent that, although the prevailing attitude in 1941 was one of disillusionment at the advent of another world war, after the bombing of Pearl Harbour people who previously did not support the war now saw their participation as a matter of Australia's survival.[29] The war effort called for unity, and most artists responded to this, although when individuals tried to reconcile anti-war views with national duty, it created anguish and conflict, which, for the most part, was only articulated well after the war was over. In public they pulled together but in private it was difficult to ignore anti-war literature. In 1941 Australia's key journal for the arts, *Art in Australia,* published an article by the influential leader of European surrealism, André Breton, who urged Australians to recognise that, in war, human thought is 'humiliated'.[30] Also confusing was the high proportion of modern artists serving in the armed forces and prepared to sacrifice themselves for a nation that, in peacetime, showed their profession little respect and support. With so many artists fulfilling their national duty, the hypocrisy was noted, and when it came time to write the catalogue for a 1942 *Anti-Fascist Exhibition* held at the Contemporary Art Society in Melbourne, it was suggested that those who 'so strongly attacked Modern Art in this country' should bear in mind how many artists were serving their country in the armed forces.[31]

The war years were a period of fragmentation when thought and action were often split and people behaved out of character, made bizarre compromises and underwent abrupt transformations as a response to the radical social changes inflicted on them. Almost overnight, for example, Frank Hinder became 'Lieutenant Hinder' in the CMF which split his life into two personalities, one civilian and the other soldier. Look through Hinder's sketchbooks and his double life is evident in the way cubist drawings jostle for space with notes about 'snipers' and how abstract sketches are interspersed with pragmatic reminders about the week's upcoming camouflage jobs (figs. 8.4 & 8.5). Hinder himself was intrigued by stories of other people shape-shifting to comply with wartime culture. He made a special note in his diary about the case of Camilla Wedgewood who was the Principal of the Women's College at the University of Sydney and who was 'a strong pacifist and would not allow any of the students to indulge in the war-like process of making camouflage nets; later she was appointed Colonel in the Army'.[32] The extraordinary events of wartime life, the surreal juxtapositions and bewildering social transformations made a deep impression on Hinder and explain why he paid such attention to his diaries, and why they tend to highlight the paradoxes and ironies of life.

In Britain, too, artists recorded stories about dramatic political and psychological changes brought on by WWII. For instance, before September 1939, Roland Penrose and Julian Trevelyan were part of the British surrealist movement, and they were also pacifists.

29 For public responses to the threat of Japanese invasion and how the war united Australians see Penglase & Horner 1992, pp. 62–63.

30 Breton 1941, p. 11.

31 See catalogue Introduction, *Contemporary Art Society Exhibition 1942*, reprinted in Stephen, McNamara & Goad 2006, p. 414.

32 Frank Hinder, Frank Hinder Papers, Sydney, Archive of the AGNSW, MS 1995.1, Box 4, Folder 10.

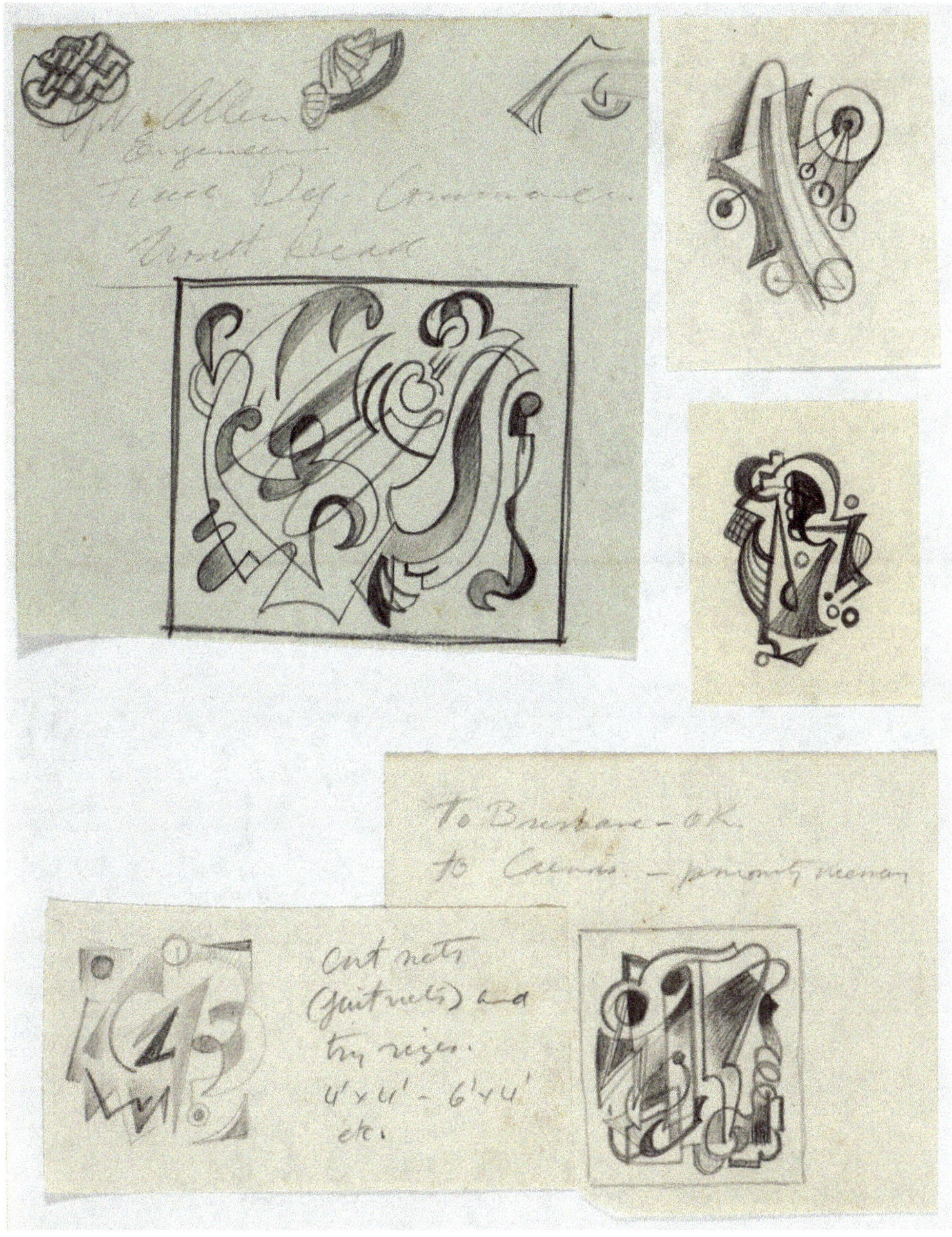

Fig. 8.4. Frank Hinder, from 'Semi Abst.', Sketchbooks 1941 and 1942. Frank Hinder Papers Collection: Archive of the AGNSW.

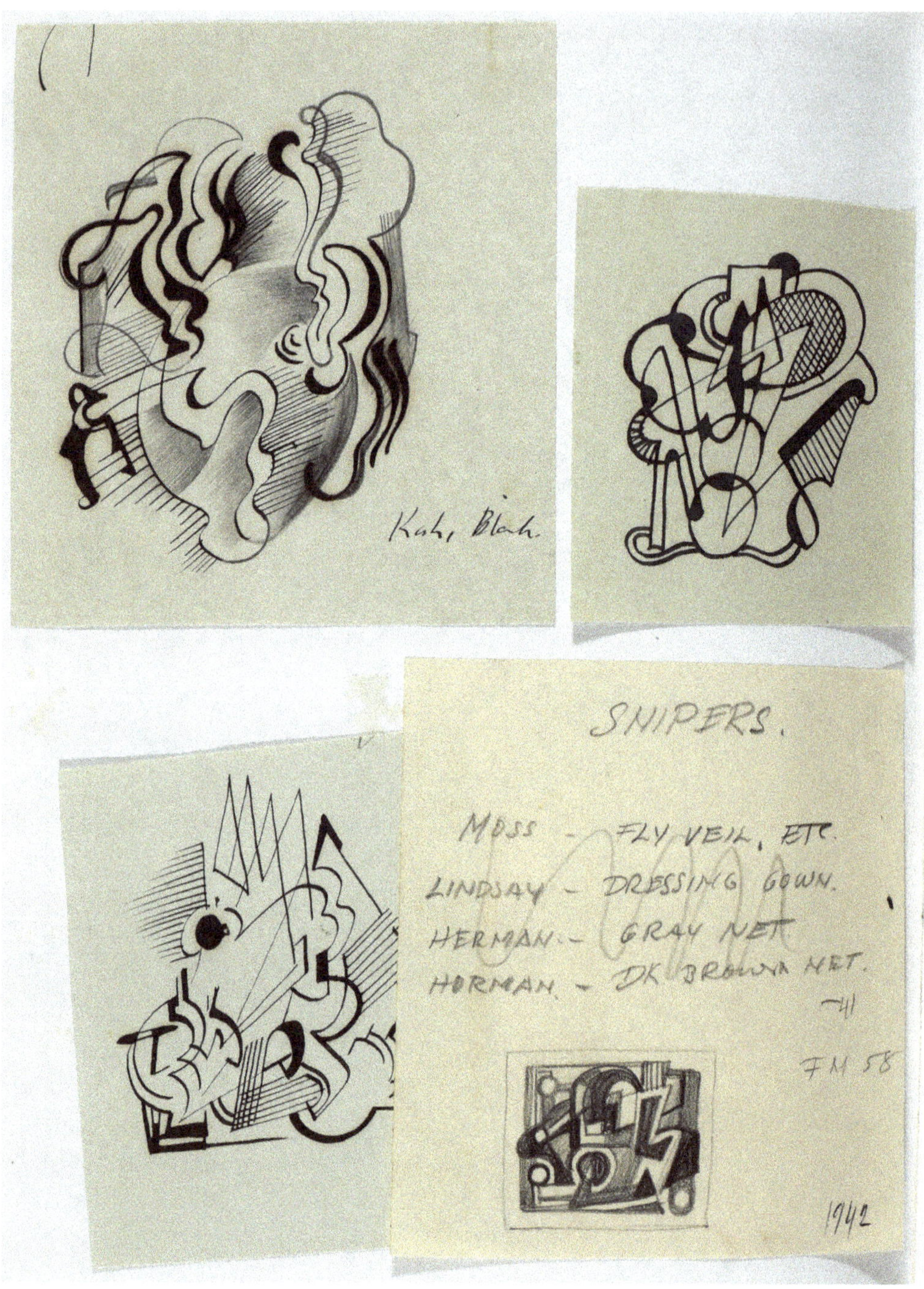

Fig. 8.5. Frank Hinder, from 'Semi Abst.', Sketchbooks 1941 and 1942. Frank Hinder Papers. Collection: Archive of the AGNSW.

Alarmed by the spread of fascism and the Spanish Civil War they marched the streets of London one May Day after 1936, waving a banner that read 'a warlike state cannot create'.[33] But with the onset of WWII they were both deployed for the Industrial Camouflage Research Unit. Within no time Penrose became a camouflage instructor and produced the *Home guard manual of camouflage* (1941). Trevelyan joined the army and served in the Middle East, but he shared his father's 'overwhelming detestation of militarism in all its forms'.[34] The sights he saw in wartime were not only horrific but also so strange and so irrational that the movement of surrealism, with its uncanny juxtapositions, lost its purpose; life had superseded art when 'German soldiers with Tommy-guns descended from the clouds on parachutes dressed as nuns'.[35]

For artists of all allied countries forced to participate in war, the longstanding argument that art is for the good and betterment of humanity had to be reconciled with the argument that war was for the same purpose, namely to eliminate an enemy in Germany and later Japan who threatened to take away freedom, including artistic freedom. In 1942 the US art journal *ARTnews*, a periodical that was read by Australians and was requisitioned by the US government in 1942 to assist war propaganda, acknowledged the ethical dilemma of artists, especially pacifists, but also felt that 'almost too much has been written about the artist's place in the war' when the real issue was not their right to freedom of expression but the risk of freedom itself.[36] Designers and commercial artists were singled out and put on notice that their contribution of skills and experience for the war effort, was expected.

War called for tough attitudes and those who did not at least try to be useful represented all that was weak about society. The ultimate mark of the warrior and model citizen in wartime was willing sacrifice if not of one's life, then one's self-concerns. The rhetoric of sacrifice was impossible to avoid and it infiltrated all manner of communication. When Max Dupain and Bob Curtis requested transfer from Sydney to the war zone in the SW Pacific area, to be more useful to soldiers in the field and to gain experience in operational camouflage, Dupain wrote Dakin a letter inflected with a sense of noble destiny:

> Your idea was for me to go north for a short period, work in the field & return to experimental work with a fuller understanding of the general conditions up there. I discussed this with Bob Curtis & he feels that it is what we have been wanting for a long time.[37]

Very likely Dupain felt pressured to accompany Bob Curtis to the SW Pacific, and the psychological pressure of Anzac history should not be discounted. Dupain, who was born in 1911, was still very young and impressionable during WWI, and this was precisely the time the Australian national psyche was infused with the heroism of the Anzacs who died for race, Empire and nation. In fact art historian Mary Eagle argues that children of that

33 Trevelyan 1957, p. 73.

34 Trevelyan 1957, pp. 111–12.

35 Trevelyan 1957, p. 80.

36 Brennan 1942, p. 7.

37 Max Dupain to William Dakin, 10 March 1943, Bankstown experiments, Sydney, NAA, Series C1905 T1, Item 3, Box 13.

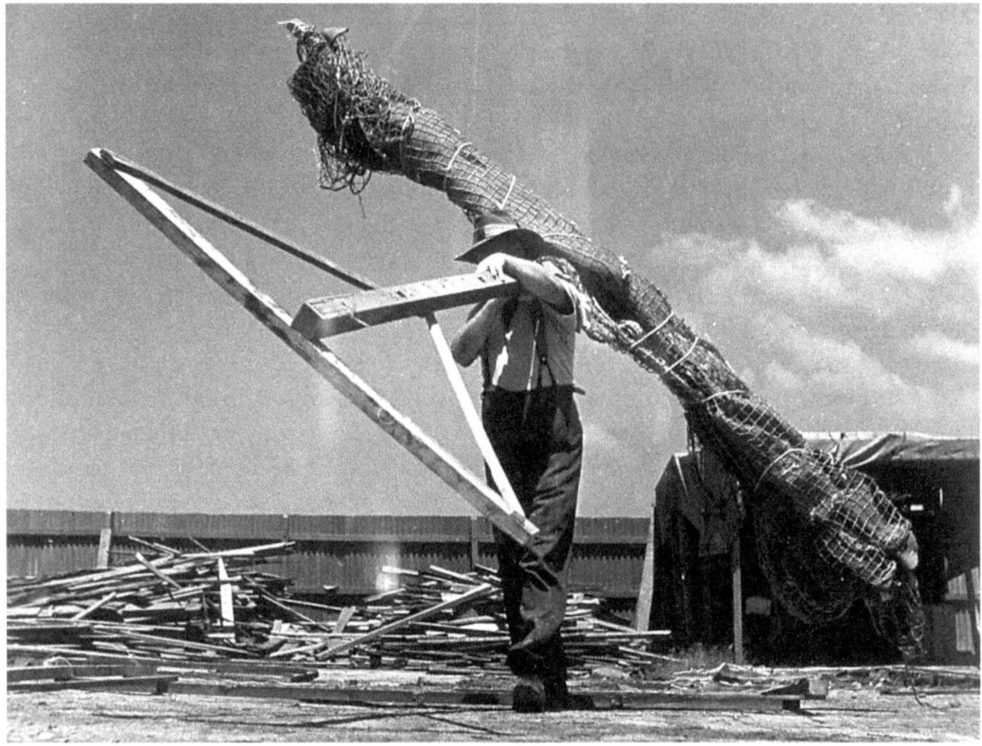

Fig. 8.6. Bankstown camouflage experiment using T-unit with garnished net, Jan/Feb 1943. From the collection of the NAA: C1905, 3. Photograph, Max Dupain.

period were not only 'burdened in childhood with an image of sacrificial death in war' but also with the responsibility in WWII of honouring the Anzac legend.[38] The psychology of this social backdrop combined with the reality that camoufleurs with the DHS were under pressure to work in forward areas, explains Dupain's wish for relocation to the war zone. Even in the DHS, where camoufleurs were non-combatants, it was a mark of bravery to volunteer for the New Guinea area and a matter of pride to pass the fitness test. And as Isobel Crombie has shown in detail through her biography of Max Dupain, moral and physical fitness were critical to his ethos as a result of having been raised in a family where fitness was understood as the eugenic ideal of a healthy mind in a healthy body.[39]

The burden of sacrifice is an interesting way to approach certain of Dupain's war photographs taken during his months serving as a camoufleur at Bankstown aerodrome, attached to the RAAF. In many he favoured a low viewpoint which, in the case of photographs of men at work, gives real nobility to their task. But the symbolism of sacrifice is overt in a photograph of a fellow camoufleur carrying a portable camouflage net in the shape of a cross (fig. 8.6). On one level it documents the practicality of a portable concealment

38 Eagle 1989, p. 15

39 Crombie 2004, pp. 14–16.

device, but on another it functions within the conventional symbolism of cross-bearing, the sign of willingness to endure a burden for a greater good. Only a matter of months later, Dupain's request to go north to the war was granted and the concept of personal sacrifice put in sharp perspective by exposure to service personnel who had been politically and psychologically transformed and damaged by the campaign in New Guinea, and who, in the words of historian John Raftery, were no longer 'ordinary men [but] dutiful aggressors'.[40] If Dupain felt initial enthusiasm at the idea of transferring to Darwin to conceal oil tanks, and of photographing camouflage schemes from reconnaissance planes in Papua and New Guinea, it was drastically dampened by the experience itself. By 1944 he could not wait to leave the north and the SW Pacific and return to what he called 'civilization'.

The clash of cultures between artists in the DHS and members of the armed forces over lack of accreditation and uniform was detrimental to their work on the mainland. But it was even more consequential in the New Guinea area. The DHS's role there was to assist the RAAF and to liaise with Canberra on conditions and problems with respect to camouflage. In 1943 a small hand-picked crew of officers from the department signed the official secrecy undertaking, and embarked on service in Papua and New Guinea. What they encountered there was not only an immense military colonisation of the area, but also an entirely foreign world of people and landscapes. They found it impossible to adapt to the physical and social world that is now called Papua New Guinea but their mission at the time was to instil a camouflage way of thinking in the minds and behaviours of jungle-fighting troops, to convince both army and airforce in the region that concealment and deception were essential for saving lives and for winning the war, and to make individual camouflage a matter of instinct and second nature.

40 Raftery 2003, p. 37.

PART 4

THE FIELD: NEW GUINEA AND PAPUA

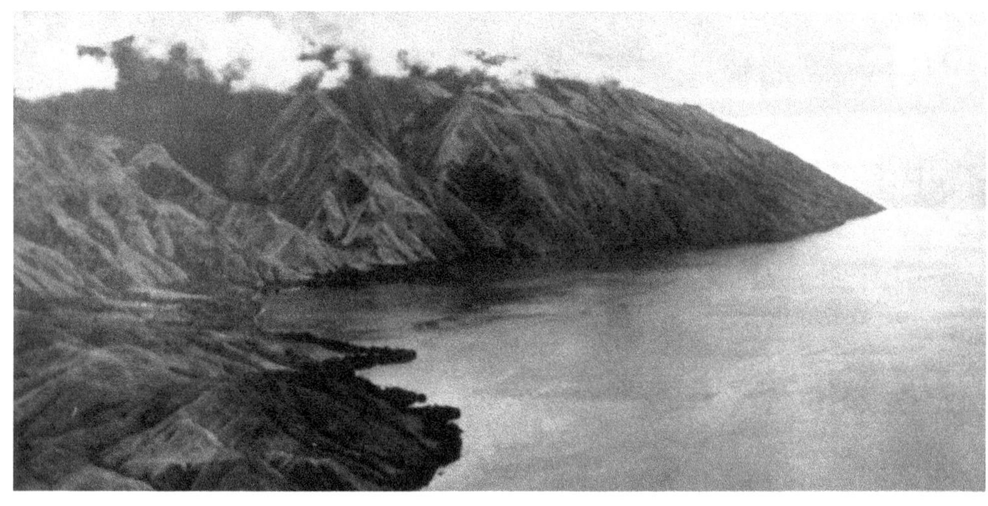

Fig. 9.1. A corner in the Milne Bay area well known to American and Australian soldiers. Hills fall sheer to the water and are covered in kunai grass from three to ten feet high'. Bridges 1945, p. 243.

Chapter 9

Jungle

Conspicuous in the 'remarks' columns of DHS staff records are the comments 'volunteer for New Guinea', 'fit for tropical service' and 'willing to serve outside Australia'. These identified men who volunteered for or requested to go forward to the war zone known variously as the New Guinea area, the New Guinea region, and the SW Pacific area. If they declined that was also noted. The pressure for camouflage officers in the department to volunteer, or accept a placement in the region, was considerable, as it was for any man in Australia who was physically fit. But the expectation increased appreciably in late 1943 when camouflage directorates were set up in the New Guinea area (fig. 9.1).

Of the 28 individuals in the DHS who did volunteer, only eight passed the fitness examination and took the oath of secrecy preventing any unauthorised discussion of the aims and objectives of their work in New Guinea:

> I … an officer of the Department of Home Security hereby undertake not to make use of and not to communicate to any unauthorised person any information concerning the contents of correspondence, plans, drawings or other documents dealt with by Home Security officers or any instructions issued with regard to camouflage or any of the methods or manners in which it is planned, arranged or carried out. I further undertake not to refer to camouflage either in my correspondence or conversation with any unauthorised person and particularly to avoid any reference to my own duties. I understand and acknowledge that any breach of these undertakings will render me liable to heavy penalties at law.[1]

Between 1943 and 1945, George Adams, Edward Billson, Victor Corlett, Bob Curtis, Max Dupain, John Robertson, Clement Seale and Colin Wyatt—peace-loving citizens at the outbreak of war working in architecture (Billson and Robertson), photography (Dupain), design, advertising, illustration and commercial art (Adams, Curtis, Corlett, Seale and Wyatt), but forced by the circumstances to take up war duties—were stationed for up to six months in the jungles of Papua and New Guinea.[2] When war descended on

---

1 'Undertaking', 1943, Camouflage—general—camouflage organisation in New Guinea, Canberra, NAA, Series A453, Item 1943/17/1510 Part 1.

2 For one source of the otherwise scant information on their exact occupational backgrounds see Camouflage—Personnel NSW staff appointments, Canberra, NAA, Series A453, Item 1942/28/2298. Adams was not the first choice of man for the New Guinea region. He was called in to replace Douglas Annand who 'did not desire, for personal reasons, to proceed to New Guinea', see A.W. Welch,

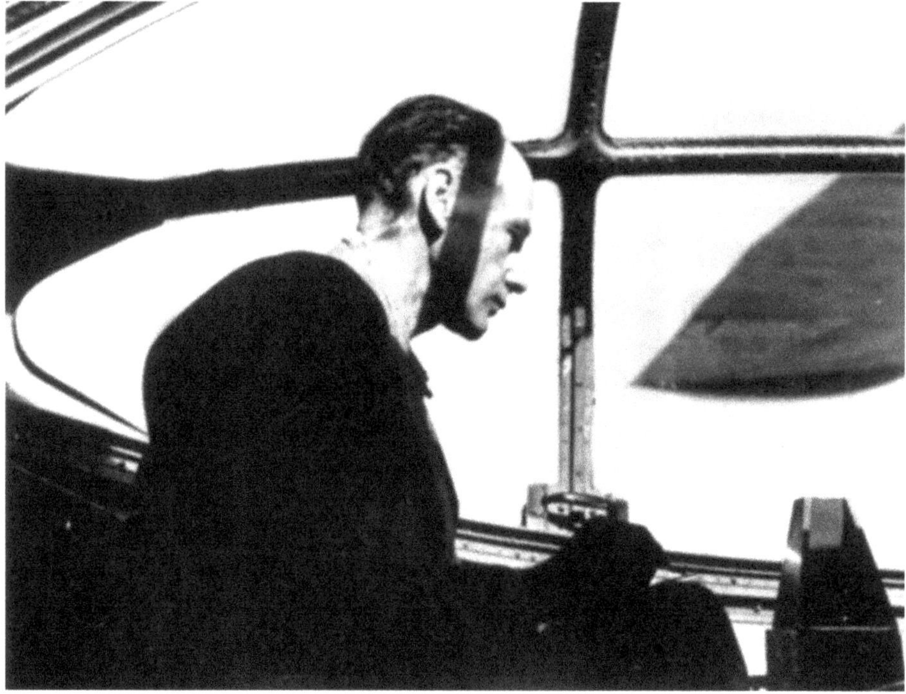

Fig. 9.2. Captain Robert Emerson Curtis sketching an aerial scene from a Catalina, Cairns, 1942. Canberra, AWM, AWM P02186.003. Photograph, Max Dupain.

the wider Melanesian region, its rugged terrain and exotic tropical rainforests and coral islands were already well embedded in the western imagination through books, stories and images created by individuals in the arts and sciences who adventured there to collect, classify and study animals, plants and people. Among them were anthropologists Bronislaw Malinowski (1884–1932) and Francis E. Williams (1893–1943), naturalists and zoologists William Saville-Kent (1845–2908) and Edward A. Briggs (1890–1969), and artists Marion Ellis Rowan (c. 1847–1922) and Emile Nolde (1867–1956). All discovered frontiers of knowledge, experience and expression in the remote Pacific. But for the majority of Australians, the region symbolised a primal, unruly world, one that only magnified the fear of jungle warfare. New Guinea, noted Donald Friend, was the place that soldiers in the AIF were 'afraid of being sent to'.[3]

In February 1942 the Australian territory of Papua and the mandated territory of New Guinea were removed from civilian colonial government and put in the military hands of the Australian New Guinea Administrative Unit (ANGAU).[4] Port Moresby became the camouflage centre of the SW Pacific area. In September 1943 after Dakin convinced the

'Camouflage in New Guinea', 13 September 1943, Camouflage—general—camouflage organisation in New Guinea, Canberra, NAA, Series A453, Item 1943/17/1510 Part 1.

3 Donald Friend, 7 December 1943, in Gray 2001, p. 436.

4 Lepowsky 1990, p. 209.

government of the need for a base in the area, Bob Curtis became Officer-in-Charge of the Camouflage Section for the DHS (fig. 9.2). He was relocated first from Sydney to the camouflage instructional unit in Townsville, Queensland, then to Port Moresby and finally to Milne Bay, New Guinea, with the RAAF's 'No. 9 Operational Group'. He was shocked by the strangeness of the isolated mountainous Milne Bay environment and only reassured by the military presence which, he said, made it 'more civilized' (fig. 9.1).[5] After further reconnaissance of the region, Curtis felt that Goodenough Island in nearby Papua was more suitable for the department's headquarters and in November 1943 he relocated once again. Goodenough offered easy connection to RAAF units, and was headquarters for US General Krueger and the Sixth Army.[6] Unlike Milne Bay it had the advantage of a large expanse of level plain, and a new airfield. He hoped the DHS would send camouflage-architects or at least people who could draw building plans and had experience of construction, photographers, trained camoufleurs, and men able to tolerate tough living conditions.[7] He requested the help of a sturdy type like Frank Hinder, but Hinder remained on the mainland.[8] He anticipated that Max Dupain would accompany him in 1943 but Dupain remained in Darwin.[9]

In August 1943 Curtis wrote to Canberra requesting men who were:

> physically fit; able to use a rifle; highly resourceful in ideas which are simple and practical; capable of quick decisions; a good mixer with officers, airmen and must have the right way with natives; ready to give a hand in building a camp, taking a share in any work which is more vital at the time than his own; self-sufficient in most things, such as gear and personal equipment.[10]

By December 1943 Colin Wyatt, George Adams, John Robertson and Clement Seale were assigned to the area primarily to work on projects for the RAAF's Survey and Design units. It was far from ideal but in Seale's case, at least, being part of the DHS, and taking up mobile camouflage duties in advanced stations, was infinitely preferable to being called

5 R.E. Curtis letter to A.W. Welch, 20 August 1943, Camouflage—general—camouflage organisation in New Guinea, Canberra, NAA, Series A453, Item 1943/17/1510 Part 1.

6 R.E. Curtis letter to A.W. Welch, 21 October 1943, Camouflage—general—camouflage organisation in New Guinea, Canberra, NAA, Series A453, Item 1943/17/1510 Part 1.

7 R.E. Curtis to A.W. Welch, 3 November 1943, Camouflage—general—camouflage organisation in New Guinea, Canberra, NAA, Series A453, Item 1943/17/1510 Part 1.

8 R.E. Curtis letter to A.W. Welch, 3 November 1943, Camouflage—general—camouflage organisation in New Guinea, Canberra, NAA, Series A453, Item 1943/17/1510 Part 1; Hinder was never stationed in New Guinea or Papua with the DHS, but made a reconnaissance trip to Rabaul in New Guinea in 1941 to report on camouflage. See Frank Hinder, 'Records of Lt F.C. Hinder Camoufleur', Canberra, AWM, PR 88/133, File 895/4/182, Item 9 of 12.

9 R. Curtis letter to A.W. Welch, 14 September 1943, Camouflage—general—camouflage organisation in New Guinea, Canberra, NAA, Series A453, Item 1943/17/1510 Part 1.

10 'Memo to the Technical Director', 31 August 1943, Camouflage—general—camouflage organisation in New Guinea, Canberra, NAA, Series A453, Item 1943/17/1510 Part 1.

up for military service.[11] All were fully accredited with the RAAF as officers: Wyatt was Assistant Officer and assigned field duties; Adams was Camouflage Officer attached to 41 Radar Wing on field duties and demonstrations; Robertson was Camouflage Officer attached to 62 Works Wing as architect and pre-planner; Seale was Camouflage Officer on field duties.[12] Max Dupain left Sydney for Goodenough Island on 22 February 1944. En route he photographed operational jungle camouflage at Finschhafen and Nadzab, the Admiralty Islands and New Britain, from the air. The images were intended for instructional purposes.[13] Goodenough Island was their base, but their service entailed constant movement with the RAAF's advancing operational units into environments where, according to Curtis, 'change is the only constant state possible'.[14] Consequently in December 1943 they were variously stationed at Finschhafen, Nadzab, and Kiriwina Island to undertake radar camouflage and to advise on individual concealment.[15]

As a special squad with a mission to protect Australian soldiers, they were keen to serve in forward areas. They were trained to handle guns and grenades. Nevertheless they were non-combatants working with only basic equipment: torch, waterproof matches, mirror, soap, tin and bottle openers, nails, rope and string, knife, bottle of ink and snake-bite kit, sandshoes to avoid hookworm, camp equipment and mosquito net, saw and wire-cutters.[16] If situations demanded additional equipment, officers were expected to improvise. Victor Corlett, who was stationed in the region from 1944 to 1945, wrote that they were ill-prepared for the conditions they met, and further that the Department of Defence in Canberra seemed uninterested in their welfare. Canberra, he said, 'did not seem to care how we fared in these forward areas and strongly feel they have let us down badly on many points'.[17] And he was not the only one who felt this way. George Adams reported on the regrettable gulf between what had been taught about camouflage back home with the department, and the reality on the ground in the New Guinea area:

---

11 Seale was expecting to be called up in 1942. See John D. Moore letter to DHS, 12 November 1942, Camouflage personnel NSW Camoufleurs staff appointments, Canberra, NAA, Series A453, Item 1942/28/2298.

12 R.E. Curtis memo to Air Officer Commanding, 27 December 1943, 'Camouflage Establishment. Advance Headquarters 9. Operations Group', in Camouflage—general—camouflage organisation in New Guinea, Canberra, NAA, Series A453, Item 1943/17/1510 Part 1.

13 R.E. Curtis letter to The Secretary Department of Home Security, 2 April 1944, Camouflage—general—camouflage organisation in New Guinea, Canberra, National Archives of Australia, Series A453, Item 1943/17/1510 Part II.

14 R.E Curtis letter to A.W. Welch, c. 15 December 1943, Camouflage—general—camouflage organisation in New Guinea, Canberra, NAA, Series A453, Item 1943/17/1510, Part 1.

15 R.E. Curtis memo to Air Officer Commanding, 27 December 1943, 'Camouflage Establishment. Advance Headquarters 9. Operations Group', Camouflage—general—camouflage organisation in New Guinea, Canberra, NAA, Series A453, Item 1943/17/1510 Part 1.

16 R.E. Curtis, 'Requirements of Camouflage Officers entering a forward area: New Guinea and Islands', 31 August 1943, Camouflage—general—camouflage organisation in New Guinea, Canberra, NAA, Series A453, Item 1943/17/1510 Part 1.

17 Victor Corlett, 'General report on Operations: SW Pacific', June 1945, Camouflage—general—camouflage organisation in New Guinea, Canberra, NAA, Series A453, Item 1943/17/1510 Part II, p. 10.

> All that the department has said about operational camouflage is perfectly alright in theory, but the practical difficulties of working in an U.S/R.A.A.F combination, with an American C/O on an operational mission involving perhaps 200 personnel of all types [making] it impossible for us to apply our specialized knowledge just at the time when it may be most wanted.[18]

Their many reports dispatched to Dakin complained about heat, mosquitoes and rain, and never failed to mention the low priority that RAAF personnel gave camouflage. But they shied away from the subject of personal emotional strain and rarely provided insight into the barbarism of war. Victor Corlett's report is an interesting exception. He was first employed by the Labour Squad on aerodrome camouflage in Sydney before transfer to Queensland and then New Guinea in 1944. When he wrote a 'General Report on Operations' from New Guinea and recounted how his radar-siting team got 'slight enjoyment' from burning a Japanese corpse, the conditions of wartime life for camouflage officers in New Guinea—and insights into the mentally unstable atmosphere of hatred and revenge that made the New Guinea campaign infamous—were registered for the first time by the DHS.[19]

The physical hardship of undertaking camouflage in the New Guinea area was overwhelming. Officers with the DHS had not anticipated insufficient supplies of paint to camouflage white tents, and rapid deterioration of linoflage and hessian in the humid tropics where Milne Bay's annual rainfall was 105 inches (2667 mm) and the skies in the Trobriand Islands and Gasmata were almost always thick with mist or heavy rain.[20] Little had prepared them for the complexity and difficulty of building and disguising radar stations in this terrain, of placing scrimmed camouflage net over the radar machine's revolving base while allowing all parts of it to move and keep balanced. Ensuring that the radar's outline was properly concealed from aerial view, as well as from ground view, was no easy matter (fig. 9.3).[21]

Victor Corlett noted that in some cases, Allied personnel were their own worst enemy. They blithely cleared large areas of jungle and kunai grass with bulldozers, making it impossible to hide ground activity from aerial view. At Nadzab and Kiriwina the RAAF did not observe 'blackout' at night leaving them visible and vulnerable to regular nightly enemy aerial reconnaissance. They made military objects conspicuous rather than invisible, with badly designed static camouflage (fig. 9.4). And Corlett noticed that Australian airforce personnel preferred to wear traditional khaki—which showed up too loudly against the

18 G.H. Adams, 'Extracts taken from field reports by G.H. Adams', 22 December 1943–12 January 1944, Camouflage—general—camouflage organisation in New Guinea, Canberra, NAA, Series A453, Item 1943/17/1510 Part II.

19 Victor Corlett, 'General report on Operations: SW Pacific', June 1945, Camouflage—general—camouflage organisation in New Guinea, Canberra, NAA, Series A453, Item 1943/17/1510 Part II, p. 9

20 R.E. Curtis, 'General Comments on Camouflage noted in Milne Bay, Goodenough Is, Port Moresby', August 1943, Camouflage—general—camouflage organisation in New Guinea, Canberra, NAA, Series A453, Item 1943/17/1510 Part 1.

21 G.H. Adams, 'Extracts taken from field reports by G.H. Adams', 22 December 1943–12 January 1944, Camouflage—general—camouflage organisation in New Guinea, Canberra, NAA, Series A453, Item 1943/17/1510 Part II.

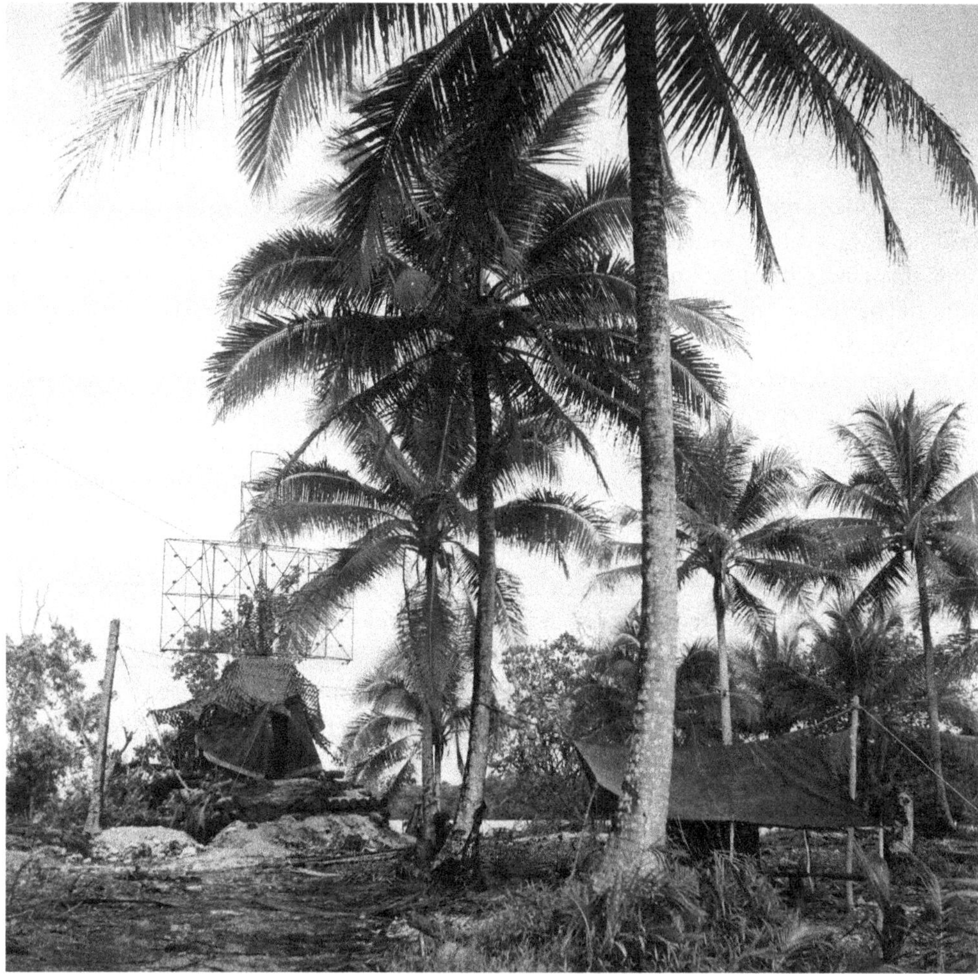

Fig. 9.3. RAAF radar station with camouflage, New Guinea, c. 1943–44. The collection of the NAA: C1905, 6.

dark green vegetation of the SW Pacific—rather than the new green uniforms. He referred to the dark uniforms as 'Black Widows' because, when worn at night, they made the wearer invisible and deadly.[22] Neither would troops wear Skin Tone Commando Cream, even though their European flesh shone under the stars at night. Being aware in the jungle of one's physical colour and shape, as well as sound and smell, was common sense but, wrote Corlett, 'will the Services ever realise that??'[23]

DHS staff were also ill-prepared for encounters with indigenous New Guineans and Papuans who offered them their labour. In anticipation of this problem Americans and

22 V. Corlett, 'Jungle Green Uniforms', 31 March 1945, Camouflage—general—camouflage organisation in New Guinea, Canberra, NAA, Series A453, Item 1943/17/1510 Part II.

23 Victor Corlett, 'General report on Operations: SW Pacific', June 1945, Camouflage—general—camouflage organisation in New Guinea, Canberra, NAA, Series A453, Item 1943/17/1510 Part II, p. 10.

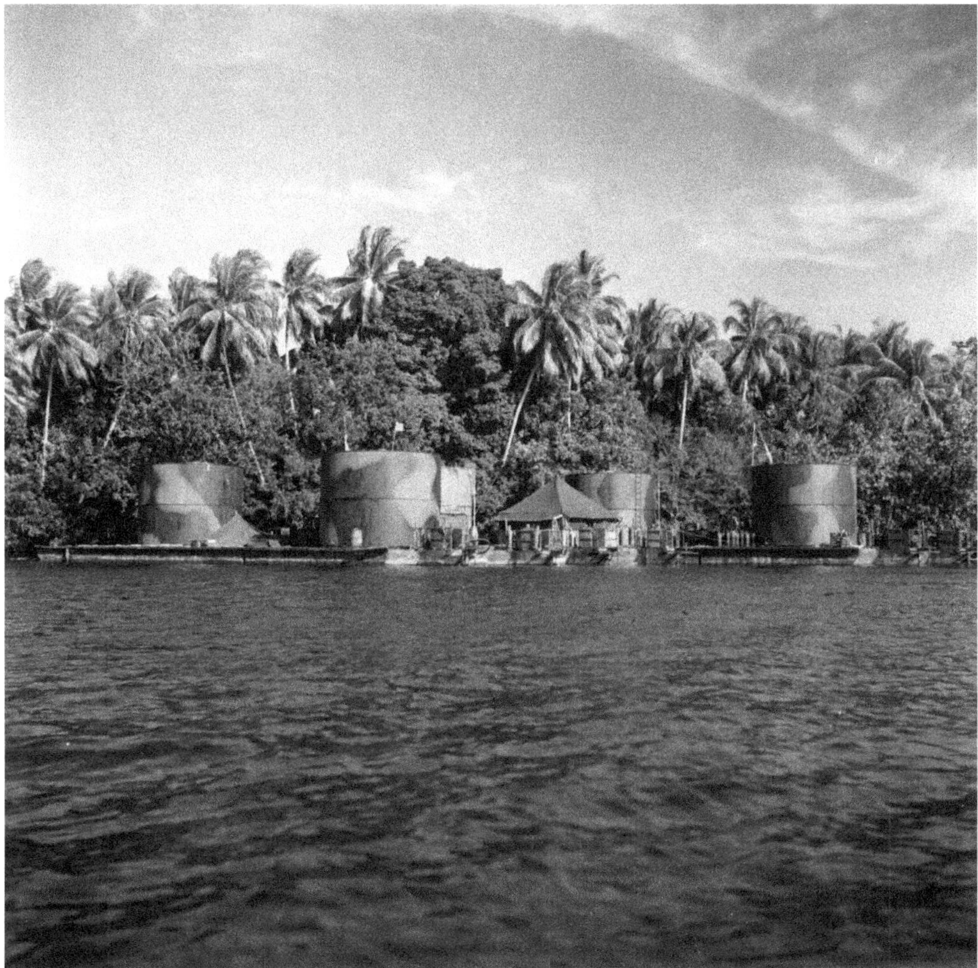

Fig. 9.4, Camouflaged tanks and vessels, New Guinea, c. 1943. The collection of the NAA: C1905, 6.

Australians were issued with pocket-book guides to local people. Australian troops read *You and the native* by Lieutenant F.E. Williams, who was also the government anthropologist in Papua during the war. He warned that 'natives are not children, are not animals, but peoples with their own definite cultures and ways of living'.[24] Americans read *Guide to New Guinea* but, when reflecting on the shock of this completely new cultural encounter, one American asked: 'how much could GIs really learn about the complex tribes of New Guinea?'[25]

Nonetheless, Colin Wyatt was grateful for knowledge of plants and bushcraft which he learnt from indigenous workers and he criticised RAAF personnel for underestimating the intelligence of local labourers. At the same time he was not convinced he could trust the

24 Lieut. F.E. Williams, 'General Errors in Attitude', Geographical Section South West Pacific Area: Relations with Natives of New Guinea in Wartime, AWM Collection, Series AWM 54, 506/8/2, 1942, p. 1.

25 Schrijvers 2002, p. 29.

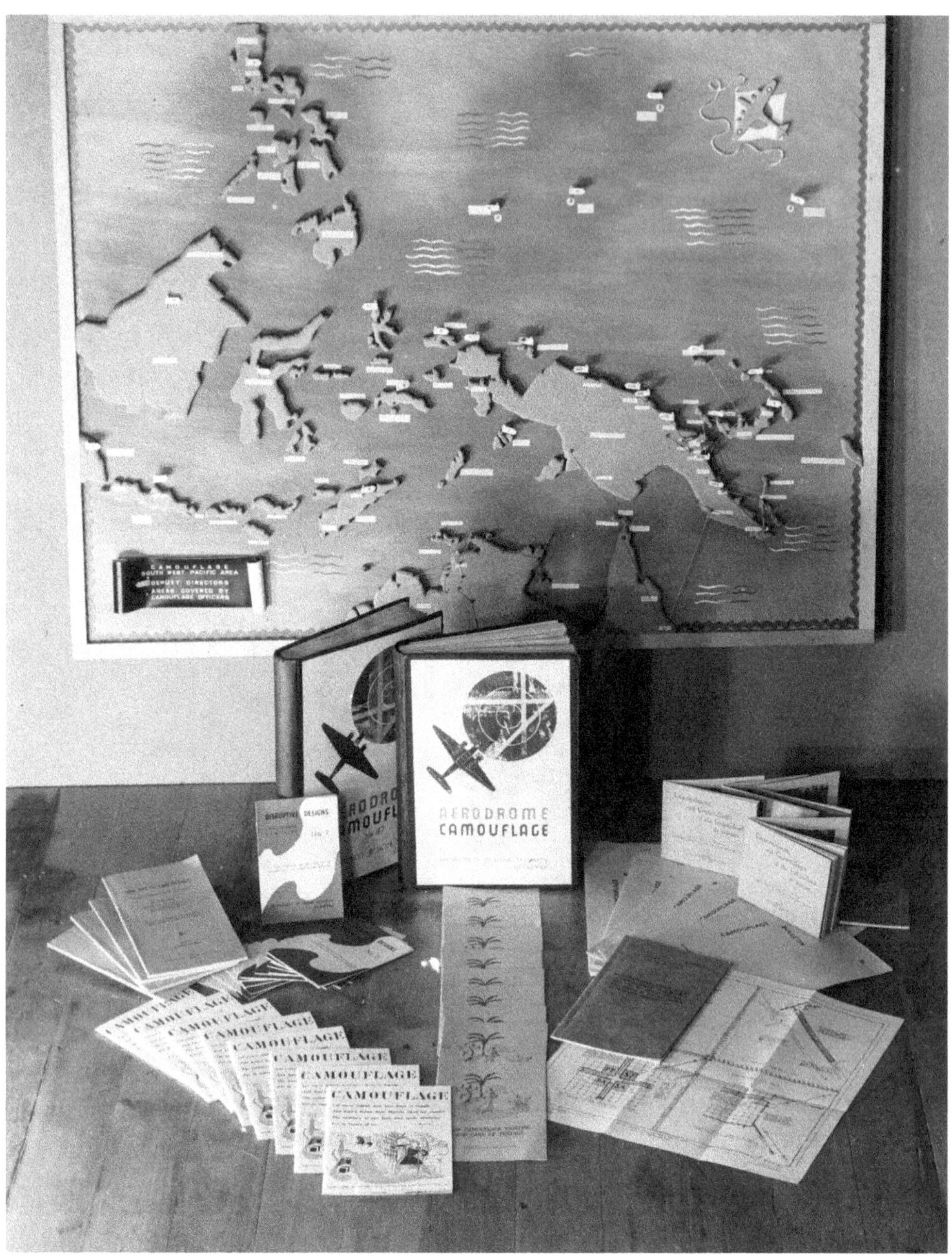

Fig. 9.5. Department of Home Security map and bulletins for camouflage in the SW. Pacific, c. 1943. The collection of the NAA: C1905, 10.

local people to keep secret about the military's camouflage installations.[26] His ambivalence was typical of Australian personnel who came from a society that justified colonisation of the region on the basis that New Guineans and Papuans were not capable of self-determination, and oscillated between romantic notions of people in touch with nature, and hostile attitudes to 'savages' who could easily trade secrets with the enemy.

Dakin, however, had a special interest in what he called 'natives' and what their behaviours could teach Europeans about camouflage. As people who belonged, in his view, to an earlier stage of human development, 'natives' had plenty to offer on camouflage instincts. He hoped the New Guinea area and its primitive tribes would rekindle in military personnel the lost knowledge and art of natural concealment and deception that had 'died out in present day man'.[27] He hoped the jungle would bring this long-buried knowledge back into consciousness and that Australians would once again feel fear and respect for shadows. In particular he hoped they would discover the primitive within. It struck him that the Japanese were more in tune with nature. Yet military personnel turned the reputation that Japanese developed for sophistication with camouflage, against them. For some it was evidence that the enemy was naturally feral, and 'the nearest thing in human form to a bush animal'.[28]

To communicate the theory and practice of concealment and deception to Australian troops in the SW Pacific, the DHS printed a selection of manuals. It was in *Camouflage Bulletin 17* that Dakin explained the concept of intuition (fig. 9.5). This, he said, was best demonstrated when 'wild animals' and 'wild natives' freeze for concealment.[29] The New Guinea and Papua regions confirmed Dakin's spurious racialised theories: black skin is a functional evolutionary adaptation to the jungle; 'natives' are at an earlier stage of evolution than Europeans and therefore more in touch with the instinct for concealment and deception; 'natives' are ruled by their physical surroundings. In this environment, he argued, it was valid and reasonable to return to a state of nature. Further, the struggle for existence was reason enough to expect soldiers to embody 'the instincts and tricks of concealment into all his activities'.[30] Not only did he caution troops about the colour of the white person's skin but he encouraged them to use all their senses, for while sight is acutely needed, 'the ears are the most important in the very silent jungle, and the sense of smell may easily detect an enemy'.[31]

---

26 Colin W. Wyatt, 'Notes on New Guinea Area', 1 March 1944, Camouflage—general—camouflage organisation in New Guinea, Canberra, NAA, Series A453, Item 1943/17/1510 Part II.

27 W. J Dakin, 'The modern position and value of camouflage in regard to war in forward and base areas', Air intelligence reports, Sydney, NAA, Series C1707, Item 36, p. 5.

28 Kneale (VX21257) 1943, p. 121.

29 W.J. Dakin, 'Camouflage bulletin no. 17', 22 December 1943, Camouflage report 1939–1945 (appendix O), Canberra, AWM, Series 81 [77 Part 4], 1947, p. 11.

30 W.J. Dakin and the Camouflage Directorate, 'Concealment, and camouflage of the individual in warfare', Canberra, DHS, March 1944, in W.J. Dakin, Camouflage report 1939–1945 (with appendices K–N), Canberra, AWM, Series 81 [77 Part 3], 1947, p. 2.

31 W.J. Dakin and the Camouflage Directorate, 'Concealment, and camouflage of the individual in warfare', Canberra, DHS, March 1944, in W.J. Dakin, Camouflage report 1939–1945 (with appendices K–N), Canberra, AWM, Series 81 [77 Part 3], 1947, p. 40.

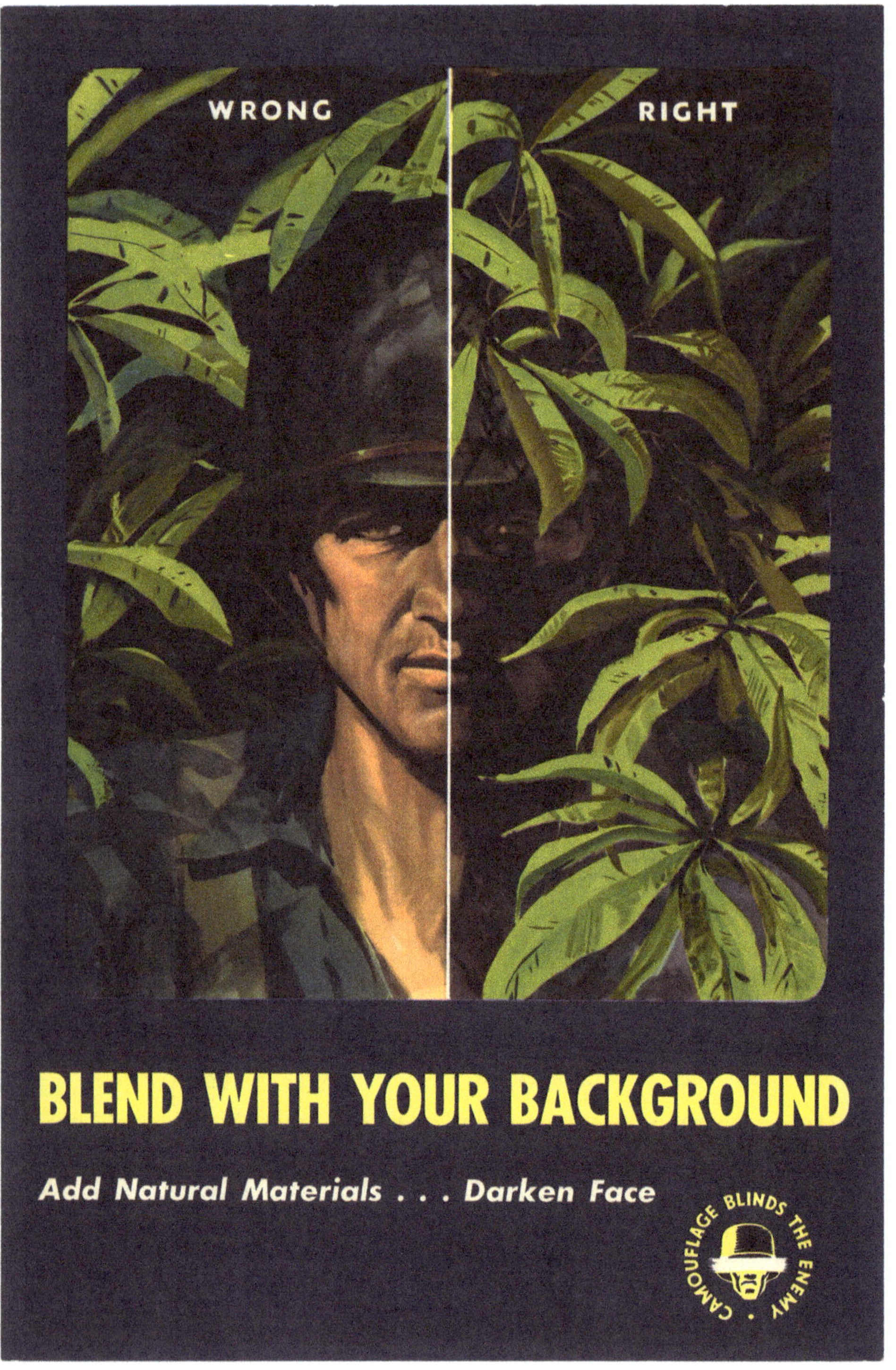

Fig. 9.6. *Blend with your background,* Department of the Army Poster c. 1942–44. The collection of the NAA: MP222/2, 3.

The Australian army, too, stressed the advantages of primitive instincts. By the time camouflage officers with the DHS arrived in the region, Allied land forces for jungle fighting in the SW Pacific area had already compiled and distributed a pamphlet on 'Soldiering in the Tropics (SW Pacific Area)' and it advised troops to make the jungle their 'friend'. It tried to allay the myth that the term 'jungle' meant impenetrable terrain, and identified silence, invisibility and surprise as the ultimate tactics (fig. 9.6). More specifically it advised that the 'jungle soldier must fight like a native armed with more deadly weapons and advanced knowledge [and] if detected by sound or sight the jungle soldier withdraws into the cover of the jungle and seeks another opportunity to strike with surprise'.[32] AIF posters reinforced the message: 'Conceal to live: live to attack'.[33]

It was as if the environment—popularly known as a place of struggle where only the fittest survived—was made for war. Certainly, the jungle came to symbolise humanity's original, primal home. Indeed, one Australian soldier, not in New Guinea but in the Malaysian jungle, drew a picture of himself devolving to an ape.[34] In the New Guinea region soldiers photographed themselves swinging across rivers, Tarzan-like, holding jungle-vines (fig. 9.7). But in many cases, primitivistic behaviour was symptomatic of serious psychological disintegration; psychologists back home on the mainland of Australia warned families to expect returning soldiers to have undergone a 'regression to the primitive' in reaction to mental and physical strain demanded by the New Guinea campaign.[35]

But was camouflage—a form of warfare where being primitive was encouraged—itself a cause of mental regression, and did it lead to psychological disturbance? The case of Peter Medcalf, a nineteen-year-old rifleman with the 15th Battalion, 29th Brigade on Bougainville in Melanesia, suggests it could. From 1944 he spent months 'listening, smelling, peering' in the jungle.[36] He learnt to embody the instinct of concealment to such an extent that in hindsight he realised he had 'become part of this, part of the dim tracks in the wastes of rainforest, the rushing rivers with their secretive, tangled banks'.[37] Among the rivers, vines and thick vegetation, among spiders, birds and animals with amazing camouflage abilities (like the Solomon Islands frogmouth), Medcalf came to feel as if the boundaries of his body had merged with the jungle. But then the war ended, and the thought of returning to life as it was before in suburban Australia—where distinction counted immeasurably more than invisibility—led to a crisis of identity. A popular term for this anxiety is 'going troppo', but whatever the name, it is widely documented that returned troops from WWI and WWII experienced feelings of radical estrangement from home.[38] Psychological disturbances were manifest in camouflage behaviours where they tried to become invisible in the world they returned to.

---

32 Allied Land Forces, 'Soldiering in the Tropics (SW Pacific Area)', August 1942, cited in Frank Hinder, 'Records of Lt F.C. Hinder Camoufleur', Canberra, AWM, PR 88/133, File 895/4/182, Item 10 of 12.

33 'Conceal to live: live to attack', AIF poster, February 1943, MP222/2, Melbourne, NAA.

34 Bennier 1942, p. 51.

35 A.H. Martin (Institute of Industrial Psychology, University of Sydney) quoted in Garton 1998, p. 91.

36 Medcalf 2000 [1986], p. 97.

37 Medcalf 2000 [published 1986], p. 169.

38 For a discussion of estrangement from home in WWI see Garton 1995, pp. 191–92.

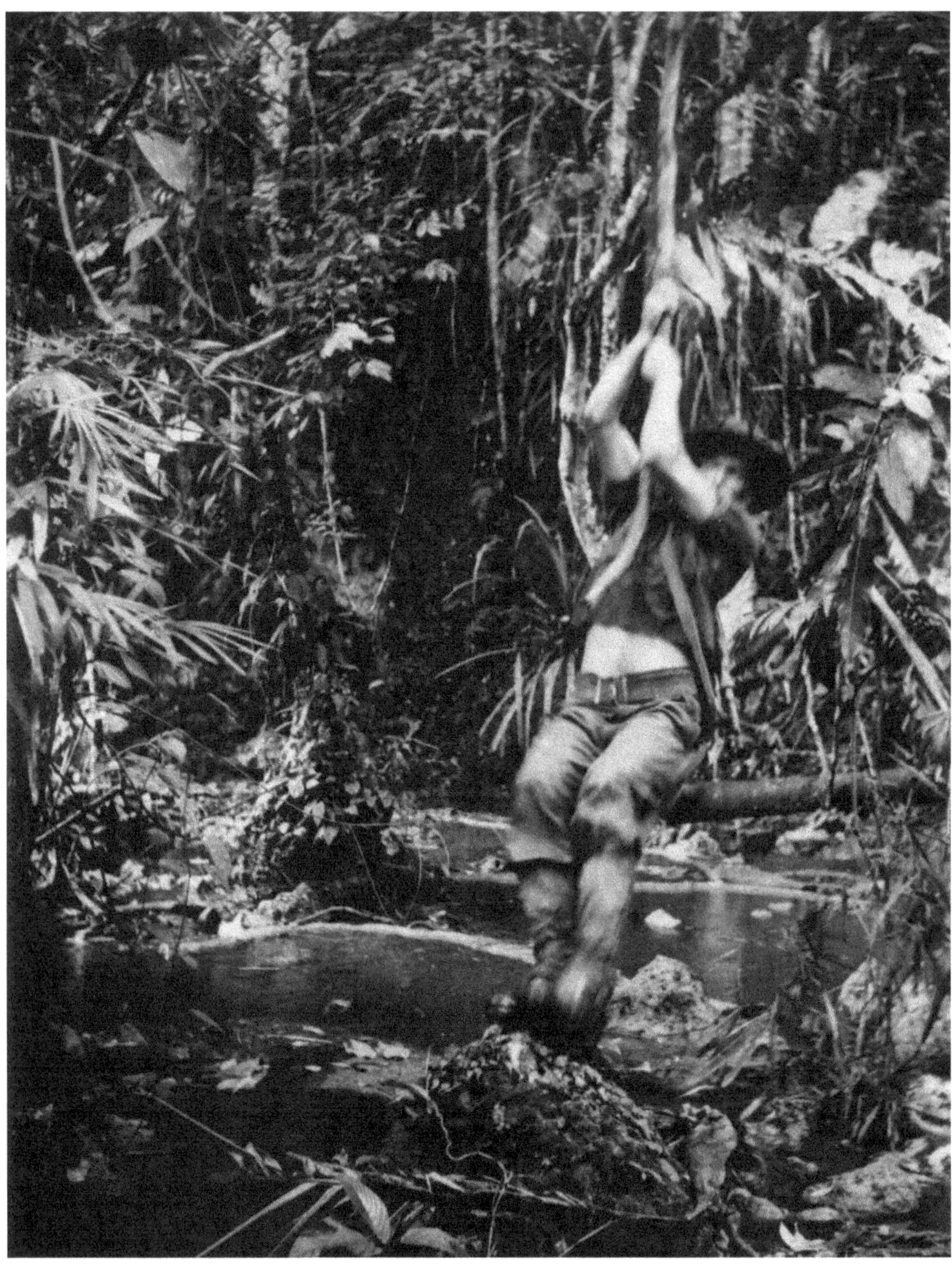

Fig. 9.7. Sergeant Spike Barer (AIF) of Parramatta swings across a stream in northern New Guinea, 22 June 1944. Canberra, AWM, AWM 017344. Photograph, Samuel James Fitzpatrick.

Camouflage, in one context, is a sign of intelligent behaviour because when used for protection and attack, it achieves military advantage. But concealment and deception in another context can be signs of dysfunctional psychological behaviour, and this was discovered in WWI as well as WWII. The action of freezing still, for example, which helped jungle-fighting soldiers to go undetected and which Dakin urged them to adopt, was symptomatic of a psychotic condition among returned troops known as 'catalepsy'. Also referred to as 'human camouflage', catalepsy was diagnosed in WWII as a psychosis induced by the shock of warfare in which individuals suffered from an urge to become invisible and 'disappear' from the world by freezing still. Comparisons were made between the behaviours of frightened soldiers and animals that use 'stillness as a defence', such as bitterns, certain caterpillars and pigeons, and other creatures that also freeze to make themselves seemingly dissolve and disappear into the background.[39]

While Peter Medcalf identified so strongly with the jungle that he did not want to be separated from it, most Australian soldiers in the New Guinea campaign feared the rainforest and undergrowth, imagining themselves watched from all directions. Most of all they worried that trees and bushes would materialise into enemy soldiers.[40] But Dakin's advice to the soldier in forward areas was to get to know the landscape intimately and:

> Study the terrain of the particular area on which he may have to concentrate and consider how its nature, 'pattern,' texture, and colouring may be used in siting any military works or concealing special objectives.[41]

If Dakin was right, Japanese soldiers in the New Guinea area were better prepared for the jungle than it was first thought. From their well-concealed positions the Japanese could see, but they could not be seen. Staying motionless when observed, feigning death, using skin pigments, garnishing their body and head with local plants, leaves and grass, masquerading in uniforms of Allied forces, and siting themselves in shadow, were standard practices. From Dakin's perspective the Japanese were:

> masters of camouflage where the concealment of the individual soldier is concerned. Careful selection of colouring in clothing, employment of netting, and full use of natural foliage cover have evidenced a complete understanding of camouflage problems and a thorough study and preparation in overcoming these problems.[42]

Japanese had better knowledge of field-craft and more thorough training for the Papua and New Guinea regions, which one soldier said was so difficult that it rendered useless many developments in modern warfare making the fight with the enemy 'nothing compared

---

39 Meerloo 1957, p. 96.

40 Cpl B.A. Harding (NX113826), 'This nerve war', in Australian Army and War Memorial 1945, p. 15.

41 W.J. Dakin, 22 September 1943, Duties of Camouflage Officers of the Department of Home Security in Forward Areas, Sydney, NAA, Series C1904, Item 4.

42 Office of the Technical Director of Camouflage, 'Camouflage in forward areas: notes from the fighting areas in which camouflage is specially mentioned', Camouflage in Pacific, Sydney, NAA, Series C1707, Item 65.

with the conquest of the land itself'.[43] In 1942 Australians learnt that Japanese soldiers in the New Guinea area were disguising themselves as Australians by wearing jungle-greens and speaking in English.[44] American soldiers wearing green to blend into the bush also found themselves in a dangerous predicament because, as indicated by US Major Allen of the 6th Army in conversation with Max Dupain in 1944, Japanese jungle suits were so similar to American ones that it was impossible to distinguish a Japanese soldier when he was wearing a captured American's helmet.[45] Bob Curtis reported that the American suits were being abandoned due to casualties and confusion among US troops.[46] One lesson to learn from this was that camouflage is not always functional and does not guarantee self-preservation. Indeed this was a criticism levelled at Charles Darwin's functionalist theory of camouflage in 1935 by the French sociologist Roger Caillois who disagreed that mimicry is an evolutionary adaptation for protection. Caillois, who wrote about mimicry in insects from a psycho-biological point of view, drew attention to the failure of mimicry to work for Phylliidae or leaf-insects, a sad case, he said, because they mistake each other for real leaves and eat their own kind.[47]

When Japanese forces, however, were pushed back from 1943 to 1945, the urgency for individual concealment receded. Consequently when DHS officers arrived in the region in December 1943, the question of individual camouflage was less urgent to airforce and army than they expected. Clement Seale felt like 'a salesman selling winter underwear in Equatorial Africa'.[48] Those fast moving theatres of war left no time for unnecessary distractions like attention to camouflage detail. Despite this setback, Bob Curtis requested the DHS become more proactive in the New Guinea area and that it pushed the limits of thinking on camouflage's conception and design. Despite the increasing ascendancy of the US in the region, he argued that:

> Each of us has to be something of a pioneer for a start. We have to promote and 'sell' the idea of camouflage and be continually exploring out new angles and improvising new methods.[49]

Japanese ingenuity in the New Guinea region was all the incentive they needed, and even camoufleurs with the DHS stationed on the Australian mainland felt a keen sense of competition and a desire to innovate. 'The Japs', wrote Frank Hinder:

---

43 (Lt) C.D. Madden (VX17681) 1942, p. 134.

44 W.J. Dakin, 'Japanese camouflage (as discovered in the South West Pacific Area)', March 1944, in Camouflage in Pacific, Sydney, National Archives of Australia, Series C1707, Item 65, p. 9.

45 Max Dupain to William Dakin, 5 April 1944, R.E. Curtis trip New Guinea, 1943, Sydney, NAA, Series C1707, Item 31.

46 R.E. Curtis, 'Report on camouflage "jungle-suit", New Guinea', Camouflage Data New Guinea Documents, Sydney, NAA, Series C1707, Item 32, p. 1.

47 Roger Caillois, 'Mimicry and legendary psychasthenia', in Caillois 2003 [1935], p. 97.

48 C. Seale letter to R.E. Curtis, 28 August 1944, Camouflage—general—camouflage organisation in New Guinea, Canberra, NAA, Series A453, Item 1943/17/1510 Part II.

49 R.E. Curtis letter to W.J. Dakin, 23 March 1944, Camouflage—general—camouflage organisation in New Guinea, Canberra, NAA, Series A453, Item 1943/17/1510 Part II,

> had the right solution in the islands. I understand an airstrip was discovered after the Japs were pushed north that had been suspected but never found. Reason, of course, was that they had cabled the tops of the palms to palms on the edge of the planned strip, cut off the trunks and levelled the area beneath. A perfect solution for that area—were we jealous![50]

Understanding Japanese methods of camouflage was an important branch of work for officers with the department stationed in the New Guinea Area. They analysed the way the enemy sited workshops, barracks, hangars, storehouses and oil-tanks adjacent to ravines, the base of mountains, and forests, to take advantage of shadows. Dakin had acquired a Japanese camouflage bulletin that warned how flat topography invites brightness and therefore greater visibility, whereas sites that are unlevel increase the variety of shadows and assist objects to blend with their surroundings.[51] Similar lessons on how to use terrain to best advantage were taught by Sun Tzu in *The art of war*, a treatise that influenced 20th-century Japanese methods.[52] But from Dakin's Eurocentric perspective the Japanese manual was evidence that their spies had stolen western methods of jungle warfare and obtained copies of instruction books published by the Allies.

Increasingly it came to light that Japanese applied extensive and inventive use of bluff, dummies and decoys in the SW Pacific area. 'This sort of thing', wrote Dakin, 'suits the Japanese temperament', but it did not suit his own.[53] He found bluff too risky. However, when the Australian army applied similar principles and methods as the Japanese to an ambitious camouflage installation on Goodenough Island, he followed with interest. Aptly code-named Operation Hackney—because by the time of its inception dummies and decoys had become a well-used and predictable ploy in the South West Pacific region—the objective was to deceive Japanese reconnaissance that Goodenough Island was strongly defended, when the opposite was true.[54] The principle was similar to one in the animal kingdom observed by British zoologist Hugh Cott who wrote that mimics and bluffers are 'typically defenceless. Here reliance is placed upon deception. These are the sheep in wolves' clothing.'[55] Nevertheless, the deception scheme on Goodenough was much celebrated and even Dakin congratulated the army for what he described as a 'brilliant effort'.[56] When the DHS set up base on Goodenough Island in late 1943, the afterglow of Operation Hackney was still evident, and camouflage in the region had almost become synonymous with bluff. Bob Curtis, Clement Seale, George Adams and Max Dupain found, when they

---

50 Frank Hinder, 'Records of Lt F.C. Hinder Camoufleur', Canberra, AWM, PR 88/133, File 895/4/182.

51 DHS, 'Japanese publication on essentials of camouflage' in Camouflage—general—camouflage organisation in New Guinea, Canberra, NAA, Series A453, Item 1943/17/1510 Part 1.

52 Sun Tzu 2008, pp. 64–70.

53 W.J. Dakin, 'Japanese camouflage (as discovered in the South West Pacific Area)', March 1944, in Camouflage in Pacific, Series C1707, Item 65, p. 6.

54 Public Relations Office, HQ 5 Aust. Div., 'When a Ghost Force held back the Japanese', 20 April 1943, Canberra, AWM 54[579/7/14].

55 Cott 1957, p. 435.

56 W.J. Dakin, Camouflage report 1939–1945 (with appendices A–D), Canberra, AWM, Series 81 [77 Part 1], 1947, p. 82.

arrived at Goodenough, a taste for adventurous approaches to concealment and deception, and for a while they were unsure whether to focus on individual concealment or on grander schemes. Yet from Dakin's point of view their role on Goodenough was simply to provide the RAAF with 'all that science and art can offer' on a minute island in Melanesia.[57]

57 WJD to Air Commodore, 14 December 1943, Camouflage—general—camouflage organisation in New Guinea, Canberra, NAA, Series A453, Item 1943/17/1510 Part 1.

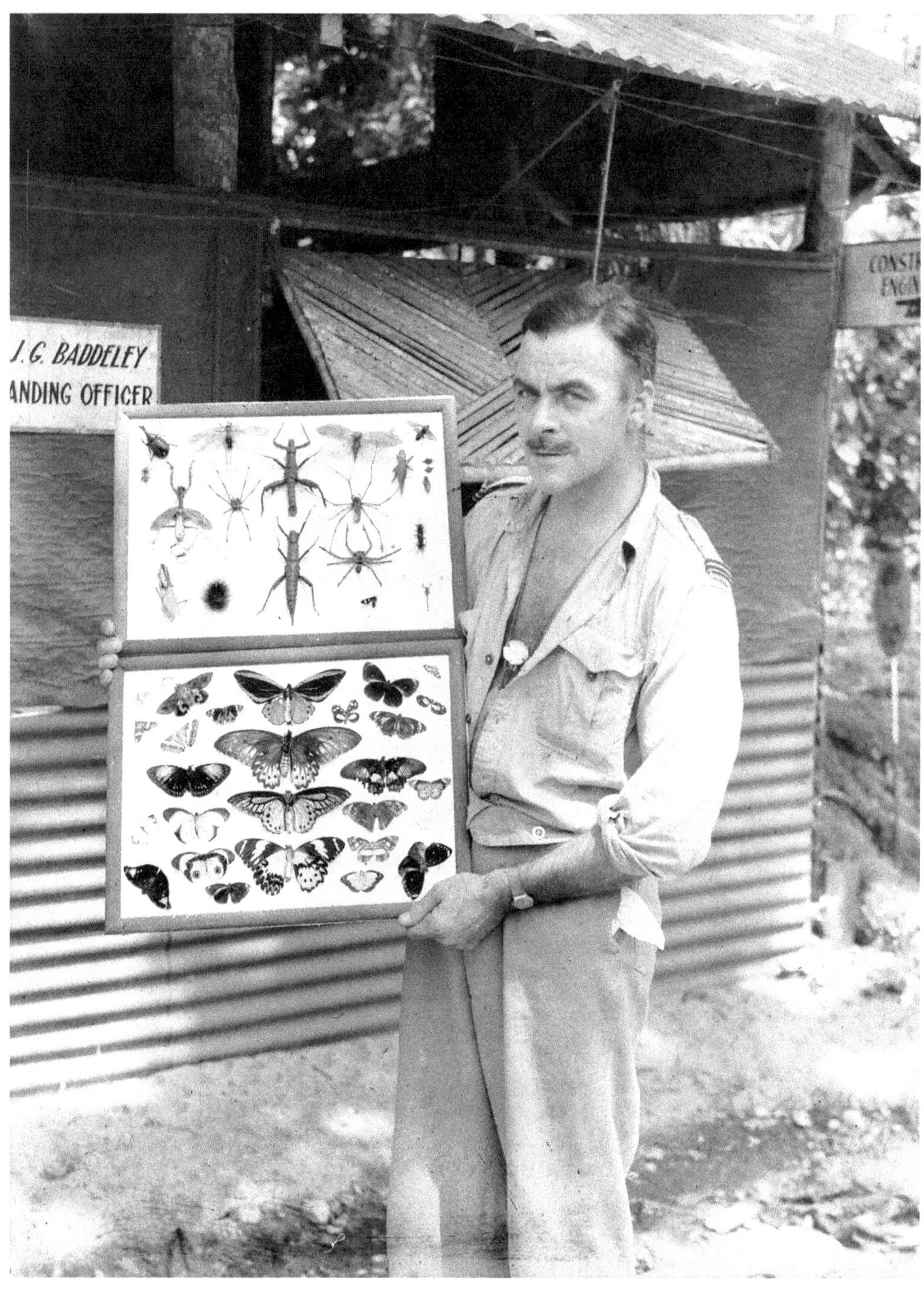

Fig. 10.1. RAAF Flying Officer J.R. McEniry, entomologist, Vivigani, Goodenough Island, 23 September 1943. Canberra, AWM, AWM OG0197. Photograph, John Thomas Harrison.

Chapter 10

Goodenough Island

Tiny, mountainous Goodenough Island—strategically positioned for campaigns in Papua and New Guinea and ideally located to protect the large Allied base at Milne Bay—was occupied first by Japanese, then Australian and American militaries between July 1942 and July 1943. The island is jungle-clad, and while only 40 kilometres long and 24 kilometres wide, has mountains that rise to 2500 metres. Today the island is a tourist attraction, especially for divers. But in 1942 Americans and Australians found it anything but a South Seas island paradise.[1] Military entomologists seized the opportunity to collect the island's exotic fauna and further their science. But to the majority of men and women stationed with military personnel on Goodenough Island, the spiders, scorpions and centipedes were horrifying, especially their size (fig. 10.1). What is more, the dimness of the island's rainforests made military staff feel claustrophobic and depressed. So melancholic did American troops find the jungle, they felled large canopy trees to make the forest floor light.[2] Indeed one of Max Dupain's memories in the New Guinea area was the awesome spectacle of US heavy military machinery as it 'rolled on the roads day and night', crushing large tracts of jungle.[3]

Max Dupain arrived on Goodenough Island, which was code-named 'Ginger', in February 1944 after a treacherous assignment photographing camouflage installations in the New Guinea area from the air. He was en route to his next assignment, photographing projects at Buna, Oro Bay and Port Moresby. For four months he stayed on Goodenough where Headquarters for the DHS had been newly established. There he processed film, annotated 180 negatives, and forwarded his work to William Dakin in Canberra.[4] As Dupain arrived, George Adams was set to leave the island. Adams had dengue fever and was exhausted after 18 months' work on isolated radar stations in north western Australia and with the RAAF's No. 9 Operational Group in the New Guinea area.[5] The difficulty

1 Two useful sources of poems and stories are: Australian Army and War Memorial 1943; Australian Army and War Memorial 1944.

2 Schrijvers 2002, p. 120.

3 Max Dupain, 'Max Dupain interviewed by Hazel de Berg', Canberra, National Library of Australia, 3 November 1975, Oral DeB 1/874, unpaginated transcript.

4 Max Dupain to W.J. Dakin, 5 April 1944, R.E. Curtis trip New Guinea, 1943, Sydney, Series C1707, Item 31.

5 R.E. Curtis letter to Secretary Department Home Security, 13 April 1944, Camouflage—general—

of his work at Finschhafen, on the eastern coast of the main island of New Guinea, was exacerbated by continuous rain. On Kiriwina, the largest of the Trobriand Islands to the north of Goodenough, concealment of military hardware demanded all his wits, especially 'when the fighters are down and the moon is up'.[6] Also assigned to radar camouflage on Kiriwina was Clement Seale. When Dupain landed on Goodenough, Seale was convalescing after a bout of malaria. He was given sedentary work making instructional posters for operational units to serve the department's ongoing campaign to educate the troops and make them camouflage-minded.[7]

On Goodenough Island camoufleurs were confronted with the social realities of the war. For the Islanders the greatest impact of WWII occurred in 1943 when 10,000 American and Australian airforce personnel descended on their population of 9000.[8] Vivigani became a large, busy airbase and many indigenous inhabitants were moved to neighbouring Fergusson Island to make way for runways, camps and war infrastructure. Men who stayed on the island were employed as manual labourers to move oil drums, build huts for personnel and carry out non-traditional work including laundry.[9] Australian anthropologist, Michael W. Young, who lived and researched on the island in the 1970s, explains that by WWII Goodenough Islanders were used to working for European masters as indentured labourers, but when war descended they 'were forced into labour by the Australian New Guinea Administrative Unit (ANGAU)'.[10] Max Dupain's photograph, *Early morning scenes, Goodenough Is.* (fig. 10.2), is testimony to the political reality of that episode, and the shock that WWII brought to the indigenous population.

Goodenough Island had three close associations with camouflage in WWII, two involving the Australian army and predating the arrival of the DHS. The third was the establishment of headquarters by the department in late 1943, staffed by up to five people. In civilian life the department's staff were artists and architects, but on Goodenough Island they were charged with the responsibility of managing the department's camouflage information and advice centre for Allied land, air and naval forces in the SW Pacific area.[11] A brief background to the army's camouflage activities helps explain the legendary camouflage history of Goodenough that met officers of the DHS in December 1943.

---

camouflage organisation in New Guinea, Canberra, NAA, Series A453, Item 1943/17/1510 Part II.

6 R.E. Curtis letter to A.W. Welch, 4 October 1943, Camouflage—general—camouflage organisation in New Guinea, Canberra, NAA, Series A453, Item 1943/17/1510 Part 1.

7 R.E. Curtis 'Extract from Memorandum', 1 May 1944, Camouflage—general—camouflage organisation in New Guinea, Canberra, NAA, Series A453, Item 1943/17/1510 Part II.

8 The figure of 10,000 Allied airforce personnel is from Young 1983a, pp. 117–18. The figure of 9000 Papuans on Goodenough Island is stated in 'Data on Vivigani Mission', Operation Hackney: Goodenough Island deception scheme, Canberra November 1942 to June 1943, Canberra, AWM 54, File 585/3/1.

9 See AWM online collection of photographs for visual records of wartime indigenous labour on Goodenough Island, www.awm.gov.au/search/collections/.

10 Young 1983b, pp. 83–84.

11 General MacArthur, 'Camouflage directive', 23 July 1942, in W.J. Dakin, Camouflage report 1939–1945 (with appendices K–N), Canberra, AWM, Series 81 [77 Part 3], 1947, appendix L.

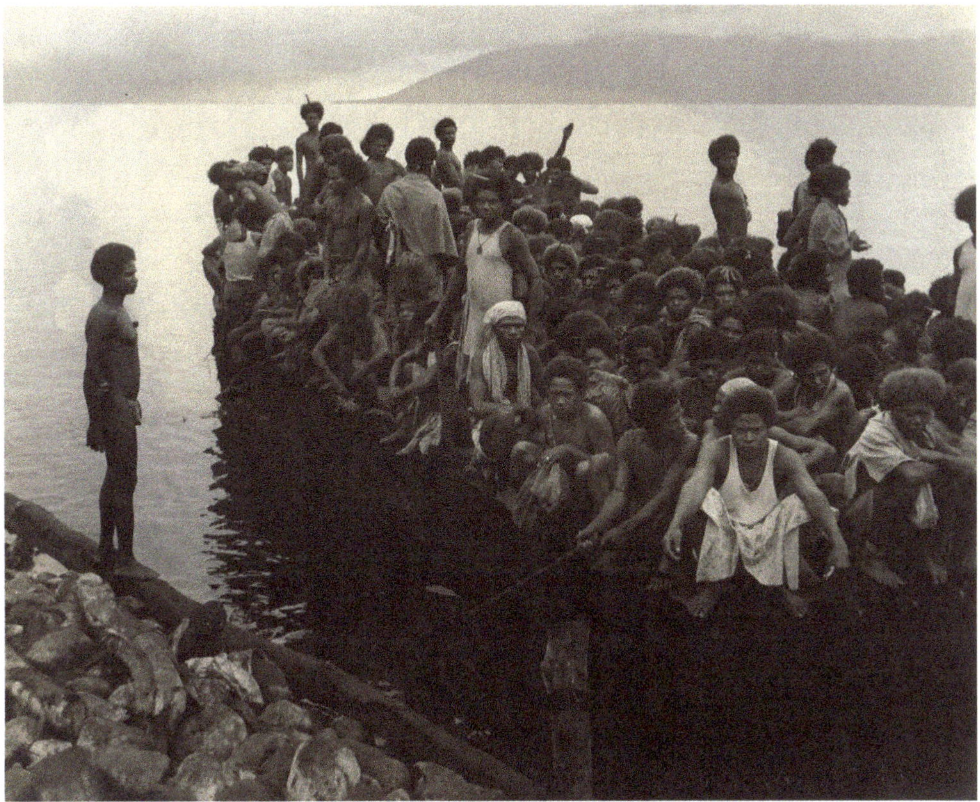

Fig. 10.2. Max Dupain, Australia 1911–1992, *Early morning scenes, Goodenough Is. New Guinea series* 1944, gelatin silver photograph, National Gallery of Australia, Canberra.

The first involved the 2/12th Battalion of the 18th Brigade of the Australian army which landed at Goodenough Island on 22 October 1942 with a mission to eradicate 350 Japanese of the 5th Sasebo Naval Landing Force. The Japanese were survivors of an Allied invasion at Milne Bay who became marooned on the island in September.[12] Australian reports claimed they were being supplied with food by Goodenough Islanders and their presence was causing Australia, the colonial administrator, to lose face.[13] Dudley Leggett, an Australian Broadcasting Commission (ABC) war correspondent, accompanied the Australian invasion party, and to him Goodenough meant only one thing: 'those 10 hours climbing through the mountains. It was the worst physical effort I've ever been asked to perform.'[14] But the psychological trauma was just as bad. As they waded through deep kunai grass the party was unaware that close by were well-concealed Japanese, waiting until they came very close. The ensuing ambush was a sharp lesson in the perils of enemy

12 The history of this event is explained in McCarthy 1959, pp. 347–49.

13 McCarthy 1959, p. 346.

14 Dudley Leggett, 'Radio talk presented by ABC war correspondent Dudley Leggett', 31 January 1943, broadcast transcript, Sydney, NAA, Series SP300/3, Item 523.

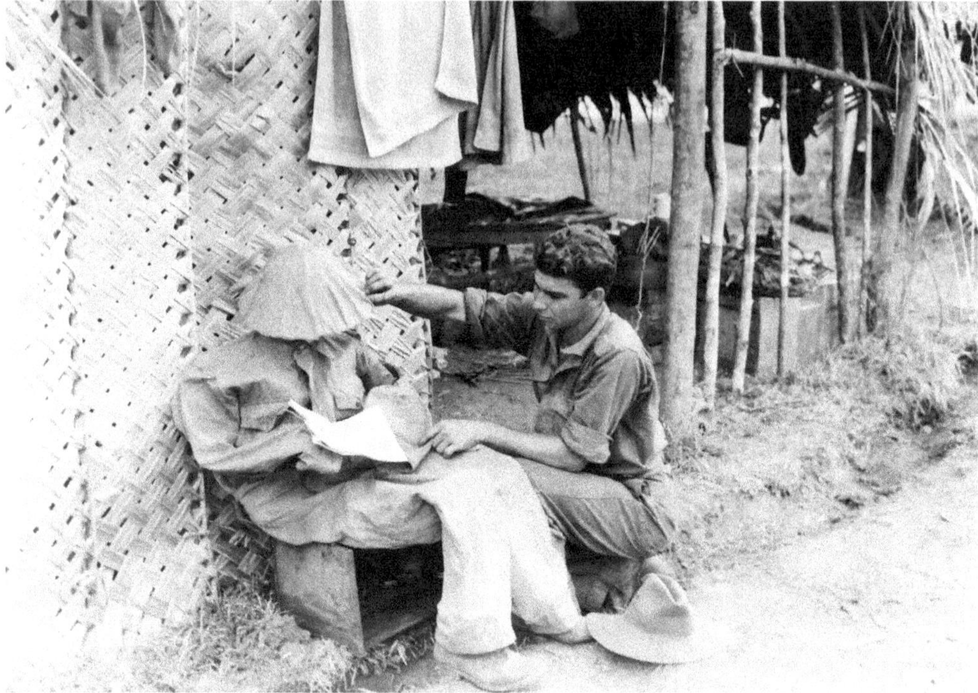

Fig. 10.3. Australian soldier arranges dummy man, Goodenough Island bluff, c. March 1943. Canberra, AWM 090202.

deception and the Australian army was forced to retreat while the Japanese escaped by barge to Fergusson Island. The circumstances were similar to the experiences of other Australians and Americans in the New Guinea area; the Japanese with their skin darkened and clothing garnished for concealment, hid silently and still before emerging with surprise. The Kokoda Track had shown that Japanese had command of the terrain and that even though their advance had been declared physically impossible, by melting into their jungle background they turned a hostile terrain into a deadly weapon.[15]

Then, in March 1943 in the aftermath of the surprise Japanese attack, those in the army who stayed to garrison Goodenough Island staged a bold camouflage scheme of heroic scale. Their numbers were perilously small and, fearing a Japanese re-invasion of the island, they staged Operation Hackney. In short, they built a fake brigade of simulated personnel and military hardware to deter the Japanese from trying to recapture Goodenough. Later it was revealed that the motive for that ambitious scheme went beyond one of deterring the enemy; it also involved a desire for 'making history in the Australian Army'.[16] The army's Public Relations Office built Operation Hackney up as a South Pacific rival for Germany's miraculous false boulevards and fake landscapes beneath which military operations

15 Reit 1978, p. 184.

16 'The address to leaders of working parties', Operation Hackney: Goodenough Island deception scheme, Canberra, AWM, AWM 54, Item 585/3/1, appendix D, p. 5.

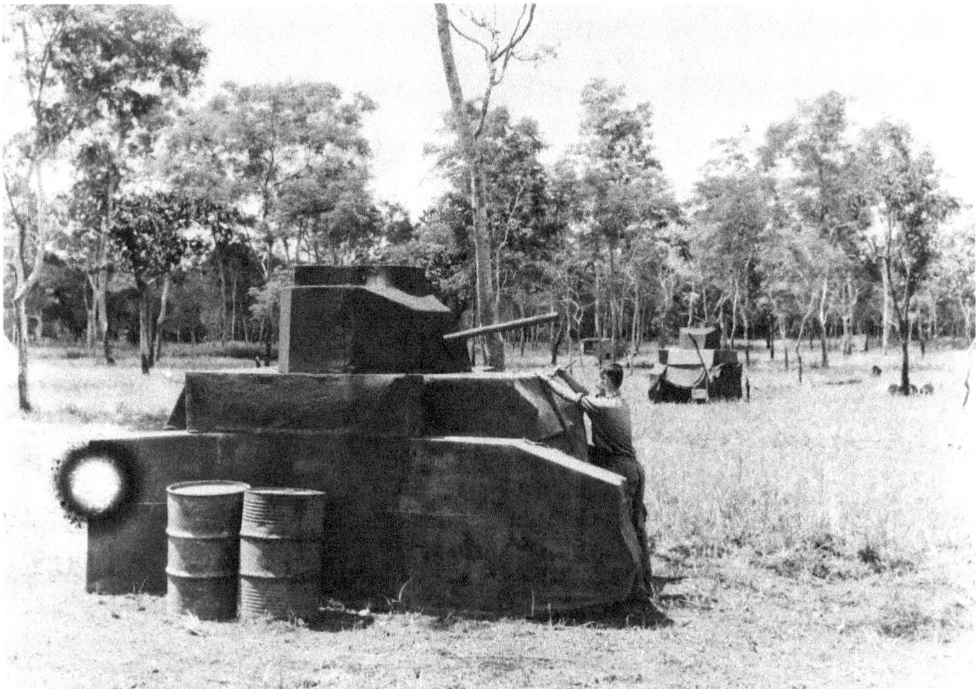

Fig. 10.4. Dummy tanks made of hessian on wooden frames, Goodenough Island bluff, c. March 1943. Canberra, AWM collection 090203.

and information were concealed.[17] Yet what was concealed behind the fake brigade on Goodenough Island was the Australian army's temporary weakness. The army seemed in no doubt that the Japanese intended to retake Goodenough, and the thought of creating its own folklore, by fooling the enemy in plain sight, was a powerful incentive to proceed with the scheme. As news spread of its success, the army called Operation Hackney 'the greatest bluff of the Pacific war'.[18]

The first stage of the operation took place on 10 February 1943 when the Commanding General of the United States Army (in charge of Combined Operational Service Command in New Guinea) ordered 12 cargo ships loaded with 'old wooden packing cases, empty petrol drums, camouflage paint, sheets of tin, hessian, timber, old tents and mosquito nets, and the many odds and ends' to be taken in convoy with fighter protection to the island. False radio reports and rumours, decoys, and sham movements helped sustain the illusion that a brigade was on the move.

Building a simulation of a brigade, in order to 'prove' its occupation, was the second stage of the project. Camouflage materials were unloaded and an array of dummies installed at Vivigani, Mud Bay, Kilia and Teleba. The details are noted in the public relations report:

17 Public Relations Office of 5 Australian Division, 'When a Ghost Force held back the Japanese: Goodenough Island Deception Scheme', Articles from Public Relations Office H.Q, 20 April 1943, Canberra, AWM, AWM 54, Item 579/7/14.

18 'Goodenough Island bluff: a Ghost Force held back the Japanese', AWM 54, 585/7/1, first page.

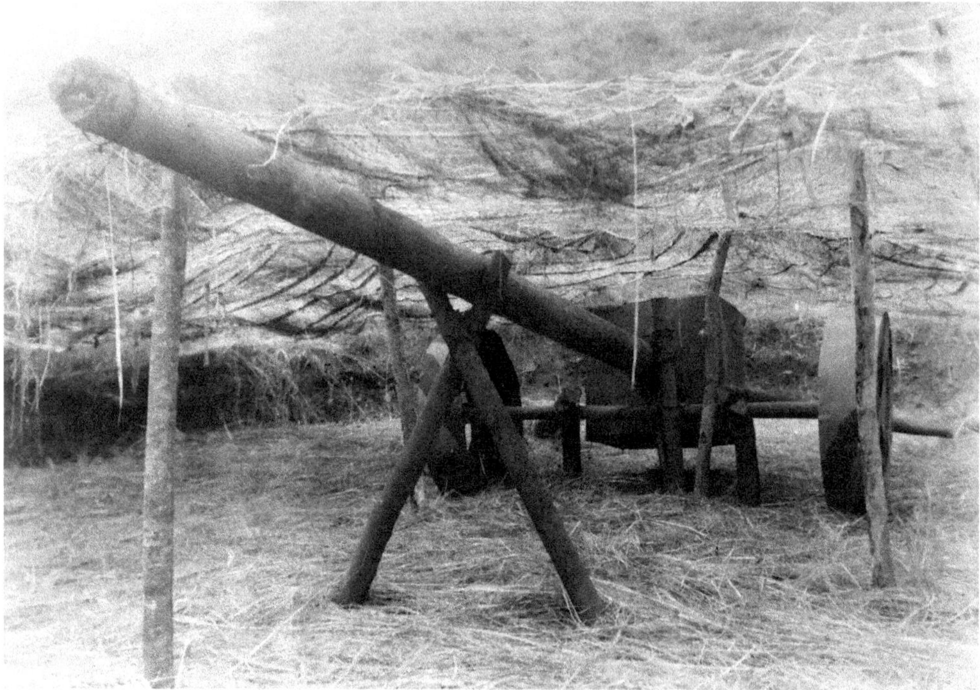

Fig. 10.5. Imitation wooden defence gun, Goodenough Island bluff, c. March 1943. Canberra, AWM collection 090200.

> Next the camps began to take shape on Goodenough Island. Areas were cleared of grass, tents were set up, kitchens were built, dumps established, slit trenches dug, tracks cleared, weapon pits rigged. Coast defence guns appeared on headlands. Bofors guns pointed skywards. Barbed wire lined the open beaches and ringed the camp areas. Smoke rose from the kitchens. Washing hung out on the lines by the tents. Apparently a full brigade had taken up its position along the eastern side of the island. But the tents were empty of men, the kitchens were useless skeletons, the dumps were old packing cases, the guns could not be fired, the tanks would not move. There was no brigade in that series of camps—just one company, scattered widely, and a few score natives.[19]

Fake kitchens, latrines, camps, tanks, guns and soldiers were rough and improvised. They were described as absolutely unconvincing when seen from the ground. But their real impact was intended from the air (figs. 10.3–10.7). The Bofors guns were constructed from bush timber and sheet iron, while the field guns and tanks were merely hessian stretched over frames. All were simple, schematic, coded pieces of visual information shaped perspectivally to increase their illusion from an oblique aerial viewpoint. The army

19 Public Relations Office of 5 Australian Division, 'When a Ghost Force held back the Japanese: Goodenough Island Deception Scheme', Articles from Public Relations Office H.Q, 20 April 1943, Canberra, AWM 54, Item 579/7/14, pp. 3–4.

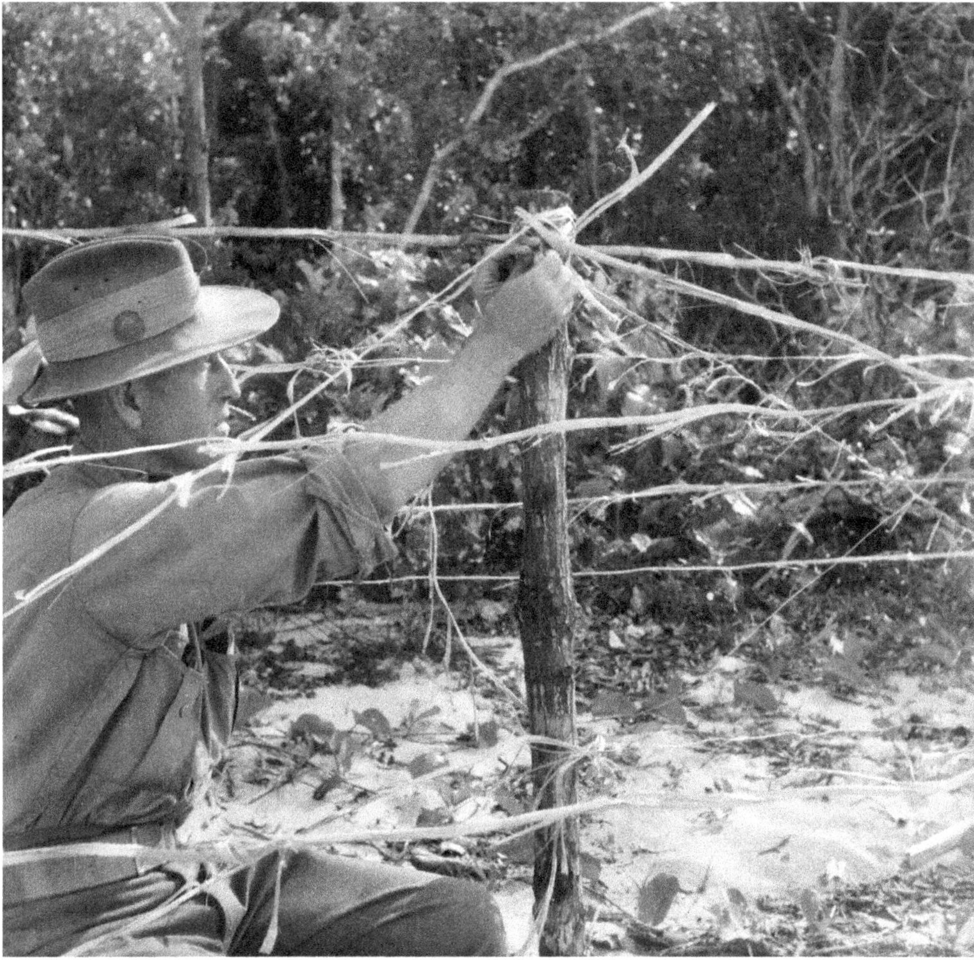

Fig. 10.6. Imitation barbed wire entanglements made from jungle vines, Goodenough Island bluff, c. March 1943. AWM collection 090196.

praised the scheme's ingenuity and saw it as an exemplary case of original, spontaneous intelligence.[20]

Yet it was less improvised than the report suggests. Use of dummies, decoys and eccentric disguises in the SW Pacific area, by Japanese as well as Allies, followed similar patterns. Japanese also built dummy guns shaped from coconut logs, and covered wooden guns with camouflage netting as ineptly as possible to provide greater realism (fig. 10.5).[21]

20 Public Relations Office of 5 Australian Division, 'When a Ghost Force held back the Japanese: Goodenough Island Deception Scheme', Articles from Public Relations Office H.Q, 20 April 1943, Canberra, AWM 54, Item 579/7/14, pp. 2–5.

21 W.J. Dakin, 'Comments on official Japanese naval publication dealing with the essentials of camouflage', Air Intelligence Reports, Sydney, NAA, Series C1707, File 36, p. 2.

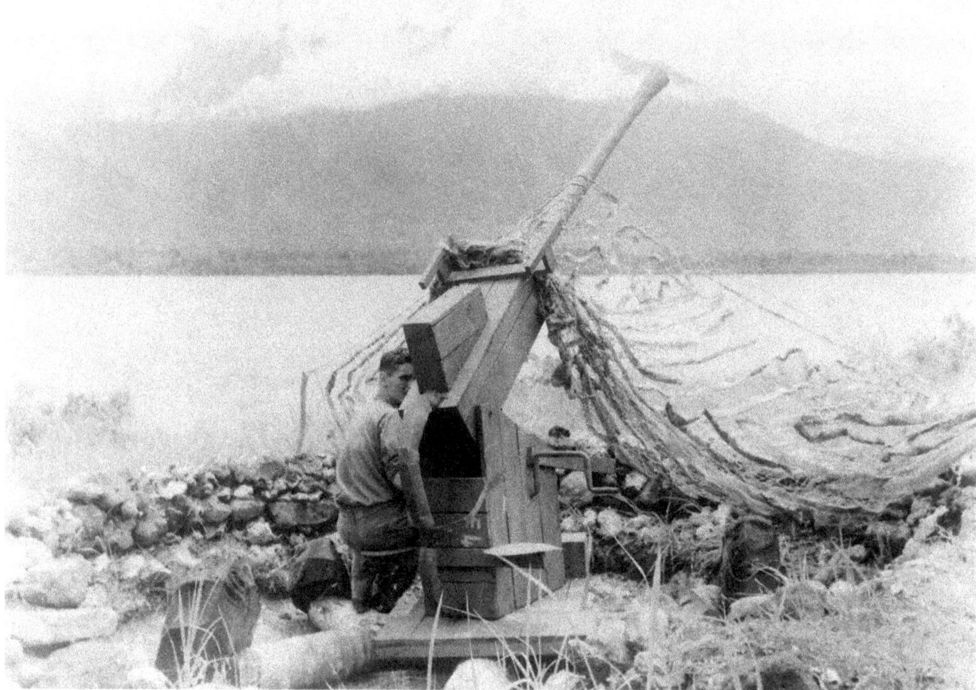

Fig. 10.7. Dummy anti-aircraft gun, Goodenough Island bluff, c. March 1943, Canberra. AWM collection 090217.

Moreover, the assortment of dummies that were utilised for Operation Hackney were prefabricated by engineers at Milne Bay.[22] The scheme was therefore a combination of the two approaches Claude Lévi-Strauss writes about in *The savage mind* (1962); engineering or scientific thought, and bricolage or untamed thinking—what the army referred to as improvisation. The intent expression of a soldier installing 'barbed wire' barricades made of jungle-vines, the engrossed look of another trying to capture the right slouch in a dummy figure made of indigenous plant matting, and another soldier who is momentarily distracted from his wooden and sheet iron Bofors gun to pose for the camera, exemplify Lévi-Strauss' portrait of the bricoleur as someone 'excited by his project' (figs. 10.3, 10.6 & 10.7).[23] For as these photographs show, the project went beyond functional necessity into a realm of satisfaction, even pleasure.

If the Pacific War was a test of intelligence and nerves, the Goodenough deception was evidence that Australians were masters of both, and magicians of a sort. Choosing its words to conform with a trend among other Allied forces, the army's public relations office spoke of how a 'Ghost Force held back the Japanese'. The celebrated camoufleurs of WWII were often likened to war magicians and beings with supernatural powers who made

22 'Outline of deception scheme—Goodenough Is', in Hackney deception scheme Goodenough Island, Canberra, AWM 54, Item 585/3/1, appendix 'A,' p. 2.

23 Lévi-Strauss 1966, p. 18.

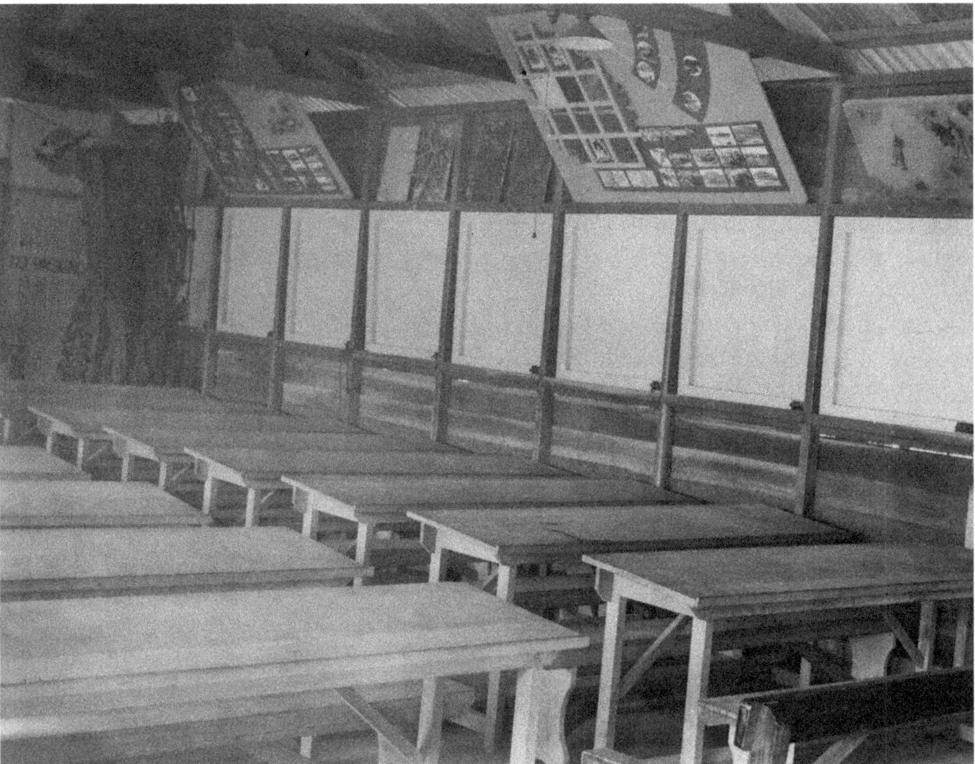

Fig. 10.8. RAAF Instruction Centre in Townsville, Queensland showing installation of portable camouflage instruction panels. Canberra, AWM, ART 91632.

things appear without themselves being seen. Two examples are the British Magic Gang, and the US army's Ghost Army that staged mock reconnaissance missions in Europe.[24] Yet the concept of an Australian 'ghost army' performing magic on Goodenough Island, a place where magic is integral to indigenous society, serves to highlight a difference between western ideas of magic and the Papuan universe that the military invaded. Michael Young explains how in the indigenous social world of Goodenough Island, secret magic is 'the most important heritable property', whereas 'magic that is known to all is of no value; it is merely folklore'.[25]

By November 1943 when Bob Curtis arrived on Goodenough, Operation Hackney's designs and methods were widely publicised in the Australian Military Forces' manual 'Camouflage in the S.W.P.A', in the hope that Hackney would act as model for future camouflage projects.[26] Curtis was officer-in-charge for the department. In November 1943,

24 Behrens 2009a, pp. 157–58.

25 Young 1993, p. 186.

26 Australian Military Forces, Engineer-in-Chief's technical instruction No. 45; Camouflage in the S.W.P.A., 30 November 1943, Frank Hinder, 'Records of Lt F.C. Hinder Camoufleur', Canberra, AWM, PR 88/133, File 895/4/182, Item 11 of 12.

with the help of Colin Wyatt, he set about prioritising camouflage for radar installations, operations and fighter quarters, tent colouration, and for concealment of aircraft and supply dumps. They thought of their work as essential ongoing support for airforce personnel who only months before they had trained in camouflage at the DHS's instructional unit in Townsville, Queensland. At Townsville, Curtis along with Gervaise Purcell, Douglas Annand and Frank Hinder, lectured to thousands of ground staff and aircrew awaiting transportation to the SW Pacific theatre (fig. 10.8).[27] But according to Dakin, what Curtis found when he entered the island war zone was 'an almost unbelievable neglect and ignorance of camouflage'.[28]

Despite this ignorance, a mood of bravado and daring prevailed, fuelled no doubt by Operation Hackney's success. Influenced by it for a brief time was Curtis himself. By August 1943 he felt divided on whether camouflage in the region should be risk-taking or conservative. But following the army's triumph with Hackney, and knowing the ascendency of the Allies in the region, he was persuaded that offensive measures suited the Papuan campaign, specifically:

> tactical movements in an advance operation, which embodies quick and temporary measures, daring conceptions in plan and materials, quick decisions and hard and dangerous work in uncomfortable surroundings.[29]

The mythology surrounding Operation Hackney quickly spread to the DHS in Canberra. A.W. Welch, secretary to the Minister of Home Security, also developed Curtis' taste for a bolder, braver approach. He confided, to a contact in the Ministry of Air, that his department's camouflage officers were on course to inspire innovations in 'tricks, decoy activity, simulating advances etc' in the New Guinea area.[30]

But in the end this was not the path Curtis followed in his capacity as officer-in-charge. Under Dakin's influence he conceded that, while ambitious schemes like Hackney were important, so too were the fundamentals of individual concealment. While decoy airstrips, mock invasion plans, fake camp sites and dummy barges all had their place, Dakin was sure it was just as important to study the terrain, interpret reconnaissance photographs, and adapt local materials to camouflage constructions. The department's officers therefore maintained their quieter campaign to convert jungle-fighting troops to intuitive camouflage thinking, a type that relied on simplicity, the individual's survival instincts, and 'jungle-craft':

> Elementary jungle craft must be learned by every jungle soldier. Jungle craft is mainly the

27 W.J. Dakin, Camouflage report 1939–1945 (with appendices A–D), Canberra, AWM, Series 81 [77 Part 1], 1947, p. 98.

28 W.J. Dakin, Camouflage report 1939–1945 (with appendices A–D), Canberra, AWM, Series 81 [77 Part 1], 1947, p. 105.

29 R.E. Curtis, 'Camouflage in New Guinea Area', 31 August 1943, Camouflage—general—camouflage organisation in New Guinea, Canberra, NAA, Series A453, Item 1943/17/1510 Part 1.

30 A.W. Welch to Department of Air, 2 September 1943, RAAF camouflage in New Guinea and Islands, Sydney, NAA, C 1904, Item 4.

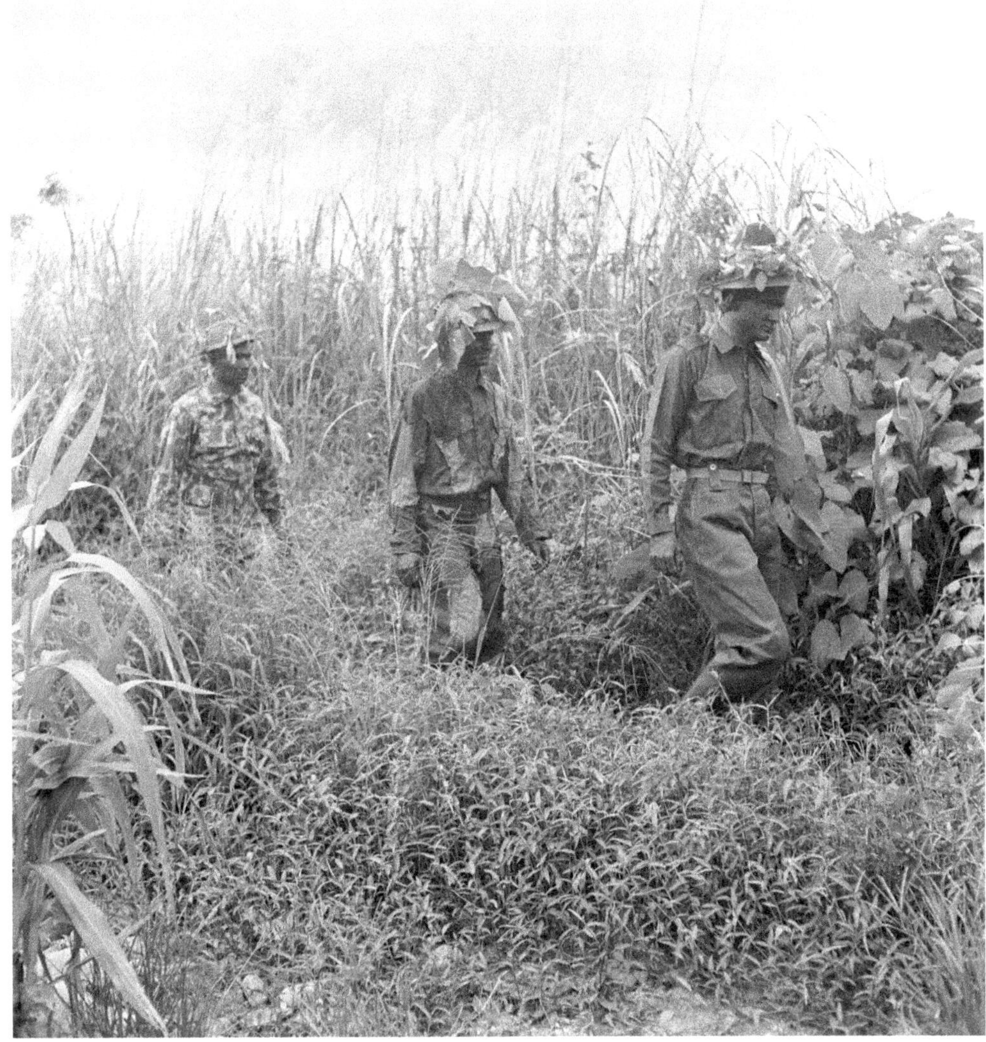

Fig. 10.9. Jungle Suits in kunai grass: American type, DHS disrupted pattern, Army drab green (Goodenough Island), c. 1943–44. The collection of the NAA: C1908, 13, Box 10.

> ability to see simple things and to know what these simple things mean. You can learn to track a man etc.—and to see animals and men in the jungle before they can see you.[31]

Instead of grand schemes, DHS officers were preoccupied with the colours and patterns of uniforms. Their experience in the field proved that the RAAF's new dark green jungle uniforms required only one shade of green, namely 'a rich, dark yellow green or

---

31 W.J. Dakin, 'Camouflage in forward areas: notes from the fighting areas in which camouflage is specially mentioned', Camouflage in the Pacific, March 1944, Series C1707, File 65, p. 2.

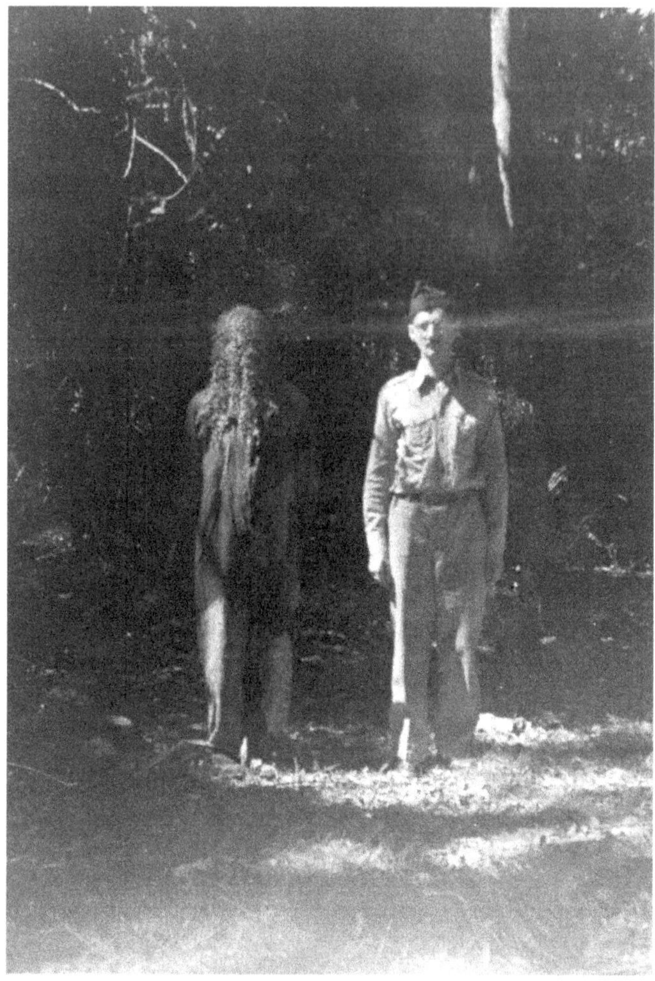

Fig. 10.10. Personal camouflage experiment, c. 1943. The collection of the NAA: C1905, T1 14 [21].

more correct "red" green … [without] any suspicion of blue'.[32] Similar in style and colour to uniforms worn by the US army and the AIF, Curtis believed they spelt 'the difference between life and death'.[33] But to reach this conclusion the department's officers tested three differently patterned and coloured suits. Smothered in Skin Tone Commando Cream and photographed against four different backgrounds including tall kunai grass (fig. 10.9), thick jungle background (fig. 10.10), creek-beds and open space, they compared 'army jungle green in plain colour, US jungle-suit with mottled colour disruption, and Home Security

32 W.J. Dakin, Camouflage report 1939–1945 (with appendices A–D), Canberra, AWM, Series 81 [77 Part 1], 1947, p. 137.

33 R.E. Curtis memo to Air Officer Commanding, 27 December 1943, 'Camouflage Establishment. Advance Headquarters 9. Operations Group', in Camouflage—general—camouflage organisation in New Guinea, Canberra, NAA, Series A453, Item 1943/17/1510 Part 1.

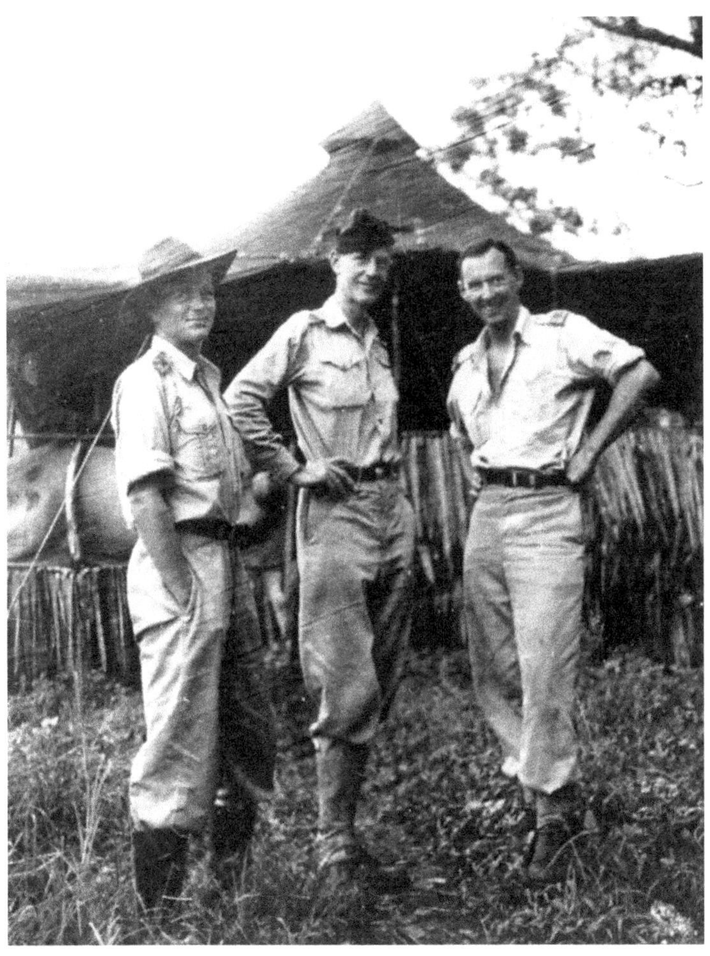

Fig. 10.11. Goodenough Island, New Guinea, 1943. Left to right: Clement Seale, Bob Curtis and Max Dupain outside their hut. Canberra, AWM collection AWM P02186.007.

jungle-suit with shirt and trousers of dark green with loosely sewn patches of yellow-green and black for disruption'.[34] They found the splotchy markings on American uniforms too small and the contrast between dark and light colours insufficient.

They tested a full-face cover because it was generally known that a human body was more difficult to recognise if the shape of the head was disguised. By de-familiarising its own shape, a human being could go undetected, even at close range. It was a lesson they had learnt in camouflage training school from lecture notes written by physicist A.D Ross who claimed that 'interpretation of what we see is based entirely on past experiences, and the camoufleur may obtain striking results if he can make the observer's experience prove his

34 New Guinea—R.E. Curtis September 1943–May 1944, Camouflage data New Guinea documents, Sydney, NAA, Series C1707, Item 32.

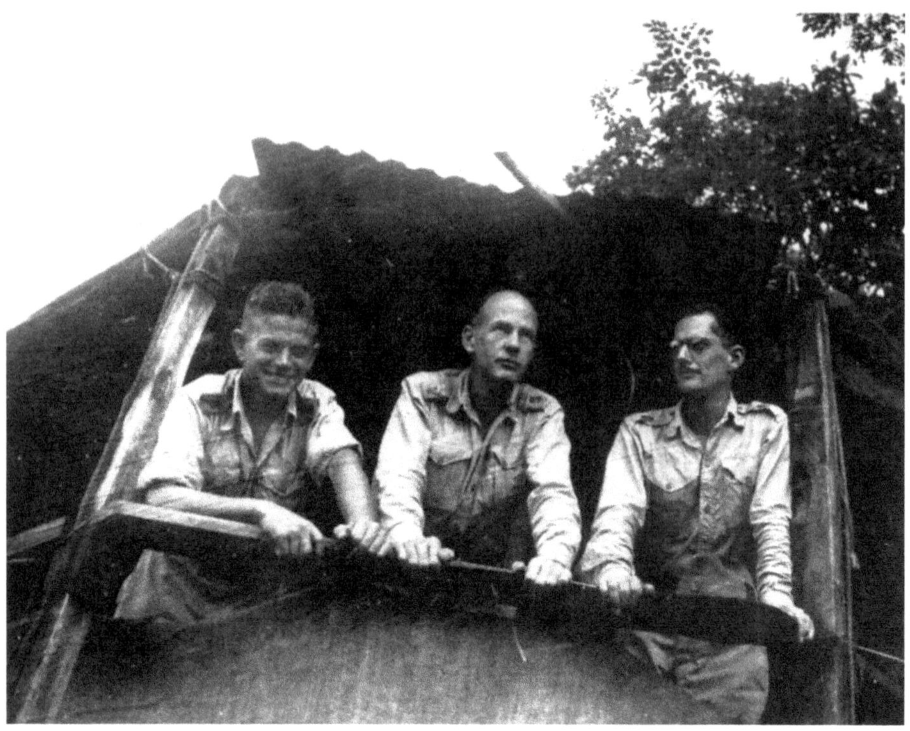

Fig. 10.12, Goodenough Island, New Guinea, 1943. Left to right: Clement Seale, Bob Curtis and George Adams. Photo: Max Dupain. Canberra, AWM collection AWM P02186.006.

own undoing' (fig. 10.10).[35] But a net over the head and neck had too many disadvantages when moving through the jungle. What the photographic tests clearly showed, however, was the advantage of wearing a darker colour. Khaki made the soldier a misfit in the jungle and therefore a danger to himself and others. The department's advice to men in the field read: 'if you are in dark clothes, keep in the shade, or keep a dark background behind you. If you are in rather light clothes, do not stand in front of a dark background, or lie on dark ground in the open'.[36]

Yet despite the disadvantages of khaki, Bob Curtis, Max Dupain, Clement Seale and George Adams appear to have worn only light colours themselves. Grainy black and white photographs capture them posing in, and beside, their 'native style' hut—as the military referred to buildings made of local materials and labour—during the brief period when they were together on Goodenough Island (figs. 10.11 & 10.12). For specialists in camouflage trying to convert troops to green uniforms, they look surprisingly conspicuous. But they

35 A.D. Ross, 'The science of light: vision', in Camouflage—general camouflage school WA, Canberra, NAA, Series A453 Item 1942/17/1957, p. 4.

36 W.J. Dakin and the Camouflage Directorate, 'Concealment, and camouflage of the individual in warfare', in W.J. Dakin, Camouflage report 1939–1945 (with appendices K–N), Canberra, AWM, Series 81 [77 Part 3], 1947, p. 4.

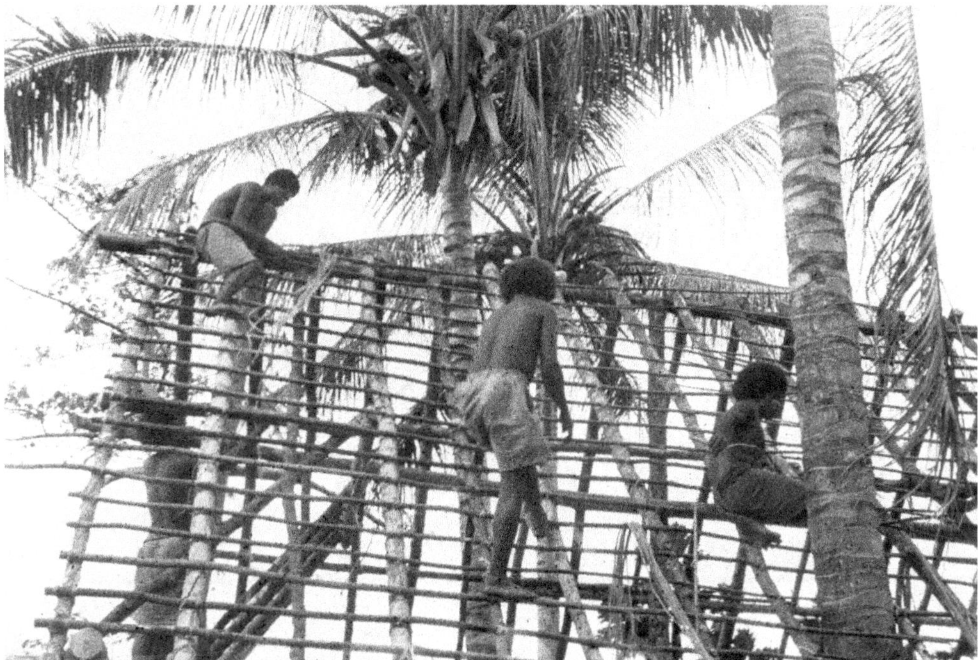

Fig. 10.13. Local men construct officers' and pilots' mess, Goodenough Island, Papua, June 1943. AWM collection P02875.147. Photograph, William H. Robinson.

also look deceptively happy for a team troubled by sickness, shock, fatigue and discontent. Curtis, at this time, was preoccupied with the grading of camouflage officers which he said was inadequate because they worked in unappreciable danger. Their entitlements and status—compared with regular troops—were inequitable:

> Camouflage officers going in with forward Siting and Installation parties, which constitute part of the second wave of assault are exposed to considerable dangers quite apart from those attendant on regular movements by barge or aircraft. To these must be added the rigours of climate and the high incidence of sickness. Further, each man must assume responsibilities for his own defence, though no course of training has been provided. He must also share in the guard duties of his unit. Briefly, the conditions under which our officers are expected to work and live and perform valuable camouflage services are, for the first few weeks at least, difficult, disagreeable and dangerous.[37]

The department's officers kept separate quarters from Allied operations units. Curtis sited their hut on a hill so they could be independent. He called it 'neutral ground', something difficult to attain in the large and fractious military community of Americans and Australians.[38] But separation was a matter of survival for a group forced to live among

37 R.E. Curtis, 'Ref. grading of camouflage officers. New Guinea Area', 24 April 1944, Camouflage—general—camouflage organisation in New Guinea, Canberra, NAA, Series A453, Item 1943/17/1510 Part II.

38 R.E. Curtis letter to A.W. Welch, 15 November 1943, in Camouflage—general—camouflage

Fig. X.

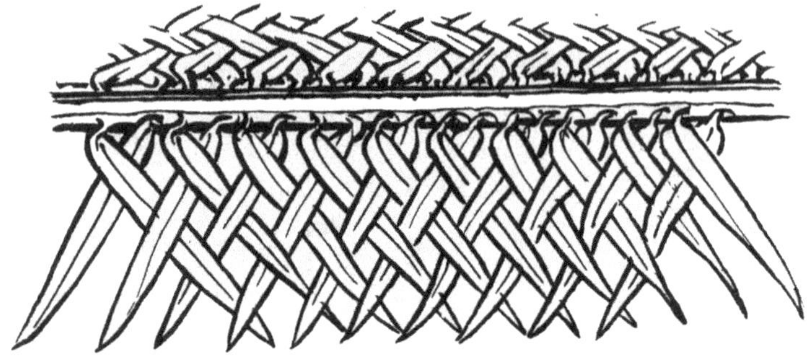

COCOANUT LEAF PLAITING, SHOWING
SPLIT SECTION OF LEAF

Fig. XI.

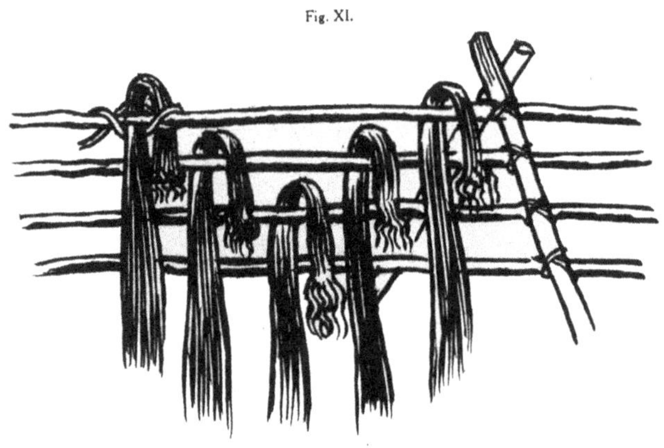

KUNAI GRASS – METHOD OF THATCHING

Design and drawing by
R. EMERSON CURTIS,

PAPUA,
December, 1943.

Issued with Bulletin No. 17.
CAMOUFLAGE RESEARCH SECTION,
Department of Home Security,
Canberra.

W. J. Dakin

Technical Director of Camouflage.

Fig. 10.14. R.E. Curtis, How to plait coconut leaf and thatch kunai grass. *Camouflage Bulletin 17*, 22 December 1943, Canberra, AWM 81 [77 Part 4], 1947, p. 19.

troops who doubted that civilians had anything useful to teach them about camouflage in war conditions. To press the point: wherever the DHS's officers went, even remote Papua, they encountered the same problem, one that Dakin called the 'fear of civilian "control" '.[39]

Island life, on the other hand, was enlightening. Bob Curtis absorbed valuable indigenous knowledge and watched with interest when Islanders were 'knocking up the framework of a hut or gathering in supplies of local camouflage material for plaiting' (fig. 10.13).[40] He remarked on the excellent camouflage that 'native style' huts provided; at the same time as sheltering personnel and hiding equipment they were inconspicuous and looked very much like houses belonging to local Papuans. He learnt the many uses of Lawyer vine, bark, sago palm, pandanus, coconut palm and kunai grass: Lawyer vine for lashing and for ropes, sago palm and pandanus for plaiting prefabricated camouflage covers, kunai grass for huts to disguise crashed planes and hide ammunition stores and petrol dumps, and coconut palm groves to cover tent installations (fig. 10.14)[41] He burnt coconut husks and coir matting to place underneath grounded planes to help them blend into the shadows of the jungle.

George Adams, Bob Curtis, Max Dupain, Clement Seale and Colin Wyatt were chosen for the New Guinea area because they had the right temperament, intelligence and attitude. Had they not been personable with servicemen and women, expressed themselves clearly, been energetic and keen, possessed practical but also creative minds, been able to make technical drawings, and been able to improvise, they would not have been selected.[42] None would have chosen war in Papua over peace in Australia, but as artists the war was a unique opportunity. Goodenough Island provided entry to mysterious Oceania. Designated primitive by westerners, Oceania represented a cultural fantasy of art and life in its purest form. To westerners it was a reservoir of primitive expression and represented the unconsciousness of civilisation.[43] Throughout the 1930s, European surrealists including André Breton and Paul Eluard delighted in the 'convulsive beauty' of the fantastic, raw energy of Oceania.[44] And at the outbreak of WWII every serious artist in Australia had either explored, discussed, embraced or rejected surrealism and other western art movements deeply influenced by what was perceived as Oceania's exotic and primitive nature.

---

organisation in New Guinea, Canberra, NAA, Series A453, Item 1943/17/1510 Part 1.

39 W.J. Dakin, Camouflage report 1939–1945 (with appendices A–D), Canberra, AWM, Series 81 [77 Part 1], 1947, p. 105.

40 W.J. Dakin, 'Camouflage bulletin no. 17', 22 December 1943, Camouflage report 1939–1945 (appendix O), Canberra, AWM, Series 81 [77 Part 4], 1947, p. 4.

41 W.J. Dakin, 'Camouflage bulletin no. 17', 22 December 1943, Camouflage report 1939–1945 (appendix O), Canberra, AWM, Series 81 [77 Part 4], 1947, pp. 1–19.

42 These skills and attributes are identified in staff records. See Department of Home Security Staff register—camouflage section, Canberra, NAA, Series A691, Item 1, accumulation dates 1 January 1942–31 December 1945.

43 For discussion of Oceania in the context of western primitivism see Rhodes 2005, pp. 9–11.

44 Discussed in Tythacott 2003, pp. 128–47.

In 1945, after five years of dedicated service to camouflage, Bob Curtis made a successful application to become an Official War Artist. The region of Papua features among his many images of the war. Max Dupain, who explored a wide repertoire of styles in the 1930s, including surrealism, was steeped in western art's perceptions of the tropical Pacific. Both Curtis and Dupain took advantage of their war service to sketch and photograph Papuan subjects on Goodenough Island.[45] In fact Dupain's time was divided between war duties involving scientific aerial photography, and private interrogations of the landscape and people. Official work, on one hand, took him into a world of logic and information, while in private he was attracted to emotive, humanist subjects. To use Les Manovich's distinction: it was the difference between being a maker of images through which the war was carried out, and being a maker of images of the war.[46]

Dupain's private work was destined for a portfolio of photographs titled the New Guinea Series. But the unfamiliar cultural environment of Papua, together with the nature of fieldwork in a war zone, contributed to what he later referred to as postwar 'shock', a feeling intense enough to persuade him (in spirit at least) to desert commercial photography forever and follow only the path of documentary.[47] While this did not eventuate and commercial photography remained important to Dupain, he did seek in documentary photography a morally decent art. And in later life, perhaps in partial reaction to a war career centred on deceptive objects and misleading visions, he also wrote that 'to fake is in bad taste'.[48]

Max Dupain's extensively documented and discussed career remains cloudy in relation to WWII, but his repeated use of the word 'shock' in memoirs and interviews indicates he was eager to declare the war episode an origin of personal transformation and change. To say that the war years were a defining episode in his life is inadequate. In fact the case of Max Dupain offers an opportunity—along with the case of Frank Hinder—for more in-depth investigation of the impact of camouflage and war on the postwar lives of camoufleurs who worked for the DHS.

45 See the collection of the AWM for drawings of Papua produced by R.E. Curtis, www.awm.gov.au/search/collections/.

46 Manovich 1993, p. 143.

47 Dupain 2007, p. 6.

48 Dupain 2007, p. 14.

PART 5

THE EDGE OF MODERNISM

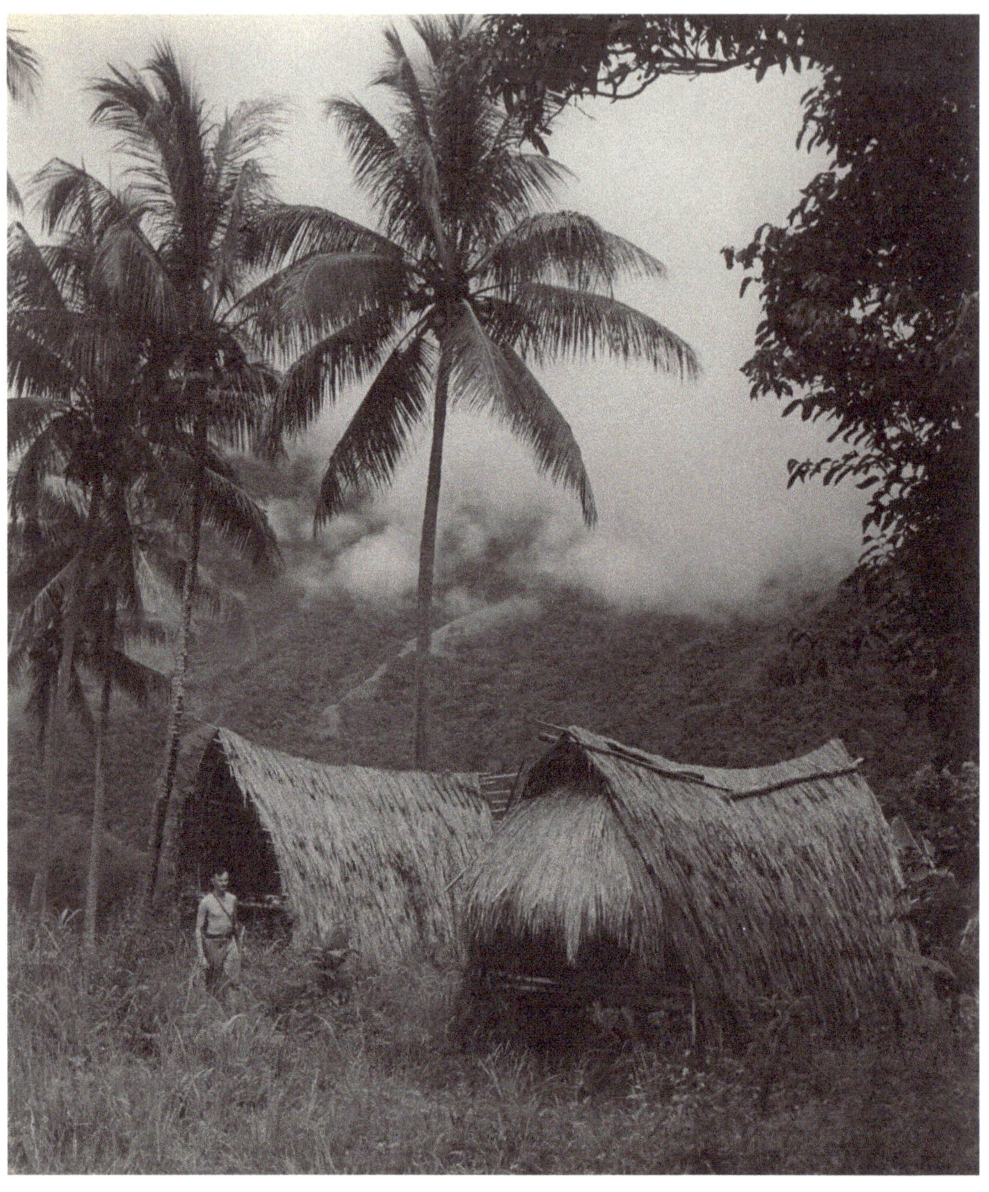

Fig. 11.1. Max Dupain, *Rain in the Mountains–Goodenough Is. (deserted native village).* New Guinea Series 1944, gelatin silver photograph, National Gallery of Australia, Canberra, 2000.58.

Chapter 11

Max Dupain

Looking back from 1986, Max Dupain (1911–1992) described how the period surrounding WWII left him in shock and took him to the brink of a psychological chasm. Unlike the men and women who returned from service in the New Guinea area and said little about the psychological effects of the war—believing it better to show mental and physical fortitude or, in the case of combat soldiers, feeling too traumatised to speak about it at all—for Dupain, the negative emotional consequences of that period became central to the narrative he constructed about his life:

> I dearly wanted to return to the studio and start a civilised life again. The unstable wartime years, the grudging adaptation to ever-changing surroundings, the thousands of impressions both good and bad of varying environments, all added up to long-term shock. I just needed a settled emotional life for a while in order to get my life and work into a new perspective. I did not want to go back to the 'cosmetic lie' of fashion photography or advertising illustration. I had seen too much of another kind of reality which probed deeper and demanded unequivocal attention—or else.[1]

Yet while Max Dupain spoke willingly and animatedly about wartime work as a camouflage officer, he was never expansive on the subject. Instead he spoke with contradictions, generalisations and innuendo so that it is difficult to know the detail of what he saw, felt, or experienced. At the outbreak of war he was a pacifist who despised brutality, but what option was there but get involved when 'even in remote Australia the war was in everybody's life'?[2] When he volunteered for the New Guinea area it was with a mixture of eagerness and trepidation, but Dupain, who always put great emphasis on physical fitness, passed the test easily where other camoufleurs failed.[3] He joined the small team of mostly younger camouflage officers in the DHS who wanted to be close to operations in the SW Pacific Area.

What the photographic record shows is that Dupain's wartime photography practice was split into two kinds. There were optical surveillance and camouflage experimentations for the RAAF at Bankstown in Sydney in 1943, but also personal projects, including a set of documentary photographs taken in 1944 on Goodenough Island. Both bodies of

1 Dupain 1986, pp. 15–16.

2 Dupain 1986, pp. 14–15.

3 Dupain 2007, p. 4.

work survive, making it possible to compare them, whereas the aerial reconnaissance photographs that Dupain took in the New Guinea area have proved impossible to identify among the vast collection of anonymous aerial photographs that survive WWII.

The New Guinea Series is a collection of photographs of sometimes gritty realism and often compelling social engagement. It marks the symbolic beginning of Dupain's intention, incited by the experience of war, to abandon what he came to see as the shallow world of commercial photography and take up the documentary genre. Although as a non-combat camouflage officer Dupain was more shielded than most from the full horror of war in the New Guinea area, in later life he intimated that the things he saw had a cathartic effect. They set him on course for a career devoted to the type of photography that 'could help clarify man's understanding of himself'.[4] For while war duties at Bankstown and later in the New Guinea area drew Max Dupain towards modern science and technology—his second passion to art—in 1944 he no longer wanted to continue with camouflage or with aerial photography and photo-analysis. Instead he wanted to change the nature of his war service and become an official war photographer.

Had Dupain's application to become a war photographer in the Pacific succeeded, it is unclear how he would have coped. Even his short time in the New Guinea area took a toll, and after four months he was very glad to leave. Before he did, he photographed the mountains of Goodenough Island with a psychological intensity that confirms the impact that 'primitive' Papua had on his sensibilities. A photograph titled *Rain in the Mountains—Goodenough Is. (deserted native village)* uses everything within the frame to convey a haunted landscape of mystery and foreboding (fig. 11.1). Magic, sorcerers and ghosts dominated the Goodenough Island society that Dupain entered in 1944, despite it being a missionised culture. Consequently, Australian military personnel were quick to conclude these were people living in a primordial state.[5] In Dupain's photograph two shadowy, deserted huts occupy the foreground of a steep and untracked wilderness while mist descends on the figure of a half-naked white man. The atmosphere is powerful and eerie. Indeed the implied primitivising narrative about isolation, magic and regression evokes Donald Friend's 1945 description of Labuan Island in Malaysia where he walked, he said, 'in the jungle in an ecstasy of fear, like a native haunted by the malignant demons of the forest'.[6]

Also photographed on Goodenough Island and part of the New Guinea Series is *Native woman cooking* (fig. 11.2), representative of how, at close range, Dupain recorded the conditions under which people on the island lived. This humanly meaningful, if conventional theme of family, work, motherhood, food and shelter anticipates the spirit of Edward Steichen's 1955 groundbreaking documentary photography exhibition at the Museum of Modern Art in New York, 'The family of man'.[7] Whether Dupain thought he was photographing a disappearing primitive culture, or simply recording the social life of

4 Dupain 1986, p. 21.

5 See Matt J. Fox discuss Papua and New Guinea as 'distant many generations of time in the primitive life of its people' in Bridges 1945, p. 240.

6 Donald Friend, 7 July 1945, in Hetherington 2003, p. 272.

7 Steichen 1955.

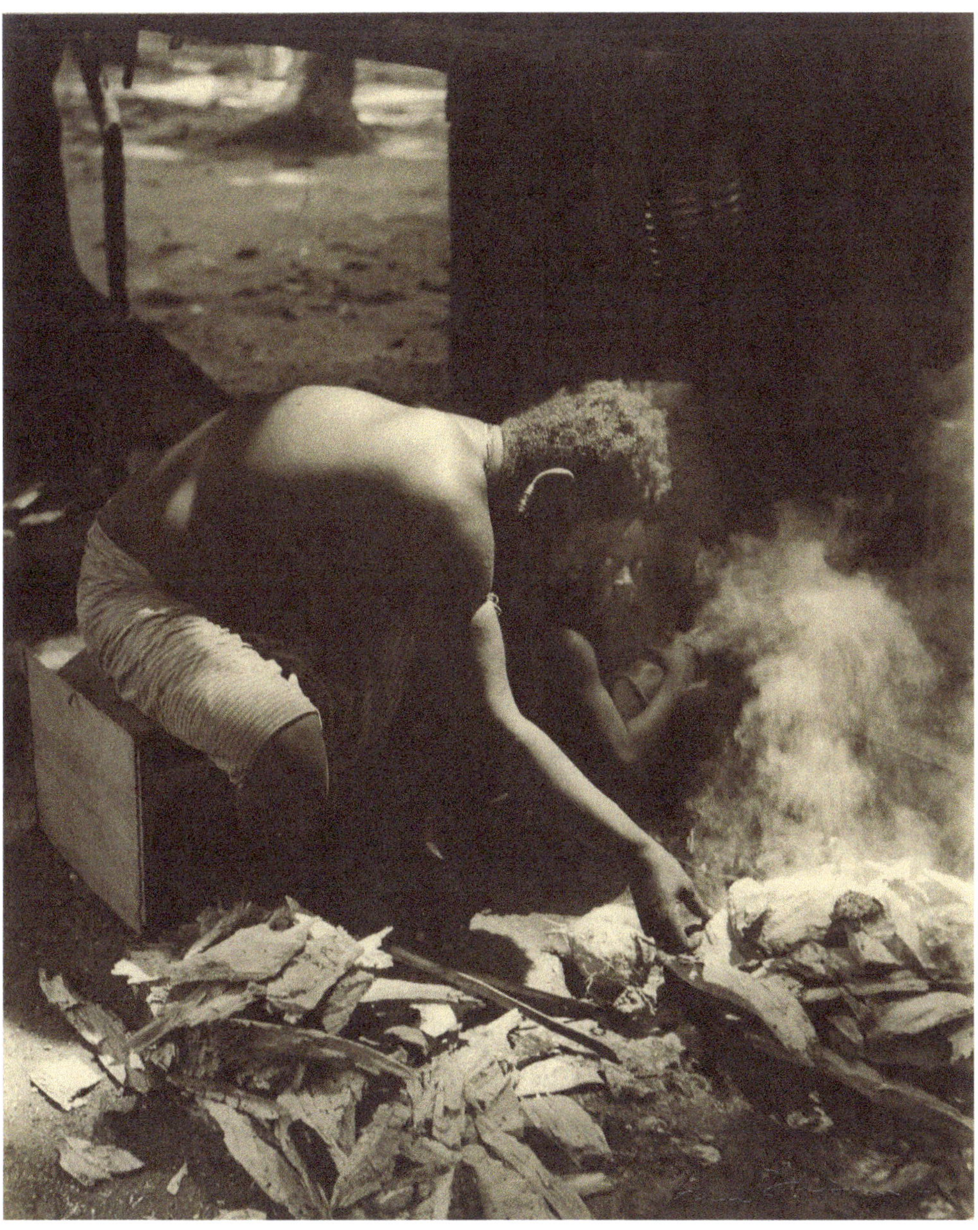

Fig. 11.2. Max Dupain, Australia 1911–1992, *Native woman cooking*. New Guinea Series 1944, gelatin silver photograph, National Gallery of Australia, Canberra.

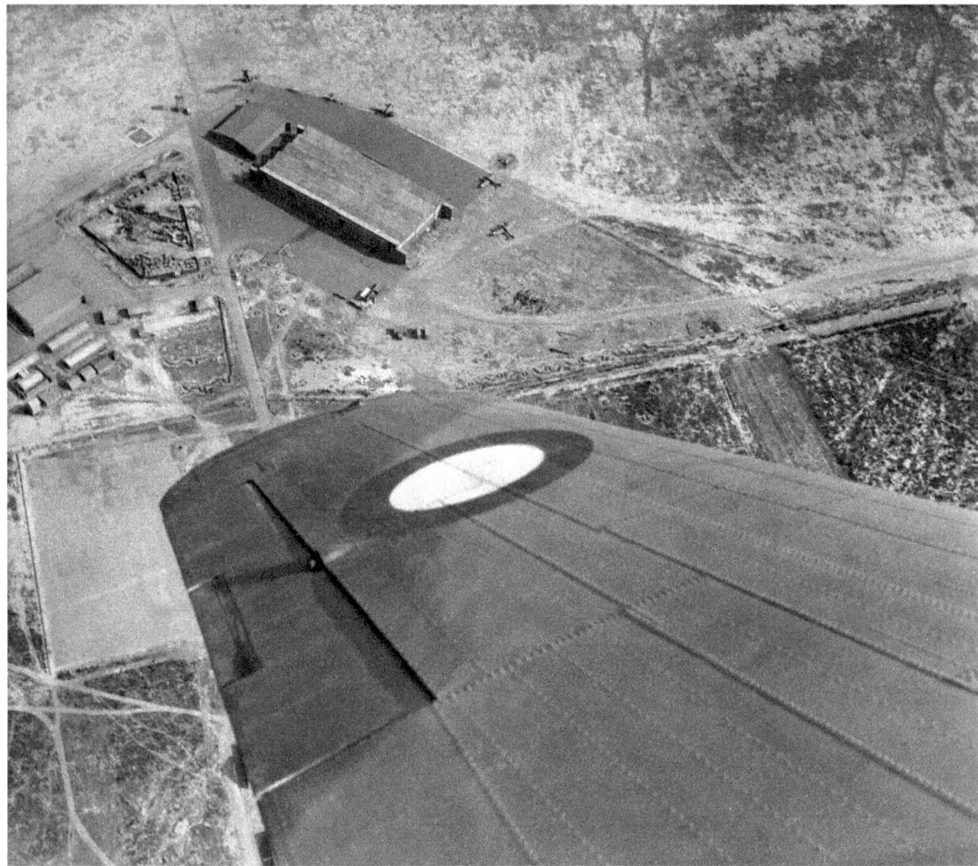

Fig. 11.3. Bankstown camouflage experiment No. 8, 1943. The collection of the NAA: C1905, 3. Photograph, Max Dupain.

the island, his intention was also to bring insight to the otherness of human experience in Papua. Gael Newton and Anne O'Hehir, the first curators to discuss the portfolio, comment that the war is hardly noticeable in these photographs.[8] Yet while it is not present as military violence in the form of decimated landscapes, bombed homes, crashed planes, dead soldiers and civilians, it is apparent here, and in other images in the portfolio, in the relationship between the photographer and his subjects and particularly through his gaze which is that of a western outsider looking in. Michael Young writes that the impact of the war for Goodenough Islanders was psychological, and less remarkable for casualties.[9] In this regard, *Native woman cooking* provides insight into the social consequences of the invasion of thousands of Europeans with guns, planes and cameras onto the island. As it transpires, the New Guinea Series is an important contribution to an otherwise slim archive recording the wartime experiences of Islanders.[10]

8 Newton & O'Hehir 2001, p. 10.

9 Young 1971, 10.

10 White & Lindstrom 1990, p. 3.

What stands out, though, is the contrast between the New Guinea Series—which exudes the artist's commitment to philosophical humanism—and the cool, scientific work Dupain undertook earlier at Bankstown. In January 1943 Dupain began intensive training in how to conduct photo-tests from the air of camouflage experiments he constructed on the ground (fig. 11.3). From aerial photographs he learnt how to read the signs of military activity, but also how to design camouflage on the ground in order to evade the enemy's attention. The question of how far photography could penetrate camouflage was vital. According to instruction material prepared by A.D. Ross, to be effective, camouflage had to make objects difficult to recognise from between one and two miles (approximately 5280–10,560 feet). In general, though, it was from a distance of 2000 feet that Dupain's experiments were designed to be viewed.[11]

Well-respected and admired for his successful photographic career before the war, Dupain was a highly valued member of the Bankstown team, and William Dakin held him up as proof that artists' minds—so used to memorising the appearances of objects and trained to create the illusion of depth on flat surfaces as well as the illusion of flatness on round surfaces—were perfectly suited for aerial photography and photo-analysis, as well as camouflage installation:

> You can't do anything without aerial photographs and without the close co-operation of men who know exactly what objects look like on the ground—not through having seen them once or twice, but whose job it is to constantly look at objects on the ground.[12]

Nevertheless, visual intelligence was nothing without training, and the only way to develop an aerial sensibility and know how objects looked on the ground was to spend many hours in the air. In later years Dupain recalled how, during practice flights his pilot flew 'over the target, turned, came back again, turned, went over the target, about a dozen times'.[13] And there was theoretical instruction in vision science to complement flying practice. Under Dakin's supervision he studied the different behaviours of light, its reflection on smooth and rough surfaces, and its absorption by dark and light objects.[14] He learnt the importance of tone rather than colour when trying to make distant objects inconspicuous. There were lessons on 'the science of light' that elucidated the physics of shadows cast by opaque bodies, that taught how to soften the edges of shadows, and how to read them from aerial photographs. Shadows were the give-away signs of a careless military that had forgotten that, even when objects and bodies are concealed, their shadows can still be visible. The perceptions of shadows, and the objectivity of shadows, were new concepts for some, but for Max Dupain they were more than familiar: they were the very underpinning of a photographic practice dedicated to space and light.

11 A.D. Ross, 'The science of light: photography', in Camouflage—general camouflage school W.A., Canberra, NAA, Series A453 Item 1942/17/1957, p. 6.

12 William Dakin, 'Notes of conference held at Premier's Department', 9 July 1940, in Notes of Camouflage, Sydney, NAA, Series SP 1048/7 Item s10/1/329, p. 5.

13 Max Dupain, 'Max Dupain interviewed by Hazel de Berg', Canberra, National Library of Australia, 3 November 1975, Oral DeB 1/874, unpaginated transcript.

14 For an example of a camouflage school syllabus see A.D. Ross, 'The science of light: photography', in Camouflage—general camouflage school W.A., Canberra, NAA, Series A453 Item 1942/17/1957.

By expanding his photographic skills into the area of military science, Dupain followed in the footsteps of revered modern American photographer Edward Steichen (1879–1973) who served in the US army in WWI as Chief of the Photographic Section, innovating aerial reconnaissance photography. In WWII Steichen was deployed for work in photographic intelligence. He referred to aerial reconnaissance photographs dispassionately as 'instruments of war' and, although he was a humanist, he also admired the science of aerial photography and appreciated the difference between ultra-modern surveillance technologies and older photographic traditions. He admired the way aerial photographs represented 'neither opinions nor prejudice, but indisputable facts'.[15]

Fresh, bold, and aesthetically challenging, aerial photography was part of the visual culture of modernity, and the aerial perspective—in the words of Arshile Gorky in the 1940s—one of 'the new vision that flight had given to the eyes of man'.[16] It captured the imagination of Alfred Steigliz, and Beaumont Newhall wrote an entire book, *Airborne camera* (1969), in response to the expansion this new genre of photography brought to depictions of the world.[17] László Moholy-Nagy, originally a teacher at the Bauhaus in Germany and later director of the New Bauhaus in Chicago, was also excited by the revelations brought by the industrial eye of the aerial camera and, like Dupain, was seconded into camouflage and aerial reconnaissance in WWII.[18] There is no question that Dupain's wartime work, in which art, science and technology were brought together, put him in illustrious company and a modern context.

The work was functionalist and utilitarian, but it was also abstract and, for a while, suited Dupain's fascination with the way optical design and industrial machinery could be integrated. But it lacked humanism in the form of direct studies of human nature, and reduced the possibilities of what Dupain later claimed was his life's aim, a perception that 'induces delight, spiritual harmony, an abundance of emotional excitement and intellectual wonder'.[19] Aerial reconnaissance and photo-analysis trained him to think like a machine, to put himself in the place of an enemy camera, and envisage his photographs as ammunition against enemy intelligence; what Paul Virilio referred to in relation to WWI as the 'logistics of military perception, in which a supply of images would become the equivalent of an ammunition supply'.[20]

The landscapes that Dupain now photographed from 2000 feet above, but which he had photographed before the war from below, where objects have depth and texture, were now abstract visual information. They were symptomatic of the schism between two opposite perspectives, one of the 'sensuous or moral experiences of space' and the other of

---

15 These quotes are from a reprint of Major Edward J. Steichen, A.S.A, 'American aerial photography at the front' originally published in *The Camera*, vol. 23, no. 7 (July 1919), and reprinted in Gedrim 1996, pp. 70–74.

16 Rosenberg 1962, p. 91.

17 Newhall 1969.

18 László Moholy-Nagy was seconded by the Mayor of Chicago for civilian defence for the City of Chicago, USA. See Moholy-Nagy 1969, pp. 182–84.

19 Dupain 1986, p. 20.

20 Virilio 1989, p. 1.

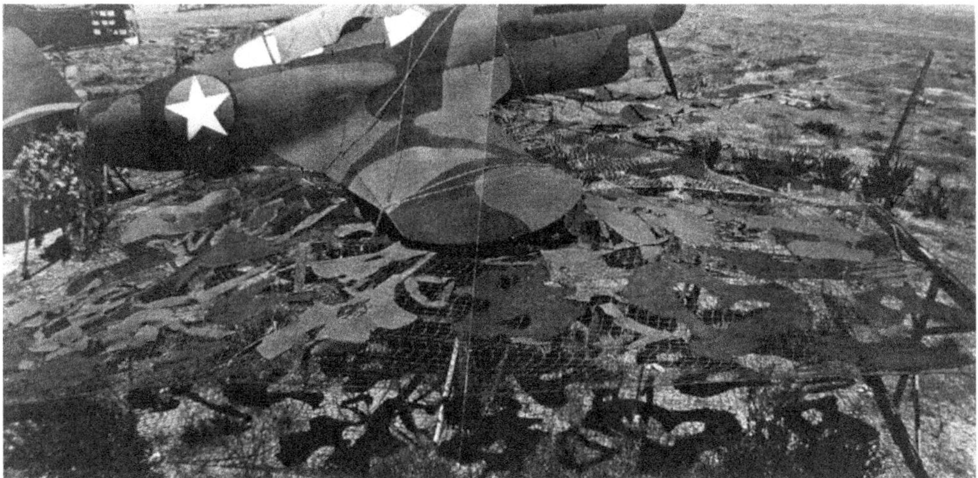

Fig. 11.4. Bankstown camouflage experiment No. 4 (ground), Jan/Feb 1943. The collection of the NAA: C1905 T1, 3[1]. Photograph, Max Dupain.

an emptying of experience from time and space created by technological perception.[21] And unlike Dupain's studio practice the work was secret and anonymous, just one small part of an enormous body of information gathered by Allied Intelligence:

> My job was to design camouflage, it might be for Beaufor guns, it might be for aircraft, a whole lot of things, and we would have this made, it might be netting, camouflage nets, it might be special mobile canopies for guns, and we'd get into the next aircraft going up over the area and photograph it. Anything that would fly in those days was what we had to rely on and I can remember the old Wirraways and the Fairy Battles and the Avro-Ansons that we used to climb in, perhaps lie down on the floor and open the bomb hatch and shoot through it.[22]

Each experiment involved measures and countermeasures to perfect camouflage on the ground, but also to perfect camouflage detection from above (figs. 11.4 & 11.5). In short, Dupain's job was to develop, through experimentation, a mental picture of objects on the ground, as they appeared to the aerial eye, and then build camouflage installations that produced incorrect mental pictures for the aerial observer. Among the paradoxes of the dual art of camouflage construction and aerial photo-tests was the reality that, when camouflage on the ground worked successfully, it stymied the penetration of aerial surveillance. Dupain therefore spent his years at Bankstown oddly trying to outwit his own photography. A typical job involved the concealment of dummy fighters made of cardboard that were covered in strip netting to conceal the plane's outline from above, and eradicate all solid shadow (fig. 11.6):

21 Hüppauf 1995, pp. 100–106.

22 Max Dupain, 'Max Dupain interviewed by Hazel de Berg', Canberra, National Library of Australia, 3 November 1975, Oral DeB 1/874, unpaginated transcript.

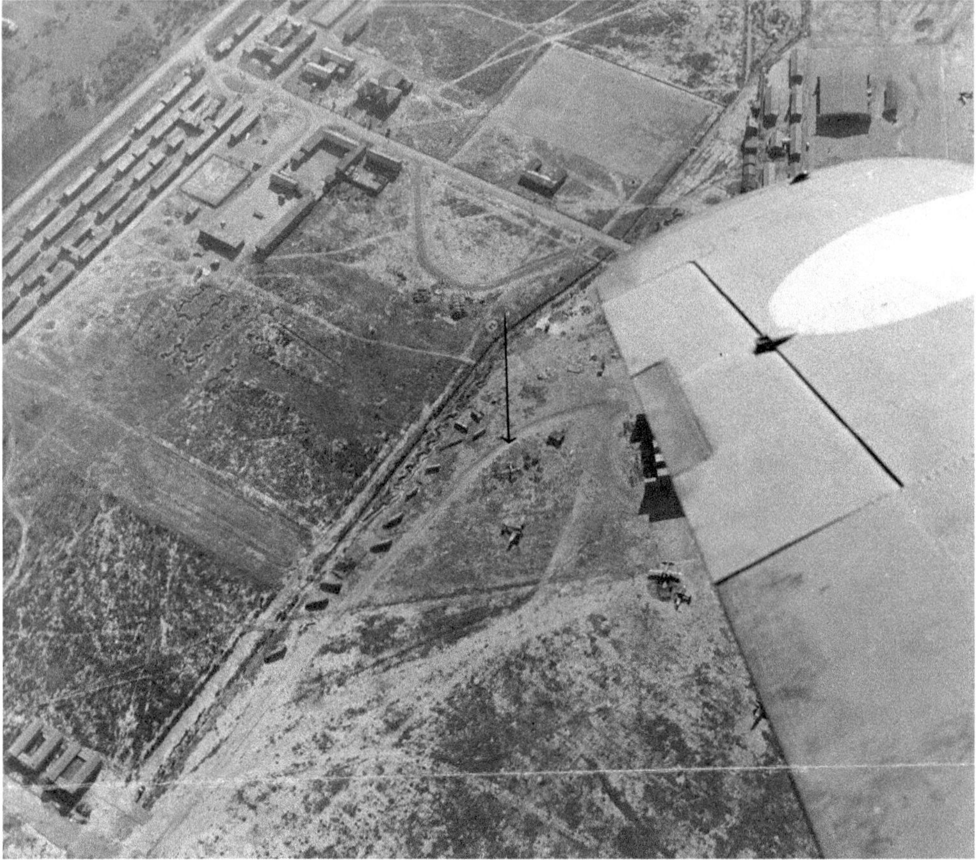

Fig. 11.5. Bankstown camouflage experiment No. 4 (aerial), Jan/Feb 1943. The collection of the NAA: C1905, Item 3, Box 13. Photograph, Max Dupain.

> Idea: to create a disruptive scheme immediately underneath two wings and part of fuselage to destroy the shadow cast thereby … and at the same time to use similar disruptive pattern on the aircraft itself so as to fuse both patterns into each other …When viewed from the air at 2000' the result looked more promising than previous experiments. The shadow was completely annihilated and the whole area broken up sufficiently to avoid suspicion.[23]

British artists also found that when building camouflage on the ground, it was critical to experience the aerial viewpoint. Julian Trevelyan, co-founder with Roland Penrose of the Industrial Camouflage and Research Unit, reflected on his early naivety as a military camouflage designer and admitted:

> It has to be confessed that we in our camouflage unit knew very little more about it than the man in the corner garage. We had none of us done much flying, and when we had

23 Max Dupain, 'Dupain's experiment no. 4', 25 February 1942, Bankstown experiments, Sydney, NAA, Series C1905 T1, Item 3, Box 13.

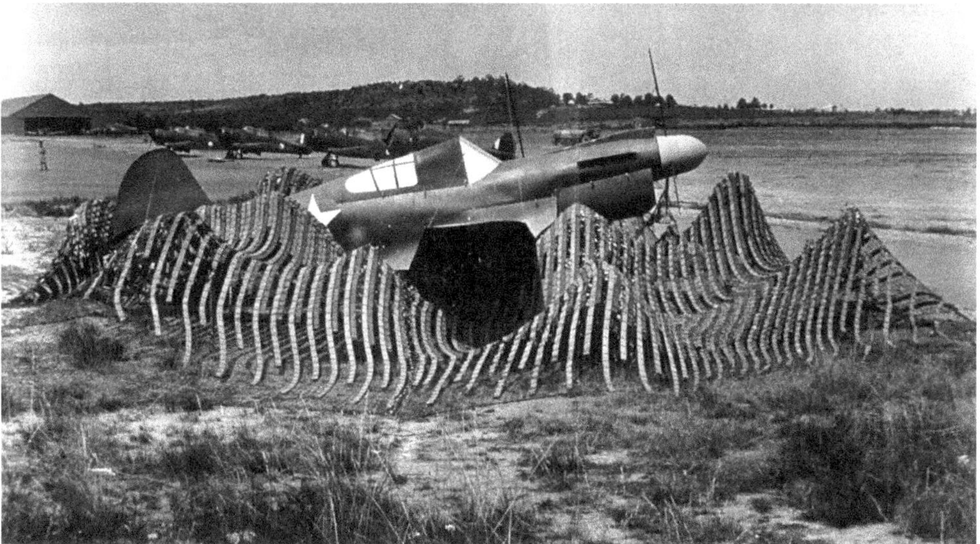

Fig. 11.6, Dummy kittyhawk and netting, Bankstown, 1943. The collection of the National NAA: C1905, 3. Photograph, Max Dupain.

> flown we had not addressed ourselves particularly to the problem of what makes things conspicuous from the air. Had we done so we would soon have realized that a lot of our assumptions were false, and that the pattern of the world from above is read very differently from the way in which we had supposed.[24]

Photo-interpretation meant that maps and aerial photographs had to be calibrated and coordinated. This was a new culture where specialists 'could look at a photograph taken straight down from an altitude of forty thousand feet, for example, and know instinctively that something had changed'.[25] But aerial photographs taken straight down did not always give information that was useful to those on the ground; therefore an oblique angle was often chosen where the camera was positioned pointing at the ground at an angle to the horizon, which helped make the landscape look familiar and objects easier to identify. A training manual for the 7 Australian Division Camouflage Training Unit described the new way of seeing needed for aerial reconnaissance as a paradigm shift—it required not just extensive training but also special mental discipline in order to attain an 'appreciation of the air view [that was] instinctive'.[26]

Dupain's interest in camouflage, indeed the fascination that impelled him to join the Sydney Camouflage Group and devote hours of training to the RAAF, makes sense in relation to his practice as a modernist photographer. He already had, it could be argued, a camouflage way of thinking. In the 1930s, for example, he experimented with

24 Trevelyan 1957, p. 113 cited in Behrens 2002, pp. 154–55.

25 William E. Burrows quoted in De Landa 1991, p. 197.

26 'The air view and air photographs', 7 Australian Division Camouflage Training Unit, Canberra, AWM, (August 1942–July 1943), AWM 52, Item 5/36/8, p. 2.

photomontage, creating two-dimensional illusions of new realities, and with rayograms (a speciality of artist Man Ray) where forms gain transparency and merge, blend and disintegrate to create, in Dupain's words, 'a new experience in light and chemistry'.[27] Both techniques put the viewer's sense of certainty with perception into doubt, the very aim of camouflage.

More directly connected to the type of camouflage photography undertaken at Bankstown in 1943, were studies made in the 1930s of shadows cast by nets onto human bodies and architecture. In 1943, when Dupain was still living in Sydney, the war was a relatively abstract concept for him, and the abstracted photographs of camouflage nets seem very much like extensions of earlier work produced as aesthetic objects without military purpose. Compare, for example, *Untitled (Jean with wire mesh)*, dated 1939, where patterns of shadow cast by netting alternately conceal and reveal the model, cut across her form and disrupt the clarity of her outline and in the process create a slight confusion of spatial depths (fig. 11.7). The results are similar to those in a photograph taken of a bomber hideout in Sydney showing light filtering through a camouflage net in preparation for the concealment of a plane by means of optical dissolution of form (fig. 11.8). Both have an element of romance and mystery, qualities that Dupain sought even in his most industrial and urban subjects.[28] Further, the steep angle that Jean is photographed from, and the unstable effect it produces, can be compared with Dupain's vertiginous aerial photographs at Bankstown where planes seem ready to slide off the surface of a topsy-turvy world (fig. 11.9). Both photographs are taken from perspectives that disorientate the viewer's recognition of familiar things and show how, with a rotation of viewpoint, the world can be made less recognisable—more so from 2000 feet over Bankstown, than from a high position like a ladder or chair used to photograph Jean, but the principle is the same.[29] These comparisons illustrate Dupain's suitability for a wartime career in visual disorientation.

When in 1944 the DHS sent Dupain to the New Guinea area, his assignment was to gather evidence of camouflage installations from planes. What an impact those visually disorientating views of the Melanesian landscapes must have had on his sense of equilibrium and fear as he travelled into war zones where it was common knowledge that planes crashed every day killing their passengers. Although Dupain had volunteered for the New Guinea area, by April 1944, when he left New Guinea, he had already spent a year living in Darwin, helping artists Ronald Rigg and Roy Dalgarno conceal radio stations and fuel storage tanks in a town where the landscape was wrecked and the heat and humidity tested his sanity. He was later to remember how the experiences of forward areas, first in Darwin and later at Nadzab, Finschhafen, Dreger Harbour, Lae and Goodenough Island, made him cynical about the role of camouflage in the war:

---

27 Dupain published a short article on the pioneer American modernist photographer and inventor of rayographs, Man Ray. See Dupain 2006 [1935], pp. 120–22.

28 See Dupain's commentary on romance and mystery in relation to *Mosman Bay at dusk* (1937) in Dupain 1986, p. 70.

29 The impact of rotation on perception, and its part in surrealist photography, is most famously discussed in Krauss 1985, pp. 31–72.

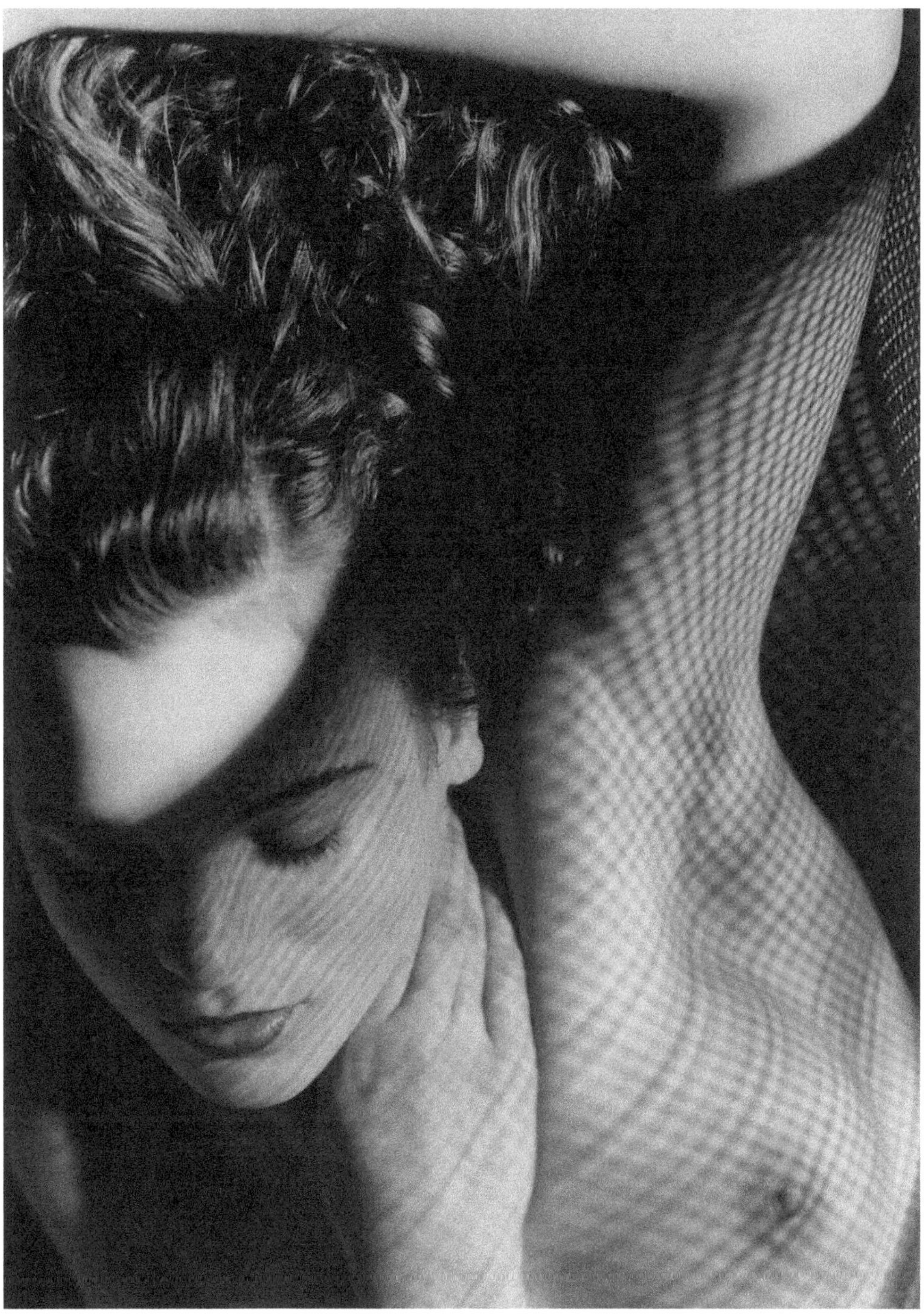

Fig. 11.7. Max Dupain, *Untitled (Jean with wire mesh)*, 1936. Gelatin silver photograph, vintage, 46.0 x 33.5 cm. Gift of Edron Pty Ltd—1995 through the auspices of Alistair McAlpine, Collection: AGNSW.

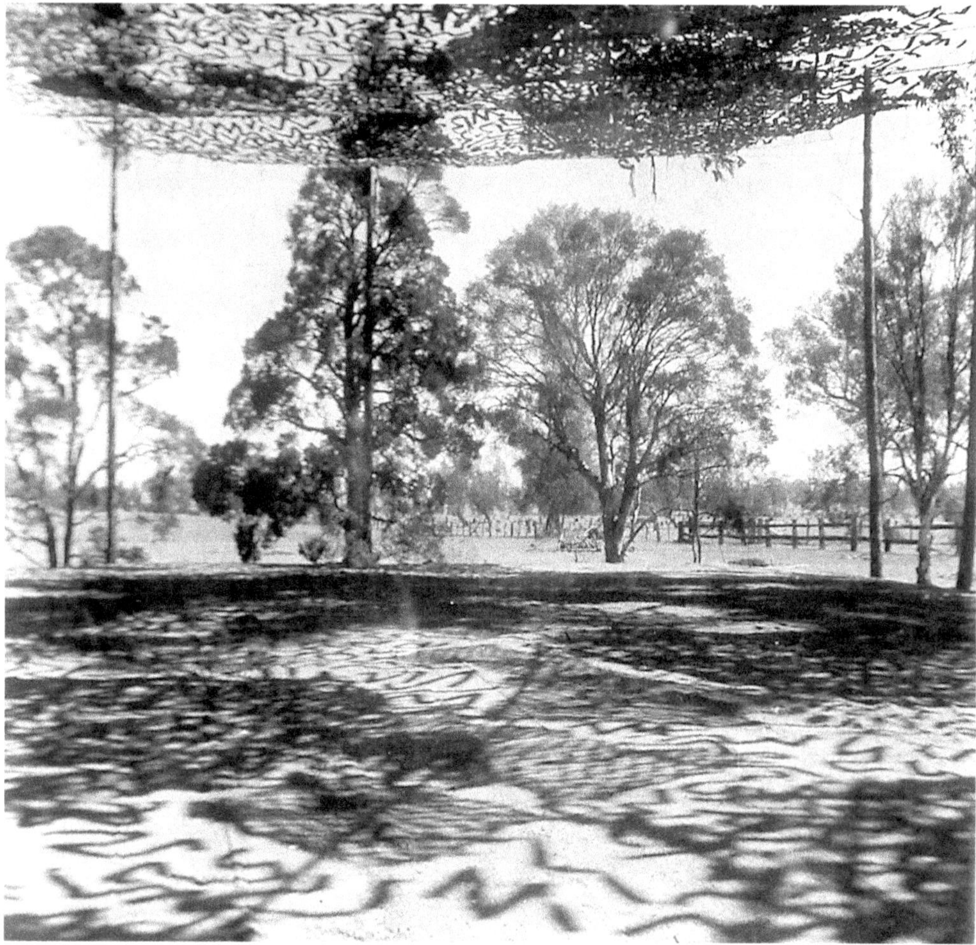

Fig. 11.8. Hideout for medium bomber with garnished ceiling and transplanted trees, Sydney aerodrome 1943. The collection of the NAA: C1905 T1, 23[1]. Photograph, Max Dupain.

> I was in the airforce taking pictures for quite a long time and I thought, oh bugger this, this place is just disintegrating (the war was nearly finished). I was going to apply for a job as a war photographer and see if I could get out in amongst it again.[30]

By 1944 Australian and American attitudes to concealment and deception in the New Guinea area had become casual, and in response Dupain became increasingly disaffected. The Americans he was sent to work with were 'not terribly concerned about camouflaging their weapons and machinery' and this made his role progressively ineffectual.[31] By late 1944 the DHS's operation in the area was wound back, and Dupain ceased working

30 Ennis 1991a, p. 23.

31 Max Dupain, 'Max Dupain interviewed by Hazel de Berg', Canberra, National Library of Australia, 3 November 1975, Oral DeB 1/874, unpaginated transcript.

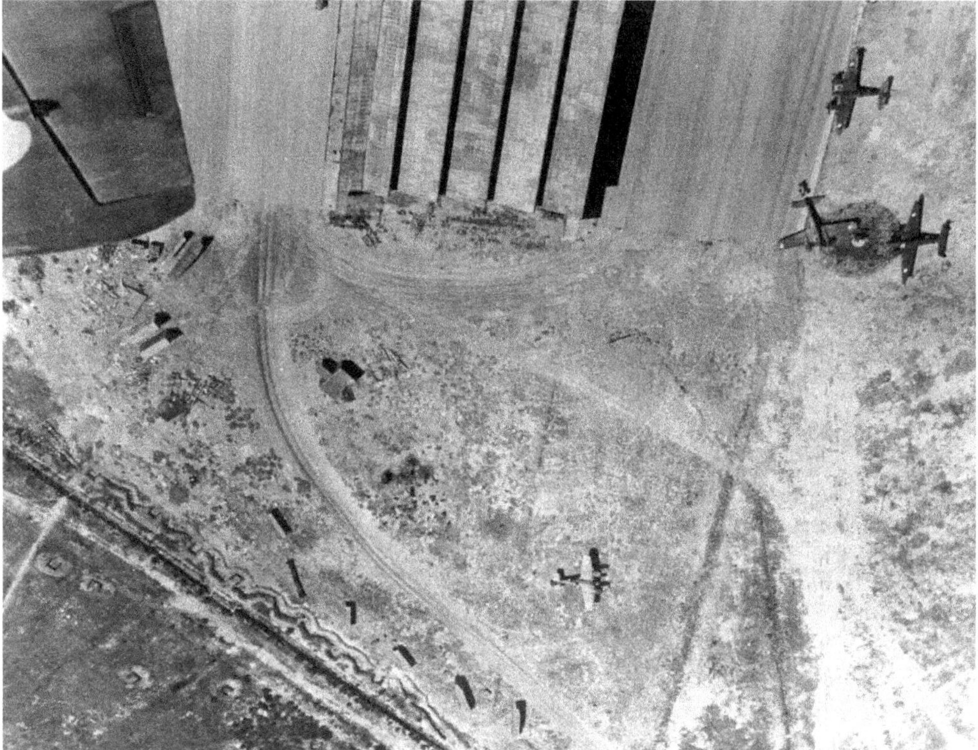

Fig. 11.9. Dupain–umbrella camouflage experiment, Bankstown, 1 March 1943. The collection of the NAA: C1905 T1, 3[10]. Photograph: Max Dupain.

in camouflage altogether. War photography appealed instead, a change of heart that nevertheless does not quite add up with a later claim in 1986 that he scorned the heroics of war. Official war photography was full of glory and war photographers like Damien Parer—who was Dupain's friend and later killed in action in the Pacific—were deployed for the heroic purpose of commemorating and remembering the decisive moments of history in the making.[32] The Department of Information, for example, a body that commissioned Australian artists to serve as war photographers, eulogised staff for recording:

> The great feats of our servicemen in a magnificent pictorial pageant—one that showed the desperate struggle to turn near-defeat into victory and told the world of the valour of Australian men and women. They tirelessly focused their cameras on soldiers, sailors and airmen as they battled in North Africa, the Middle East, Europe, India, Malaya and right on through to New Guinea. They pictured the dusty, weary desert marches, the sweaty struggle in mountain jungles, the dejection of retreat; they carried their cameras to the triumph of offence and finally to victory on land, sea and in the air.[33]

32 Dupain 1986, p. 15.

33 D.H, 'Picturing Australia: the story of the "D.O.I" photographers', *The Australasian Photo-Review*, 1948, p. 364.

Almost certainly the magnificence of struggle, and the sacrifice of men and women in war, were subjects that attracted Dupain whose work as a camouflage photographer was, by contrast, behind the scenes. War photography, on the other hand, was visible to the public and it was commonly understood that the photographer's life was one of risk and sacrifice too. But when Dupain's application for a position was turned down, he was offered a documentary assignment with the Department of Information, and this helped to set him on an altered course towards what he came to see as the moral purity of documentary photography and what Helen Ennis construes as his symbolic return to 'normality'.[34] The war made him question life's meaning and purpose and, like Lee Miller—the American photographer who, as a result of WWII, ceased her career in fashion photography and took up work as a war correspondent, photographing the barbaric realities of concentration camps, including Buchenwald—Dupain sought a similar transformation towards photography for humanitarian purposes.[35] By 1947 documentary photography seemed the only way to engage usefully and truthfully, but also simply and clearly, with the social world.[36] And several times over his life he affirmed the same belief: 'I can't get away from Documentary. (It is the school I have adopted and will stand by. Any other attitude would be phony.)'[37]

One conclusion that can be drawn from the documents and records relating to Dupain is that the character and demands of camouflage—arguably, too, a practice of the 'phony' in the sense of involving tricks and fakes—hastened Dupain's postwar determination to concentrate on the truth and clarity of documentary photography. Camouflage was mentally challenging work that was never straightforward but instead paradoxical and cryptic. It required the mind to think in doubles, entailed an aesthetic orientation to optical disintegration and delusion, and demanded that attention be given to loss of distinction rather than clarity. Further, it was a practice of deception and masquerade, and in that sense shared much in common with the fashion industry that Dupain also wanted to turn his back on:

> The war or my very limited part in it was a shock to my sensibilities. It showed me the difference between life and death, neither of which I had really considered in any depth before. This experience determined that I should not return to anything so trivial as fashion photography. It had to be photography of some sort but please God something with meat in it; not ephemeral bullshit.[38]

There is something else, though. The camoufleur's role, as we have seen, was not well respected, and from Constance Babington Smith—who wrote about photographic intelligence in WWII—we also learn that the contribution of crew who worked in allied

---

34 Ennis 1991c, p. 139.

35 See Marien 2002, pp. 304–05.

36 Dupain 2007, p. 14.

37 Dupain 2007, p. 8.

38 Dupain 2007, p. 6.

photo-reconnaissance was held in 'little acclaim'.[39] As the above quote shows, Dupain's role in camouflage affected his ability to talk confidently and without hesitation about his part in the war, so much so that he almost apologises for it. Yet for anyone reading the voluminous war record of the camouflage section of the DHS where Dupain's name is mentioned constantly, it becomes obvious that he was held in the highest esteem, and his work in camouflage was seen as playing an important part in the war's outcome.

While Dupain did not write about the richness and texture of life as a camoufleur, Frank Hinder did, and from Hinder's diaries and notebooks a picture emerges of camouflage as an intellectual and phenomenological investigation that was sustaining and satisfying. A closer inspection of Hinder's part in WWII reveals how camouflage and modern art were mutually informing. Photography was Dupain's profession at the onset of WWII and the war utilised his knowledge of photography fully. But Hinder, no less than Dupain, also focused his camouflage work on the way the camera sees. However, his unique contribution was to combine industrial, constructivist design with geometric abstraction, a passion for the impact of light on colour and form, and a flair for bricolage. Hinder was an inventor, and the war was not only a way of putting art at the service of democratic freedom, but also of putting art at the service of military design.

39 Babington Smith 1957, p. 13.

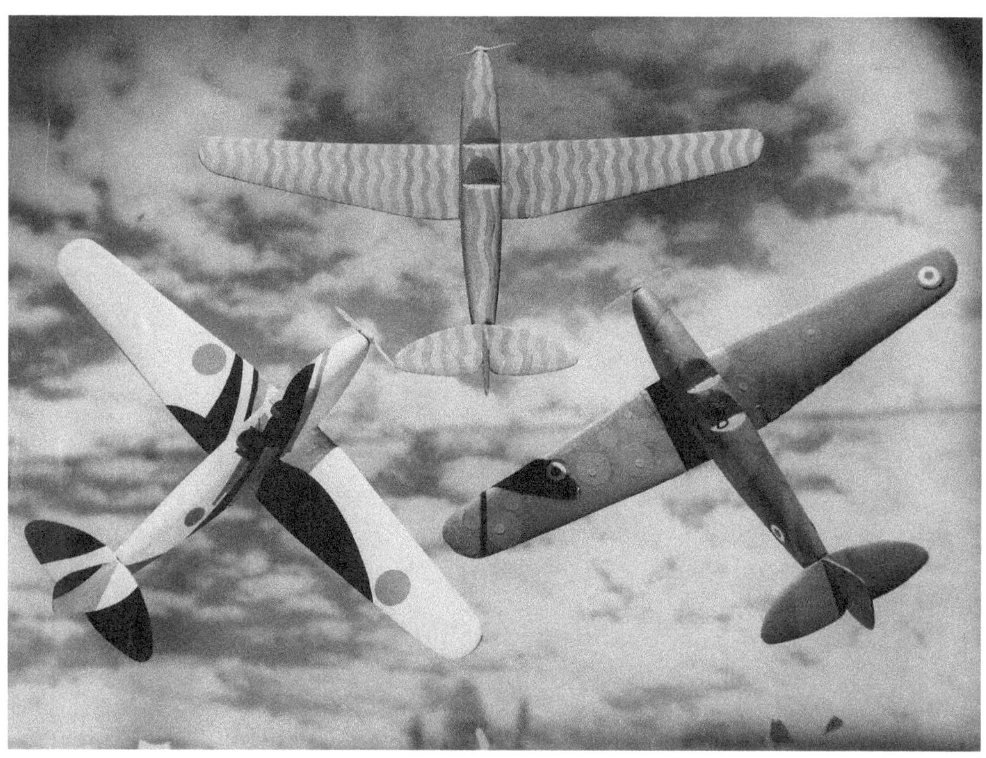

Fig. 12.1. Camouflage experiment by Frank Hinder and Russell Roberts, 'Distracting designs on A/C: dazzle colours'. AWM, Canberra, Frank Hinder Personal Records AWM 88/133 File 895/4/182. Photograph, Russell Roberts.

Chapter 12

Frank Hinder

Frank Hinder's life was like a cubist painting; it was a series of intersecting planes devoted to art, design, science and technology, all of them emerging and receding in relative importance at different intervals over time. When he became involved with camouflage during WWII, it simply meant a shift of focus, not a new beginning. In 1943, for example, he stood before a class of airforce personnel and delivered a camouflage syllabus on 'Light and colour'.[1] A decade earlier, as a design student in the US, he too listened to lectures on light and colour from teachers who inspired him to pursue a modern artistic vision through abstraction.[2] But when he lectured alongside Russell Roberts and Gervaise Purcell to military personnel in Townsville, Darwin, Sydney and Canberra, Hinder kept art's principles submerged behind the social principle of assisting men and women in wartime develop a sense of visual protection. Camouflage, like cubism and modern design, was a problem of space, light and colour, but understanding how retinal impressions of colour impact on spatial perception was only important in the war zone if it helped save lives (fig. 12.1).

Hinder (1906–1992) was an oddity in Australia in the 1930s; he was cosmopolitan in outlook, had worked in Chicago, New York, Montreal, Boston and New Mexico, and studied in the new world when most Australian artists studied in the old. He returned to Australia with a sense of being part of international modernism, and of sharing aesthetic and social philosophies in common with the vanguard artists of the 20th century including László Moholy-Nagy, Jacques Villon and Franz Marc.[3] As a student and a commercial artist in North America between 1927 and 1934 he developed wide-ranging and ongoing interests in kinetic art, light and physics, geometry, constructivism, the development of a creative society through art education, and importantly—for his future role in camouflage—the idea of furthering public life through design and technology.[4]

---

1 Frank Hinder, 'Records of Lt F.C. Hinder Camoufleur', Canberra, AWM, PR 88/133, File 895/4/182, Item 5 of 12.

2 For a discussion of Frank Hinder's education in art and design, colour and form with Emil Bisttram in the US, see Free 1980, pp. 12–13.

3 For an account of Frank Hinder's life and work (with a chronology devised by David Thomas) and for discussion of Hinder's exposure to European and American modern art see Free 1980, pp. 12–14.

4 For a list of books in the Frank Hinder Bequest to the Art Gallery of New South Wales see the AGNSW online library catalogues, www.artgallery.nsw.gov.au/lib/cat, accessed 01/10/09.

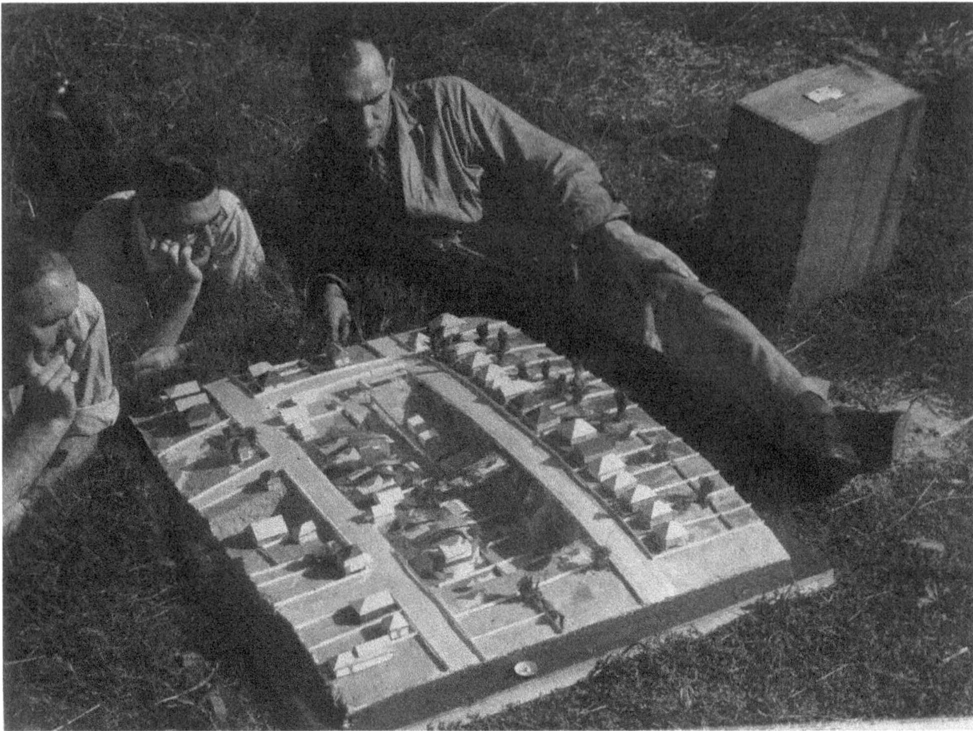

Fig. 12.2. Frank Hinder (right) with model for scheme to camouflage petrol tanks, Brisbane, 25 April 1941. Frank Hinder Papers. Collection: Archive of the AGNSW.

Frank and his wife, Margel Hinder, returned to Australia in 1934 from living in the US, and eked out an existence as practising artists, supplementing this whenever possible with commercial design. But then the war came. In 1939 Frank became a member of the Sydney Camouflage Group, and not long after was assigned to the Department of the Interior working with an architect to design camouflage schemes for bulk storage petrol tanks in Sydney as well as Brisbane, Newcastle and Port Kembla (fig. 12.2).[5] Without a doubt this collaboration between architecture and art, construction and design, remained important to Hinder throughout his life.[6] With commercial jobs scarce, he found his new wartime appointment a 'financial Godsend'.[7] Shortly afterwards he joined the CMF attached to the RAE and worked with Russell Roberts and Sali Herman in the research station at Georges Heights in Sydney, as camouflage instructor and research officer (fig. 12.1). Then in 1942 a welcome secondment from the army to the camouflage section of the DHS took Hinder

5 Frank Hinder, 'Records of Lt F.C. Hinder Camoufleur', Canberra, AWM, PR 88/133, File 895/4/182, Item 7 of 12.

6 See Hinder 2006 [1952], pp. 627–31.

7 Frank Hinder, 'Records of Lt F.C. Hinder Camoufleur', Canberra, AWM, PR 88/133, File 895/4/182, Item 7 of 12.

to Canberra to work with William Dakin, and from that point on he was exempted from military service.

The war years were a strange, incongruous episode in Hinder's life. Quite apart from the political hostilities threatening Australia, he was embroiled in an art world battle to protect the democracy of art from conservatives like the critic Lionel Lindsay, but also Sydney Ure Smith who initiated the formation of the Sydney Camouflage Group. As a member of the camouflage group Hinder had much in common with Ure Smith, but as a modern artist trying to make Australian art progressive, he—along with Margel Hinder, Rah Fizelle, Eleonor Lange, Ralph Balson and Grace Crowley—was disenfranchised by Ure Smith's circle of friends and associates. Ure Smith's circle supported establishment artists like Arthur Streeton and Hans Heysen, and tried to prevent the fine arts practices of painting and sculpture in Australia from becoming part of the movement of modernism.[8]

Hinder, though, joined the camouflage group convinced, as no doubt Ure Smith was also, that it would be impossible to modernise methods of warfare, and protect Australians from Japanese invasion, without the cooperation of modern designers. Camouflage was more than just a regional design problem; its principles were—according to the DHS—'universal'.[9] Accordingly, from 1939 onwards, Hinder kept up to date with camouflage methods overseas by reading *The New York Times*.[10] And as it turned out, he fast became one of Dakin's best recruits. Versatile, fit, knowledgeable about the military (after the compulsory military training in his youth), an improviser, an innovator and an educator, Hinder also worked easily in teams and he had valuable experience in theatre and poster design.[11] Educating military personnel, as well as camoufleurs, was an important part of his role. Among the many design-based jobs he undertook was the preparation of visual teaching aids and striking camouflage posters. The posters were clear, blocky, simple, eye-catching and blunt, like the machine aesthetic of German Bauhaus graphics (fig. 12.3).

Dakin's ideal camoufleur was an 'ordinary man endowed with imagination and fieldcraft' and Hinder stood out for ingenuity.[12] Later he received a war innovation award for an eight-legged portable camouflage net, a variation of an English design known as the 'Walgrove Spider' that he altered to make 'infinitely superior, cheap, light and inexpensive'.[13]

---

8 Ure Smith distinguished between modernism for the high arts, which he could not accept, and modernism for commercial art, which he actively supported, even employing the Hinders to design advertisements for his journal *Home*. See Underhill 1991, p. 23.

9 Written above one illustration in the 'museum of camouflage' (the portable demonstration kit referred to in ch. 5) were the words: 'the methods may vary but the principles are universal'. For photographs, by Dupain, showing how the panels of the museum were installed, see Camouflage photos not used in book, Sydney, NAA, C 1905, 10.

10 For clippings from *The New York Times* see Frank Hinder, 'Records of Lt F.C. Hinder Camoufleur', Canberra, AWM, PR 88/133, File 895/4/182, Item 4 of 12.

11 For a discussion of theatre design and Hinder's place in it see Harding 2004, p. 10.

12 W.J. Dakin, 'The outbreak of war with Japan', Professor Dakin's camouflage report, Sydney, NAA, C 1908, Item 5.

13 Frank Hinder, 'Records of Lt F.C. Hinder Camoufleur', Canberra, AWM, PR 88/133, File 895/4/182, Item 7 of 12.

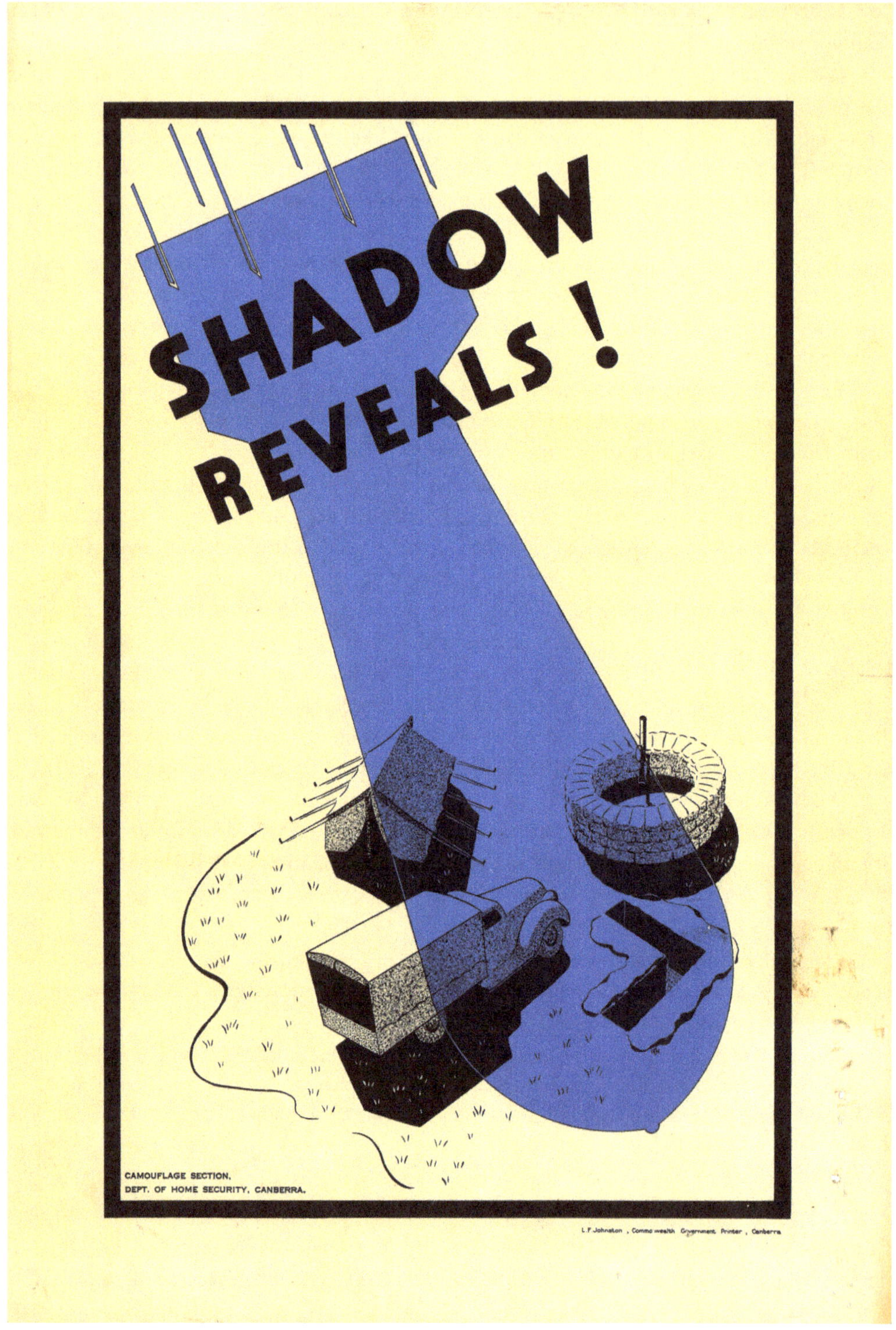

Fig. 12.3. *Shadow reveals*, poster designed by Frank Hinder for the DHS. The collection of the NAA: M222/2, 8.

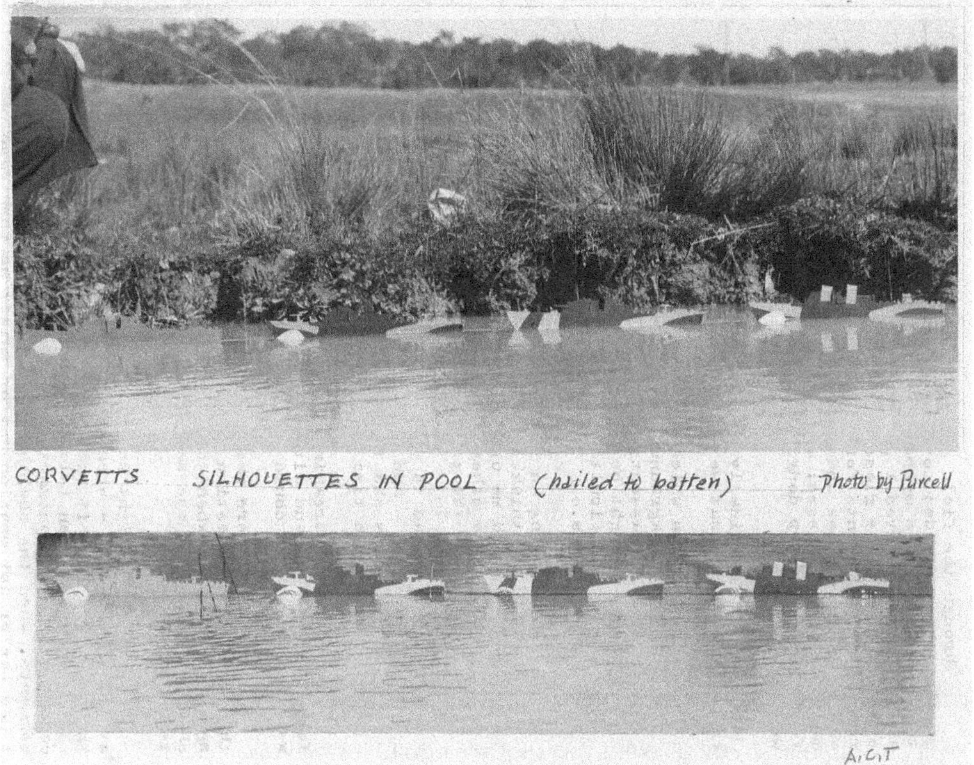

Fig. 12.4. Hinder experiment with corvette silhouettes in pool. Canberra, AWM, Frank Hinder Personal Records PR 88/133, 8 of 12. Photograph, Gervaise C. Purcell.

In response Dakin reserved a special section for the 'Hinder Spider' in *The art of camouflage*, making Hinder one of only two artists—the other John Moore—to be recognised by name or initials in that book.[14]

Even a cursory glance at the range of Hinder's production for the DHS shows the overlaps in conception and design between camouflage practice and the modern, creative field of abstraction. In the studio he investigated perception and illusion in relation to cubist designs, showing how certain figures recede against ground and others project, and called his experiments 'semi-abstracts'. The margins of his notebooks ask why figures appear nearer to the observer than ground, when both occupy the same planar space, and also proposed alternative ways of seeing and thinking: 'ground may be perceived as a surface or a space'; 'ground areas have form—the negative form of space left'; 'figure is usually perceived on top of or in front of ground'.[15]

But Hinder also interrogated the psychology of perception and the relativity of vision in relation to model war-planes, trucks and ships. His greatest concern was to make

14 Dakin 1942, p. 76.

15 Frank Hinder, 'Semi abst.', Sketchbooks 1937–1977, Sydney, Frank Hinder Papers Collection: Archive of the AGNSW, Ms 1995.1.

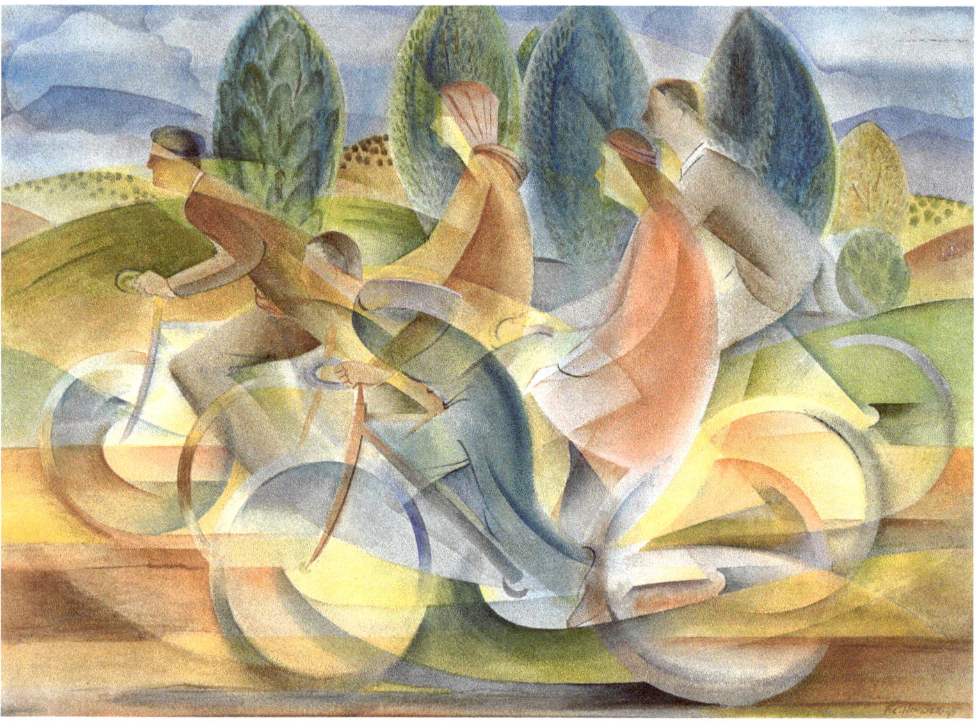

Fig. 12.5. Frank Hinder (1906–1992), *Cyclists, Canberra*, 1945, watercolour, 50.8 x 70.5 cm. National Gallery of Victoria, Melbourne, purchased 1947.

colour, tone and line useful for visual disorientation and confusion. Using the cubist principle of dissolving facts into design, he painted model ships—built by Margel—in dazzle camouflage patterns, and photographed them against landscape and water (fig. 12.4). Where the ship's dark fragmented patterns matched the tones of the river-bank, its contours all but disappeared, leaving only a pattern of nonsensical light fragments visible. The effect is similar to cubist paintings and can be seen in Picasso's portrait *Daniel-Henry Kahnweiler* (1910) where the human figure is fragmented and fractured and appears to alternately lie beneath and above an array of overlapping planes differentiated by colours and tones. Indeed, a similar illusion occurs in Hinder's landscape *Cyclists, Canberra* (1945) where overlapping planes alternately animate space but also make parts of the subject hardly visible (fig. 12.5).

In the camouflage laboratory Hinder explored a nascent curiosity with kinetic design and with animated geometric abstraction. It was to evolve, in later life, into a focused exploration of kinetic art. Among the challenges for designers of camouflage was how to conceal a moving object, but also how to make the movement of an object confusing to read. Hinder tried this out on cardboard aeroplanes (fig. 12.6). He painted them in linear patterns of contrasting colours. While they look unconvincing as military science, they compare well with the disruptive camouflage stripes painted on WWII prototypes by US

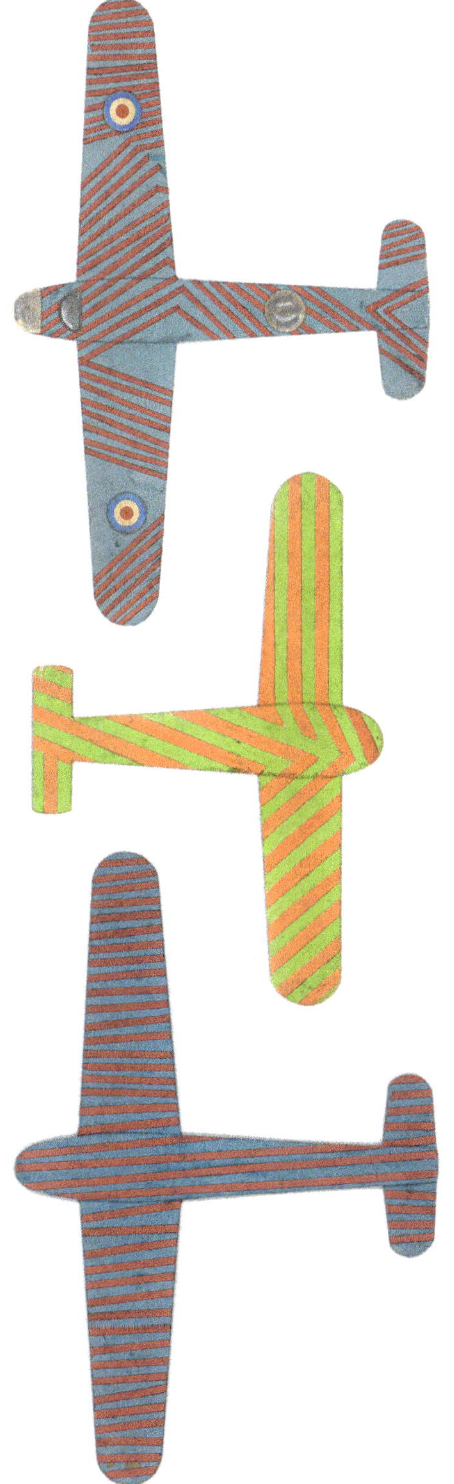

Fig. 12.6. Frank Hinder, three cardboard model planes painted in dazzle. Canberra, AWM, Hinder Personal Records PR 88/133, 4 of 12.

designer McClelland Barclay.[16] Their purpose was to make it difficult to judge the shape and location of the aircraft. Hinder carried this idea forward into the 1960s when Op Art emerged as a force in contemporary art, and he produced two kinetic works, *Red sweep* (1967) and *Blue arcs* (1967). He called them 'luminal kinetics', and they were investigations of the impact of motion on colour, but also the way the motion of colour in space alters a viewer's perception of reality.[17]

Modern art and modern camouflage were mutually informing, and this was true in relation to the work of other international modernists. More than once has Frank Hinder been compared with László Moholy-Nagy, the artist Walter Gropius said 'incessantly strove to interpret space in its relation to time, that is, motion in space'.[18] Hinder studied in Chicago before Moholy-Nagy took up residence there as the director of the New Bauhaus, but he was well acquainted with his books, including *The new vision: fundamentals of design painting sculpture architecture* (1939) which he bought in 1941.[19] But their lives were parallel in many respects, and not least because Moholy-Nagy, along with his assistant, György Kepes, developed a course on civilian camouflage and advised on defence for the city of Chicago in WWII.[20] WWII brought sudden changes to Moholy-Nagy's life, as it did to Hinder's. Virtually overnight Margel turned from sculpture to model-making and became John Moore's secretary, while Frank found himself travelling throughout Australia and the islands of the SW Pacific, including Rabaul, advising the military.

It was during the trip to Rabaul that Hinder gained his first insight into the perils of working in operational areas. In November 1941, as a lieutenant in the CMF, and equipped with only 'one water colour box and colours, one water colour pad (block), two water colour brushes, one sketch book, one set of camouflage colour cards' he flew with the RAAF on a camouflage mission, first to Port Moresby, and then Rabaul to inspect the aerial visibility of the Vulakanau airstrip.[21] But the plane, a Lockheed-Hudson, crashed, then burned, and Hinder was lucky to survive. His sometimes illegible diary entry for Sunday 23 November 1941 captures the terror of the moment in comical understatement:

> Roared up runway—seemed to take long time to get the tail up—seemed to be getting close to end of field—finally lifted from ground but bumped back again—lifted again but tipped and swerved to one side —Cpl clutching at seat with eyes popping—me the same way—plane swung straight and crashed—shot me towards end minced up with [?]—lot of noise and racket tried to protect head and face, waiting for next bump and crash. Came to rest and flames shot up all round—Pilot and F.O came racing back—Cpl tried to open door but it jammed. Maj. Pitt joined us—Cpl & Pilot heaved and bashed door—jammed.

---

16 For images of McClelland Barclay's prototypes and more information see Behrens 2009a, pp. 40–41.

17 For a reproduction and description of *Blue arcs*, see Harding 2009, pp. 142–43.

18 Gropius 1969, p. viii.

19 For a list of books in the Frank Hinder Bequest to the Art Gallery of New South Wales see the AGNSW online library catalogues, www.artgallery.nsw.gov.au/lib/cat, accessed 01/10/09.

20 Moholy-Nagy 1969, pp. 183–84.

21 W.J. Dakin, 'New Guinea and Islands Area', Professor Dakin's camouflage report, Sydney, NAA, C 1908, Item 5.

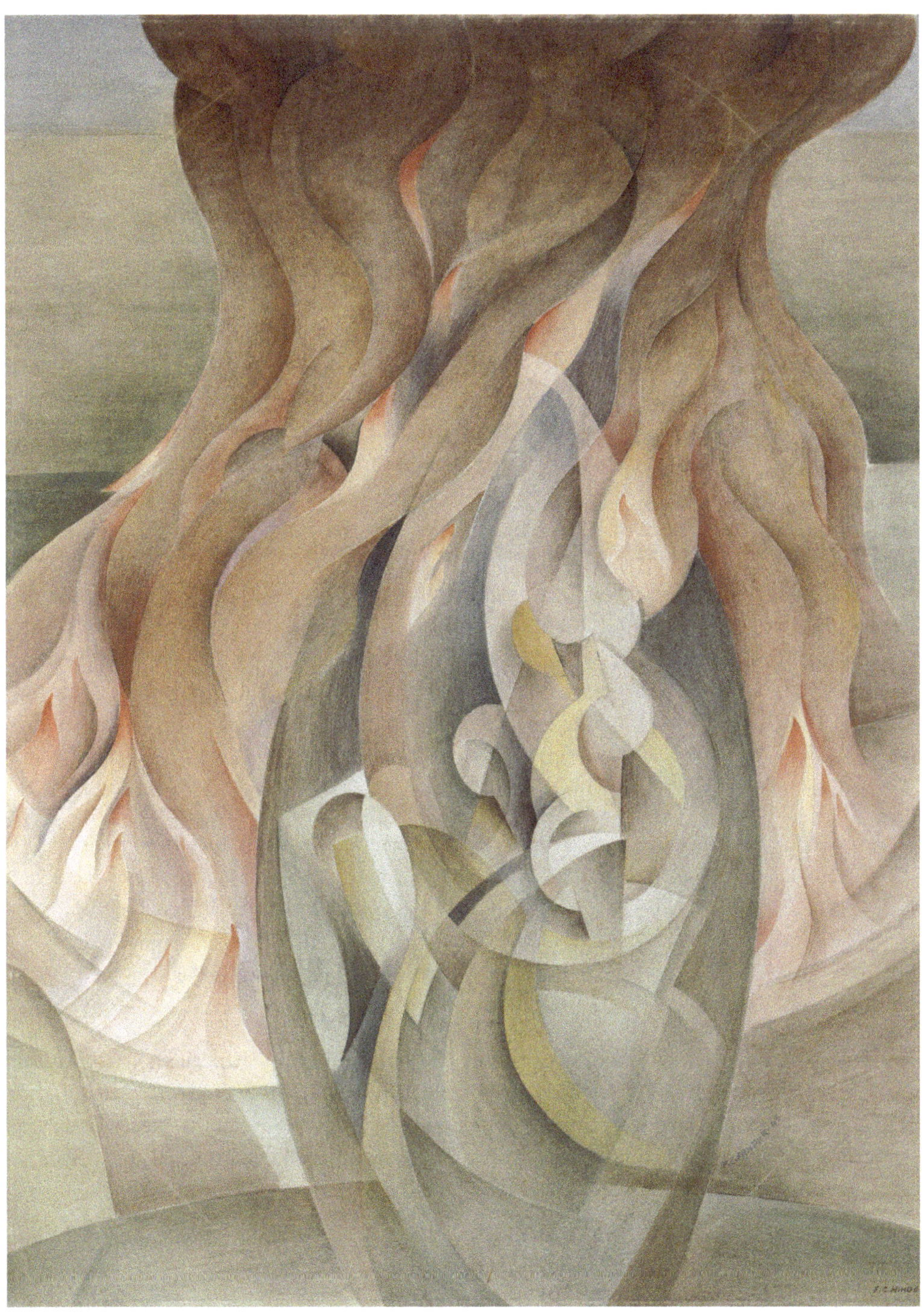

Fig. 12.7. Frank Hinder, *Bomber crash*, 1943, tempera on paper on paperboard, 54.9 x 40.0 cm. Purchased 1967, Collection: AGNSW.

> Flames shooting up all round—red glow inside—flames in cockpit—got my back to wall and kicked at door—flames outside but slightly clear to one side ... Before door opened wondered what it would be like to be incinerated—thought of Margel and her reaction to news. Peculiar feeling did not seem possible that we would actually be burnt up, although it looked like it when the door jammed. Plane blazed well—ammunition and rockets going off. Metal & plane burnt well.[22]

Later Hinder painted two impressions of the crash and inferno (fig. 12.7). One impression, *Bomber crash* (1943), has been compared with the work of the Italian futurists whose manifesto (1909) celebrated war, speed and technology.[23] But unlike Filippo T. Marinetti, leader of the futurists, Hinder was not a lover of war and violence, although he comes across as a fetishist of modern technology since, as Renée Free points out, the subjects of planes, aerodromes, and flight (literally and metaphorically), occupied his imagination for a lifetime.[24] *Bomber crash* is not a celebration of war but, as Hinder himself said, a speculation on incineration. It is also a study in dynamism and abstraction, and in this it shares a similar visual language to that of the futurists, one of fracture and fragmentation, wholly suited to the subject of explosion.

For all its terrifying moments, Hinder never complained about the war. On the contrary he was prone to highlight its benefits, and downplay its horrors. Unlike Max Dupain he kept detailed diaries, and in 1982 turned these into a lively memoir. The illustrations, photographs, and commentary that make up his personal files in the AWM, are unique records of the contribution of artists and designers to camouflage in WWII. Like Max Dupain, Hinder was very willing to answer questions about wartime, and from interviews, as well as diary entries, we learn that life in Canberra with the DHS was positive and surprisingly secure compared with the hardships of the preceding Depression years:

> As far as I was concerned, it was really a very pleasant war, because it's nice to work with your family and nice surroundings. From there I made occasional trips to the islands and various other places, looking at aerodromes and so on, and as I say, it was most helpful, most constructive, and I met a lot of very interesting people. At the end of the war I went back to commercial art again ... I think possibly having a regular salary in the Army upset most of us—it was very nice not to have to worry about it.[25]

But he was by no means alone in finding the positive in war. American artist Thomas Hart Benton, for example, assigned to camouflage in WWI, later reflected on the stimulation it brought his art practice by temporarily taking him away from what he described as 'all my grooved habits, from my play with colored cubes and classic attenuations, from my

---

22 Frank Hinder, 'War diaries 17 June 1941–25 March 1942', Frank Hinder, 'Records of Lt F.C. Hinder Camoufleur', Canberra, AWM, PR 88/133, File 895/4/182, Item 3.

23 Wilson 2002, p. 92.

24 Renée Free discusses Hinder and flight in Free 1980, p. 13, and on her website, Renée Free, Frank Hinder, www.frankhinder.com.au, accessed 23 September 2010.

25 Hazel de Berg, 'Interview recorded with Frank Hinder', Canberra, National Library of Australia, 20 July 1963, Oral TRC 1/45, transcript, pp. 560–61.

aesthetic drivelings and morbid self-concerns'.[26] Closer to home, Stella Bowen, on return from service as an Official War Artist, wrote that her appointment in WWII was a great experience and opportunity, one that took her into a world that 'civilians usually don't get a glimpse of'.[27]

While Hinder wrote vividly about camouflage, in 1963 Hazel de Berg taped his most revealing comment. Even before the war, he told her, he had 'always been interested in camouflage, or the results of camouflage'.[28] He certainly made this clear in a diary in 1938 when he wondered what part light would play in camouflage, in the event of war.[29] But looking at Hinder's personal and professional history, other connections with camouflage also become clear. In childhood he was passionate about natural history, and this instilled a life-long interest in animal studies, including the camouflage habits of birds. He attended Newington College, a school that introduced him to military culture and modern art concurrently. And from the 1930s onward he studied Jack Burnham, Jay Hambidge and Roger Caillois, three intellectuals concerned with similar questions to designers of military camouflage, but in relation to cubism, bricolage, and the conceptual space between reality and its representations.[30]

From Jay Hambidge's theory of 'dynamic symmetry', Hinder learnt about the mathematical foundation and symmetry of nature. He applied the ideas to studies that blend organic and geometrical perspectives on space.[31] The connection with camouflage was almost logical. If the underlying mathematical structures of nature could be learnt and understood, they could also be distorted and denatured. Especially relevant is a quote from Hambidge dated 1914, one that Hinder jotted down in a notebook; it is a definition of cubism that explains it as 'the attempt to dissolve facts entirely in design'.[32] There in a nutshell was also a definition of military camouflage: a practice intended to alter the specificity of an object in relation to its surrounding space. This was proof of why the rise of modernism in western art is said to have occurred at the same time as the modern development of military camouflage.[33] It explained why—when Picasso and Gertrude Stein watched a camouflaged truck on the boulevard Raspail in Paris in WWI—Picasso shouted 'yes it is we who made it, that is Cubism'.[34]

26 Thomas Hart Benton quoted in Cork 1994, p. 193 and cited in Behrens 2009a, p. 52.

27 Stella Bowen to Tom Bowen, 27 September 1944 (private papers) quoted in Wilkins 2002, www.awm.gov.au/exhibitions/stella/article.asp, accessed 27/07/10.

28 Hazel de Berg, 'Interview recorded with Frank Hinder', Canberra, National Library of Australia, 20 July 1963, Oral TRC 1/45, transcript, p. 560.

29 Frank Hinder, 23 September 1938, the Grosvenor Galleries Diary, Frank Hinder Papers, Sydney, Archive of the AGNSW, MS 1995.1.

30 Frank Hinder, 'Extracts from philosophers, artists and scientists', notebook, Frank Hinder Papers, Sydney, Archive of the AGNSW, MS 1995.1, Box 1.

31 Yuill 2005.

32 Jay Hambidge from 'The ancestry of cubism', *Century Magazine*, 1914, p. 8, quoted in Frank Hinder notebook, 'Extracts from philosophers, artists, and scientists', Sydney, AGNSW, MS 1995.1, Box 1.

33 Developments in modern art and modern camouflage are the subject of Behrens 2002.

34 Gertrude Stein cited in Behrens 2009a, pp. 99–100.

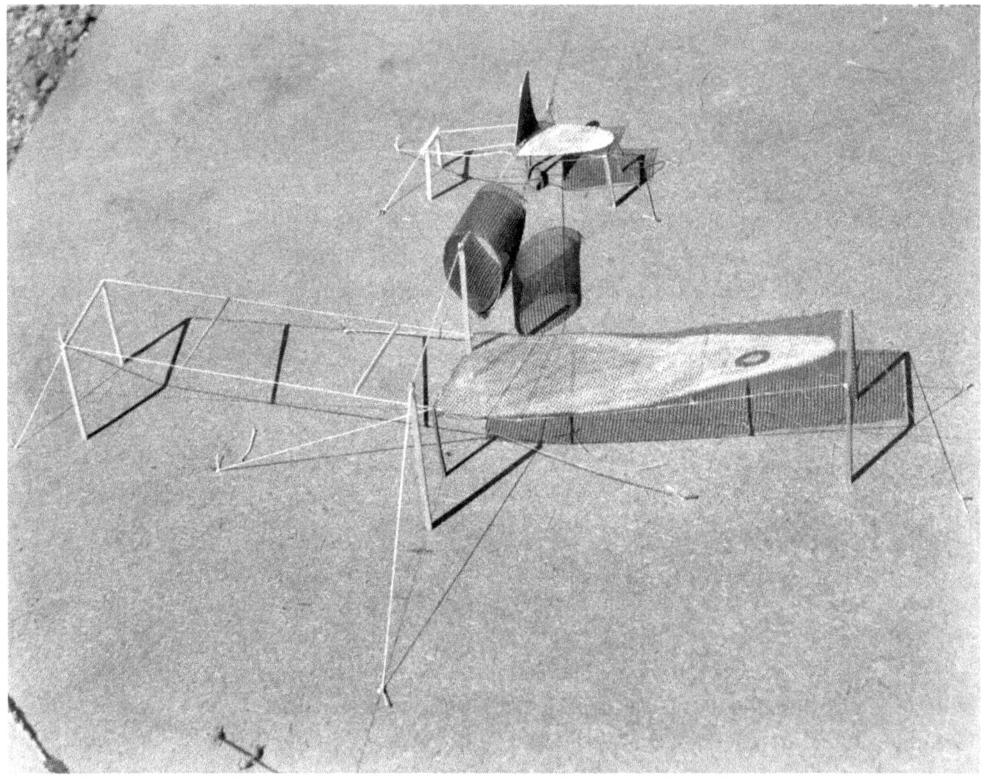

Fig. 12.8. Frank Hinder, decoy plane, Canberra. Canberra, AWM, Hinder Personal Records PR 88/133, 6 of 12. Photograph Gervaise C. Purcell.

From Jack Burnham's book *The structure of art* (1971), Hinder discovered and noted the term 'bricoleur', that refers to someone with an intuitive ability to make useful objects out of whatever is at hand, and who is different from an engineer because, as noted by Burnham, while the engineer 'gives form to function [the bricoleur] gives meaning to form'.[35] The bricoleur makes do, as Hinder himself made do in WWII by creating his inventions from scraps and waste. The decoy aircraft he designed were brilliant, simple, abstract assemblages of recycled wire, rope, metal and mesh.[36] Schematic and reductive rather than imitative, they were designed to deceive when seen from high in the air (figs. 12.8 & 12.9).

Quotes from Roger Caillois in Hinder's notebooks stand out because this one-time associate of the European surrealists has captured the imaginations of generations of readers in relation to camouflage ever since his essay, 'Mimicry and legendary psychasthenia',

35 Jack Burnham quote from *The structure of art* (1971) cited in Frank Hinder's notebook, 'Extracts from philosophers, artists, and scientists', Sydney, AGNSW, MS 1995.1, Box 1.

36 The back of the photograph reads 'dummy a/c, 27/3/43, wire netting on wire frame. Shape with hessian or new light material' with acknowledgment of photographer Gervaise C. Purcell. See Frank Hinder, 'Records of Lt F.C. Hinder Camoufleur', Canberra, AWM, PR 88/133, File 895/4/182, Item 6 of 12.

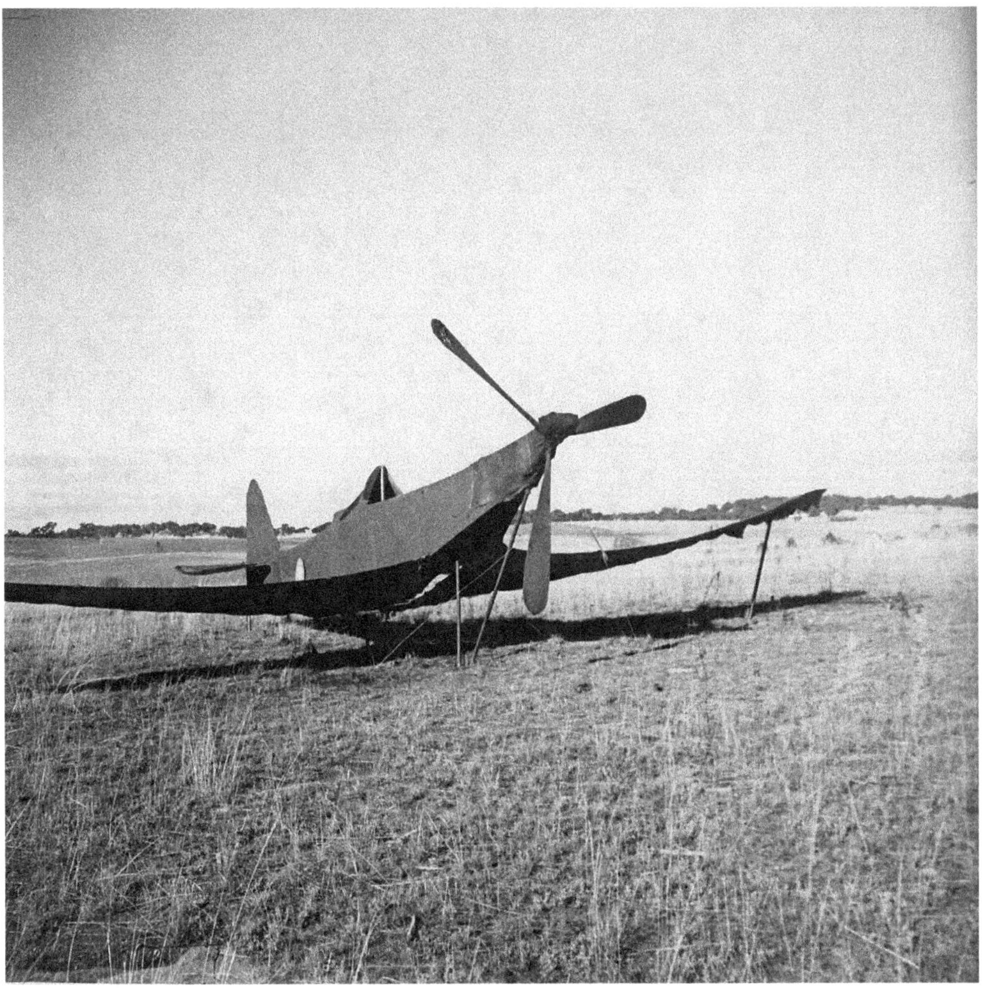

Fig. 12.9. Dummy aircraft, 'The Hindup', designed by Frank Hinder and Dale Upton. Canberra, AWM collection, Hinder Personal Records PR 88/133, 7 of 12.

was first published in *Minotaure* journal in 1935. It is a social analysis of insect mimicry. What caught Hinder's attention, though, was Caillois' broader interest in the subject of reality. Referring to the 1960s movement New Realism (Nouveau Réalisme) and the work of Jean Tinguely and Arman, who made sculptures from recycled industrial and urban objects, scrap metal, or other items of modern detritus, Caillois was struck by a reality apparently unmediated by representation. His focus on the question of the real opened up a philosophical field that placed the idea of objective reality under question. By its very nature, camouflage refutes the concept of objective reality. And it was in wanting to understand more about the deceptive game of representation, and the nature of reality, that Hinder was attracted to Caillois.

Hambidge, Burnham and Caillois captured his imagination, and their ideas overlapped with his own thinking on camouflage. But the origin of his curiosity with camouflage

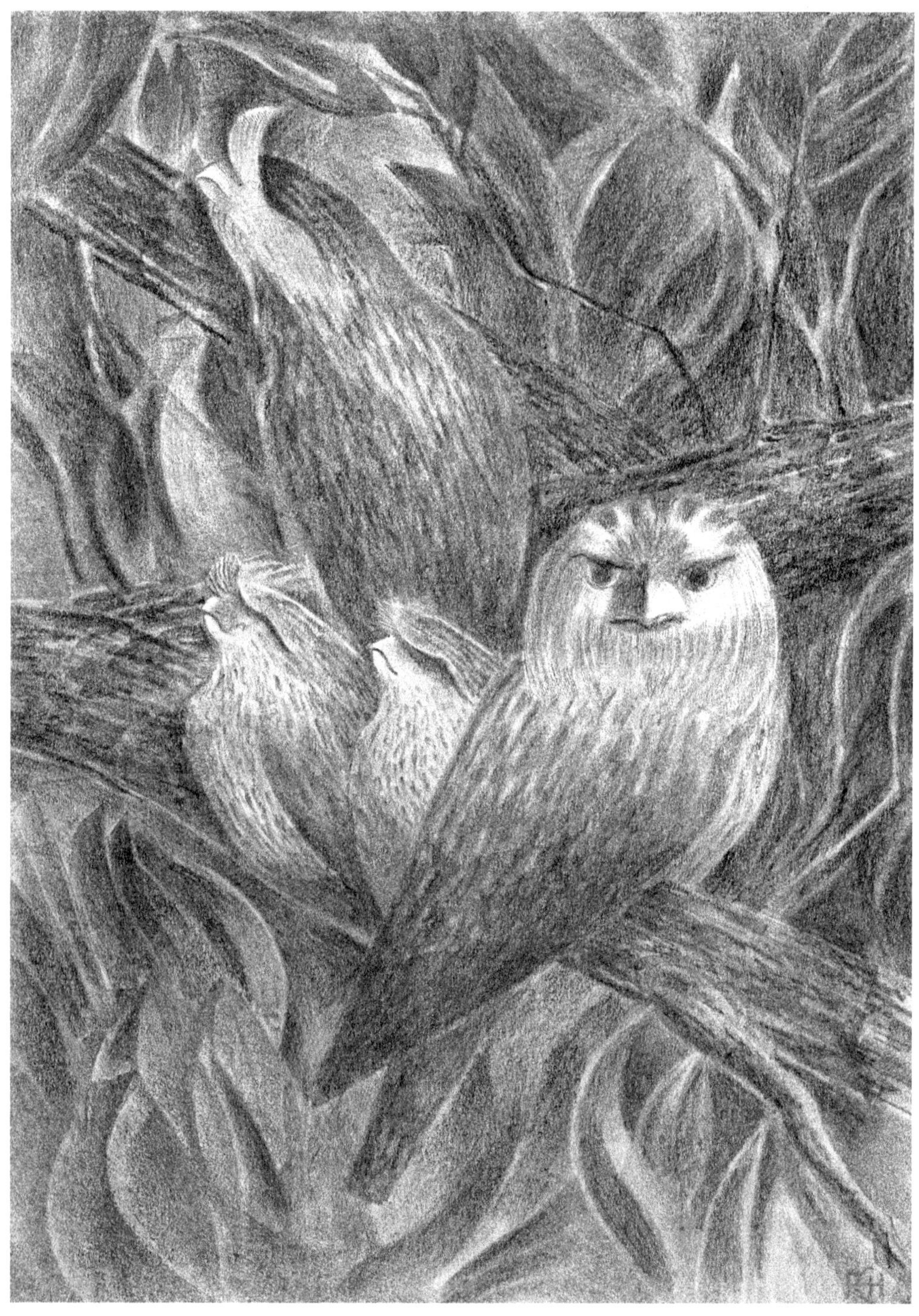

Fig. 12.10. Frank Hinder, *Frogmouth family*, 1948, lithograph, 29.6 x 21.6 cm. Private Collection.

Hinder attributed to 'an early interest in scouting and bushcraft'.[37] They taught him survival strategies, enthusiasm for reconnaissance and investigation, and how to solve problems with a limited range of tools and materials. His father, Critchley Hinder, was knowledgeable about natural history, especially animal behaviour.[38] During WWI Frank kept notebooks on insects, naming each different species (stick insect, scorpion and so on) with its popular and scientific names, dwelling on butterflies.[39] His entomological notes refer to methods of mounting and preserving dead insects, although not to insect camouflage. But later, as an artist, he collected photographs of camouflaged animals from newspapers and magazines: the American bittern that elongates its body to resemble reeds; the hellgram mite that mimics leaves; the Sumatran leopard that frightens would-be attackers with ocelli, or false 'eyes'; and pronghorns mottled in white and brown for invisibility in snow-covered hills.[40]

In the 1940s Hinder made a series of prints of well-camouflaged animals, including *Frogmouth family* (1948), a lithograph that recalls the work of German artist Franz Marc (1880–1916) who also specialised in animals. Marc, like Hinder, blended the forms of animals into broader designs through pattern and colour. Both artists found this a perfect language for communicating the concept of nature's universal laws. Franz Marc was also deployed to work in camouflage (although in WWI), before he was killed in action.[41] Hinder captures the uncanny sight of these night birds as they perch together in groups so still they are difficult to distinguish from bark (fig. 12.10). The tawny frogmouth (*Podargus strigoides*), found Australia-wide, is 'so close to being invisible that only the keenest eye will spot it', even at close range.[42] Cryptically coloured in marbled plumage, the tawny frogmouth also proves the soundness of Thayer's Law of concealing colouration; the darkness of its topside and the lightness of its underside camouflage the animal by making it seem flat.[43] It was an aesthetic, and behaviour, that Hinder himself tried to copy during experiments conducted with Dakin in the bush surrounding Canberra. Dressed in the colours of his surroundings, with skin darkened to hide reflections, and wearing textured clothing to increase the play of light and shadow (for breaking up his body's form), Hinder crouched still and silent among the trees while Gervaise Purcell took photographs (fig. 12.11).

The many philosophical books that Hinder read reinforced his belief in the philosophy that design is labour for improving human life and is, of necessity, a pairing of art and scientific knowledge.[44] Likewise his colleagues in the DHS, most of whom were architects

37 Frank Hinder, 'Lt F.C. Hinder Personal Records', Canberra, AWM, PR 88/133, Item 9 of 12.

38 Information on Critchley Hinder is from Hinder's daughter Enid Hawkins. Phone interview 12 October 2009. She recalls Hinder's fascination with books by Robert Ardrey. See Ardrey 1971.

39 Frank Hinder, 'Memorandum book: nature study notes, 1917–1918', Frank Hinder Papers, Sydney, Archive of the AGNSW, MS 1995.1, Box 1.

40 Frank Hinder, 'Records of Lt F.C. Hinder Camoufleur', Canberra, AWM, PR 88/133, File 895/4/182, Item 6 of 12.

41 See Cork 1994, p. 111.

42 Hollands 2008, p. 223.

43 Pizzey & Knight 2007, p. 316.

44 See, for example, Neutra 1954, p. 381. This book is among those bequested by Hinder to the AGNSW. See the AGNSW online library catalogues, www.artgallery.nsw.gov.au/lib/cat, accessed 01/10/09.

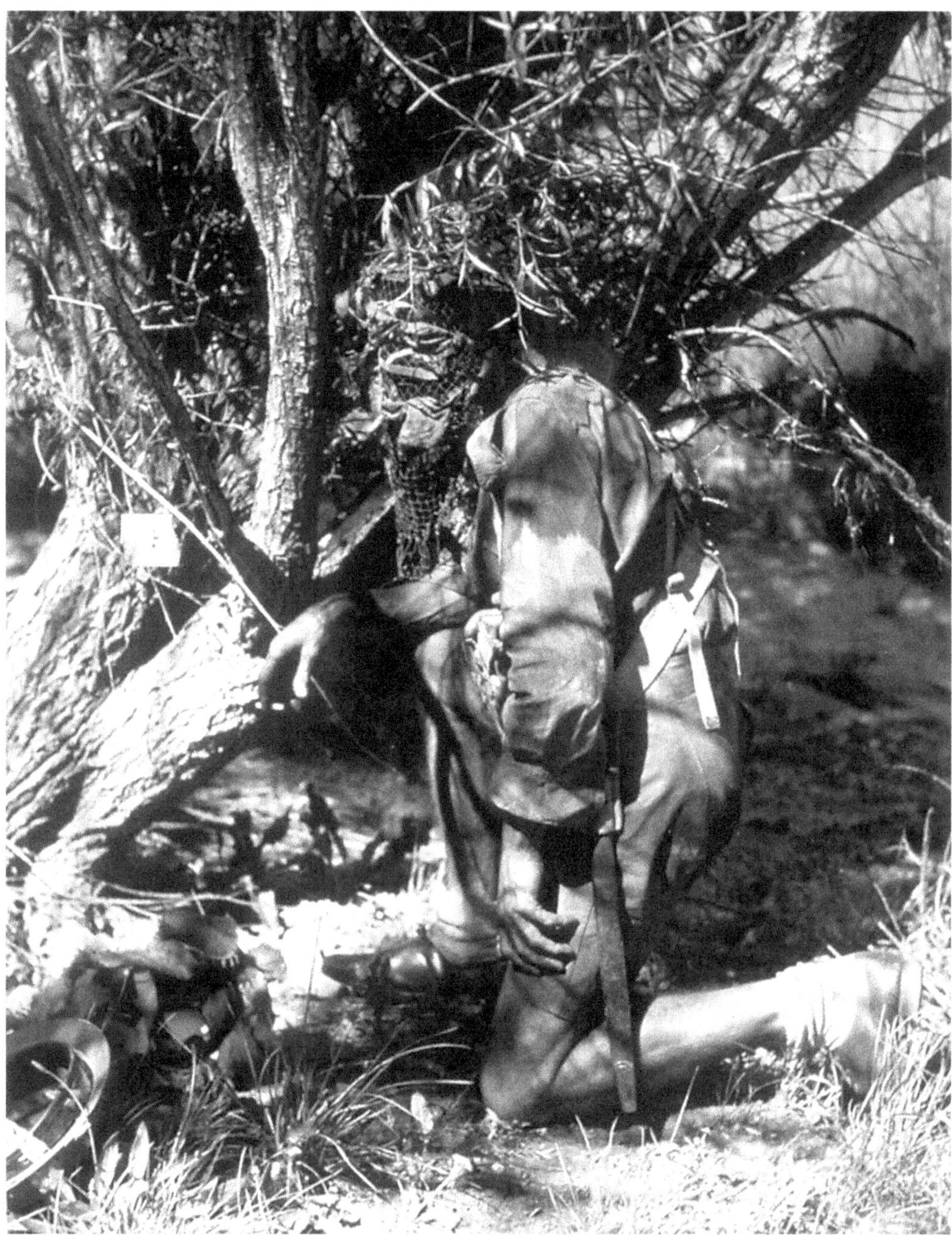

Fig. 12.11. Frank Hinder conducting an experiment with personal camouflage, c. 1943. The collection of the NAA: C1905 T1, 1 [17]. Photograph, Gervaise C. Purcell.

and designers, entered into wartime duties hoping to bring about a better, more enlightened world. John Moore, for example wrote in 1927 about the symbiotic relationship of modernity and modern art.[45] In 1941 he hoped that in the postwar reconstruction of Australia the modern architect would 'take his place in the forefront' to shape the everyday living of a new generation in a more democratic society.[46] But design and modernism at the service of war? It seemed antithetical to modernism's belief in freedom of expression, peace, life, vitality and beauty.

The debates that ensued after the war about art's autonomy were important to Hinder. He was naturally reflective of the ongoing relationship between science and art after the social catastrophes of WWII, the millions dead, and the onset of atomic warfare. On the inside cover of a notebook titled 'Extracts from philosophers, artists and scientists', which he first assembled in the 1930s, Frank Hinder pasted an undated but intriguingly titled newspaper article: 'Time that art snubs science as dictator'.[47]

Postwar, Hinder never failed to be amazed and intrigued by scientific advances on the macro scale in space exploration, and on the micro level with molecular science. But the science-directed war years had created a social hiatus in Australian society, and he became its unwitting victim. When Hinder tried to go back to commercial art after the war, he found that a younger generation had usurped his own, leaving him to feel:

> very much out of touch with things, and much to my surprise the men I had worked for before looked at my work and said "The trouble is your work is oldfashioned". I'd always considered myself rather modern.[48]

Yet, as André Breton said of that period of history, WWII side-tracked the artist's spirit.[49] It was time to get back to investigations of semi-abstraction, to explore kinetic art, to capture the idea of a universal energy, and to visualise the abstract rhythmical 'sounds' of life and the sensations of universal interconnection.[50]

45 Moore 2006b [1927], pp. 69–71.

46 Moore 2006b [1941], p. 556.

47 Frank Hinder notebook, 'Extracts from philosophers, artists, and scientists', Sydney, AGNSW, MS 1995.1, Box 1.

48 Hazel de Berg, 'Interview recorded with Frank Hinder', Canberra, National Library of Australia, 20 July 1963, Oral TRC 1/45, transcript, p. 561.

49 Breton 1941, p. 12

50 These points discussed by Henshaw 1978, pp. 12–13.

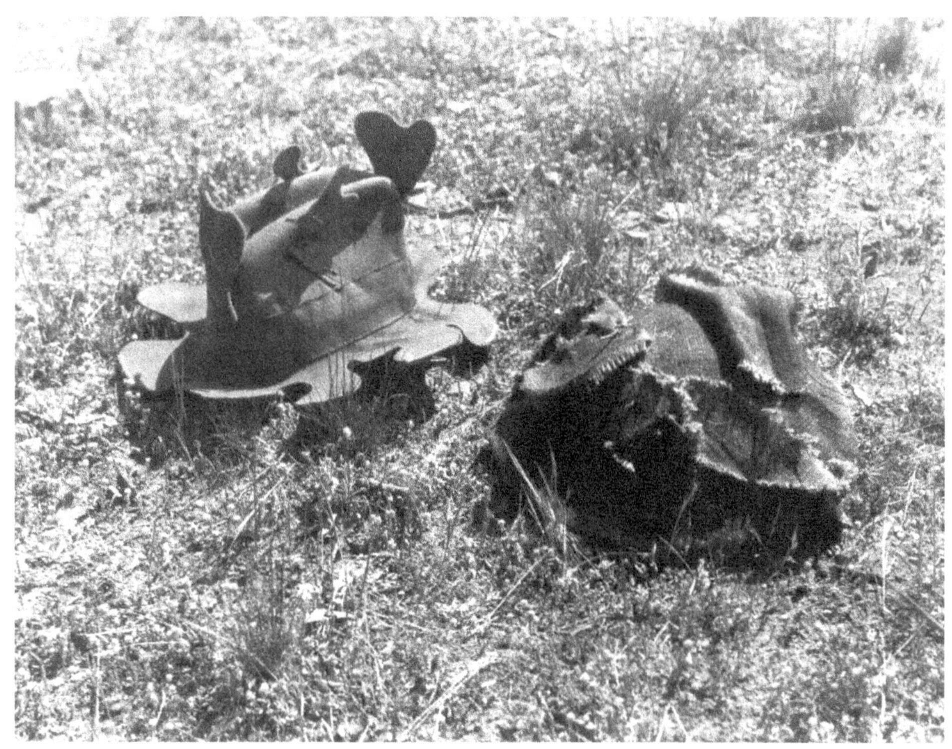

Fig. C.1. DHS hat camouflage experiment, c. 1943. The collection of the NAA: C1905 T1, 1 [15].

Conclusion

The opening pages of this book claimed that when Margaret Preston introduced Frank Hinder to William Dakin in March 1938 it marked the symbolic beginning of the story of camouflage in wartime Australia. Hinder, a fellow modernist, shared Preston's interest in analytical and scientific methods in art as well as in the organic geometries of nature. But his destiny for the next six years was to apply abstraction to camouflage design while Preston's was to spend that time devising nationalistic imagery and protesting the 'madness' of national mobilisation.[1] Yet their aesthetic concerns were not so different for, while Preston's wartime paintings include indictments of military life, others, like *The camp, Hornsby, NSW* (c. 1942), were painted in a cubist style already recognisable as a military aesthetic via the appropriation of modern art for camouflage.

In postwar culture, however, where fascination with the retinal impact of the colours and forms associated with camouflage is matched by the allure of its power as a conceptual tool for cultural critique, the relevance of camouflage to Margaret Preston takes another turn. Throughout the war she painted flowers and landscapes—*Bush walk* (1942) is one example—interweaving cubist fragmentations of colour and form with naturalistic plays of light and shadow. But she also appropriated Aboriginal cross-hatching and patterning from shields and bark paintings and in 2005 Djon Mundine referred to Preston's use of Indigenous motifs as a 'veneer' of Aboriginality, while Hetti Perkins compared it to an act of 'assimilation'.[2] These are criticisms that underscore how in contemporary culture the term 'camouflage' and its wide-ranging tropes including 'veneer' and 'assimilation' readily engage with the deconstruction of social processes.[3] Consequently, we live in a time when contemporary cultural criticism, and contemporary art, readily utilise the inherent binary nature of camouflage—that of revelation and concealment—to address the workings of the human psyche but also social circumstances that lead to the cultural visibility or invisibility of certain histories, events and people. The reason this book was written, after all, was to redress the near-invisible history of Frank Hinder, William Dakin, and their circle of civilian artists and scientists who worked in camouflage for the DHS in WWII (fig. C.1).

In 1950 William Dakin was gravely ill with cancer and in his last book, *Australian seashores,* he reflected on the modern world. He worried about a future where nature study

1 Margaret Preston and the war years discussed in Edwards & Peel with Mimmocchi 2005, pp. 176–86.

2 Djon Mundine and Hetti Perkins quoted in Edwards & Peel with Mimmocchi 2005, p. 10.

3 This point was raised by Hsu 2006.

might no longer matter. He feared that the beguiling realm of nature, and the plants and animals that had so much to offer human knowledge, including military defence, had 'fallen on evil days in this epoch of such diversities as atom-splitting, swing music, and sport'.[4] He also worried about the future of camouflage defence in the wake of the atomic bombing of Japan. It made him question whether camouflage as he knew it—a phenomenon first honed to deceive the eye, then the camera, then radar—had a place in the future. What, he asked, is 'to be the position with the advent of the Atomic Bomb?'[5] Since then the atomic bomb itself has come to be seen as an object of camouflage, ushering in perpetual, 'virtual' war but under the guise of a deterrent.[6]

Dakin needn't have worried about the demise of nature study for future generations, and he was wrong to think that camouflage designs would become redundant. Twenty-first-century art is proof that nature study has reached new levels of significance in culture, and that camouflage has been culturally and aesthetically recast and reinvigorated. Artists of a generation that succeeded the painters, sculptors, designers and photographers who worked alongside Dakin in the DHS continue to explore the sensations and ideas inherent in the play of doubles in camouflage, particularly its embodiment of concealment and exposure, invisibility and visibility, stability and disorientation. Where they differ from the circle that worked for Dakin lies in the distinction raised in the early pages of this book: artists in history worked with camouflage to participate in the production of war, while contemporary artists investigate camouflage to make art.

The intertwinement of art and science, so critical to the development of camouflage during World War II, is a stronghold of contemporary art in the twenty-first century, including the creation of work specifically about camouflage. Maria Fernanda Cardoso, for example, is both surrealist and entomologist. She collects insect specimens, dead and alive, and builds physical but also dream-like spaces for them to inhabit. Unlike the insects themselves she flaunts their powers of mimicry. She puts on exhibit the butterfly's ability to be visible and invisible in a split second, and the leaf-insect's capacity to blend with its surroundings but 'with the option to be loud and attractive' (fig. C.2 & 3).[7] Indeed the eye she brings to nature is similar to that brought by Dakin who once compared the foreshores of the coral reefs of Australia to the beauty and strangeness of surrealism.[8] Yet where Dakin used animals as models for camouflage defence in WWII, Cardoso, true to the cultural paradigm in which she lives, utilises natural history for philosophical purposes, seeing in animal mimicry a perfect analogy for the strategic concealments and revelations in human social behaviour, including forms of mimicry that can lead down a darker path to dispossession, disempowerment, assimilation and invisibility.

A second example of contemporary art that contradicts Dakin's concerns about the future demise of camouflage design, is the work of Debra Dawes whose geometric abstract paintings take up the aesthetic of disruptive patterning, wrenching it from its military

4 Dakin 1953 [1952], p. 73.

5 W.J. Dakin, Camouflage report 1939–1945, AWM, Series 81 [77 Part 1], p. 152, point 15.

6 Virilio 1994, p. 67.

7 Britz 2009, p. 27.

8 Dakin 1955, p. 13, and p. 68 respectively.

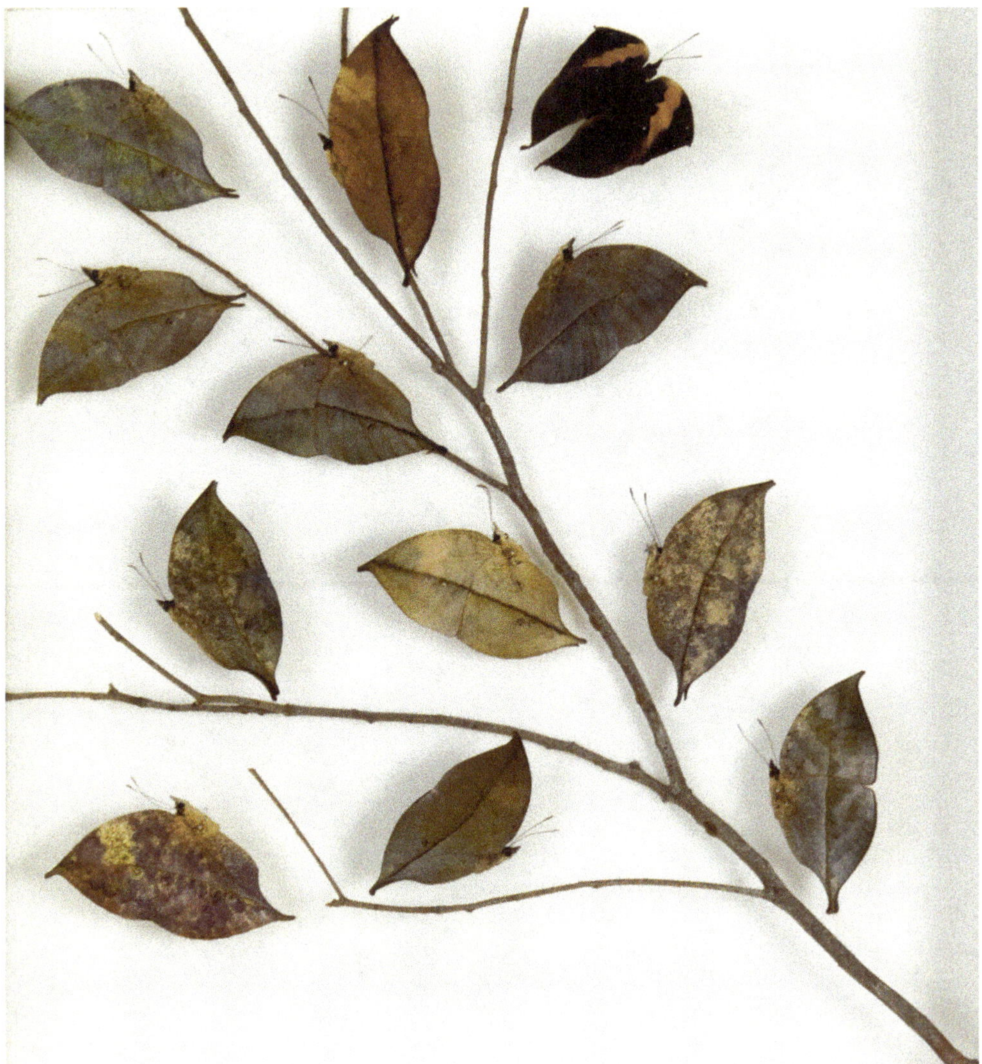

Fig. C.2. Maria Fernanda Cardoso, *Brown and green leafs, dead leaf butterflies*, 2010. Courtesy Grantpirrie Gallery, © Maria Cardoso.

history, and using it to create a personal critique of deception and disguise in the political landscape. For readily apparent in the dazzle patterns of warships, including the *Zealandia* discussed in the first chapter of this book, is a reality where perceptions are easily warped.[9] Consequently Dawes uses optical means to put the observer's powers of perception in peril while at the same time engaging metaphorically with political manipulations of public perception, and lack of disclosure and cover-ups in the public realm (fig. C.4).

The geometric abstractions of Debra Dawes have a context in the postwar paintings of Frank Hinder and among their similarities are agitated surfaces, optical confusions and

9 Millner 2010.

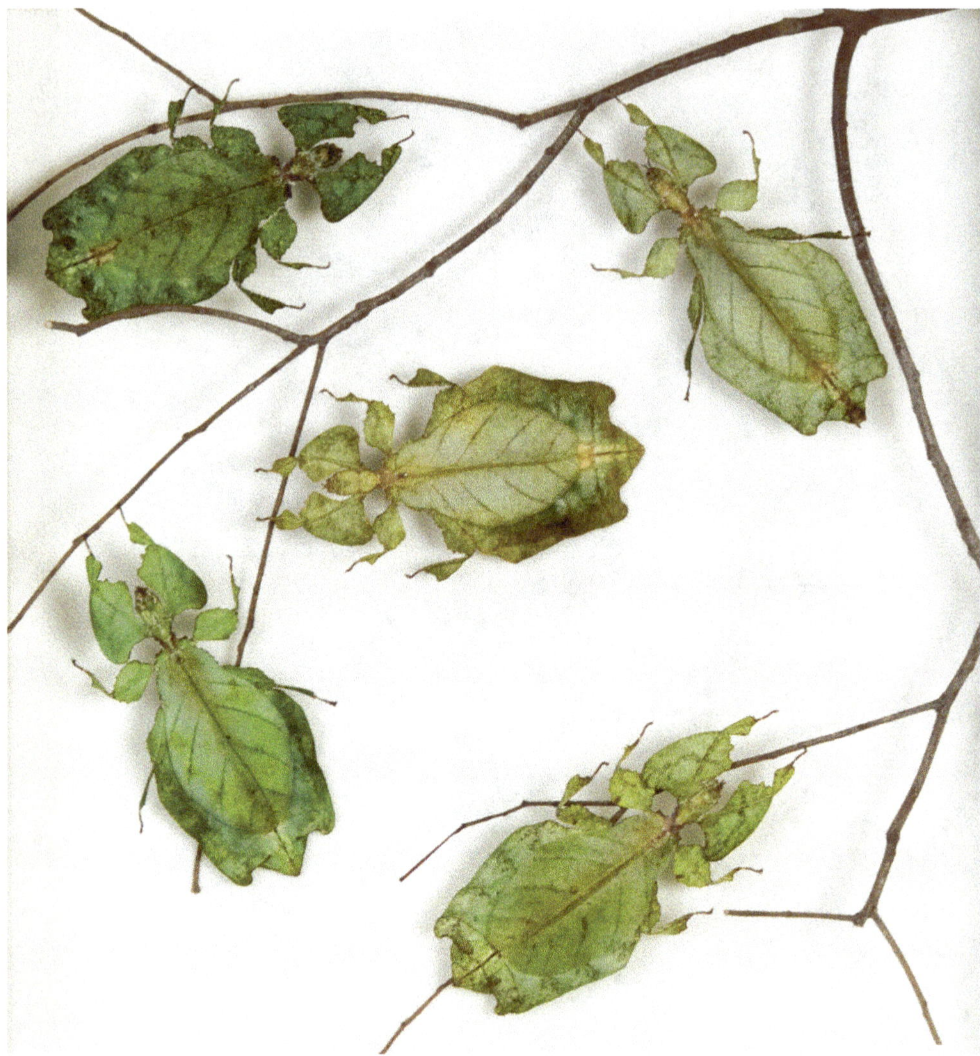

Fig. C.3. Maria Fernanda Cardoso, *Brown and green leafs, Linnaeus leaf insects*, 2010. Courtesy Grantpirrie Gallery, © Maria Cardoso.

spatial disorientations caused by light bouncing and flickering over patterns (fig. C.5).[10] Frank Hinder, who did not live long enough to witness the proliferation of national military camouflage patterns throughout the world or of the co-option of disruptive patterning for street fashion, never lost his fascination for military camouflage and in later life looked forward to the day when camouflage clothing would incorporate the 'chemical or nervous qualities of a chameleon'.[11] Strangely enough, seven years after his death, in 1998, the *SMH* reported that British marines had developed a 'chameleon suit', an item of stealth clothing

10 This description used for Hinder by Yuill 2005, p. 8.

11 Frank Hinder, 'Lt F.C. Hinder Personal Records', Canberra, AWM PR 88/133, Item 9 of 12.

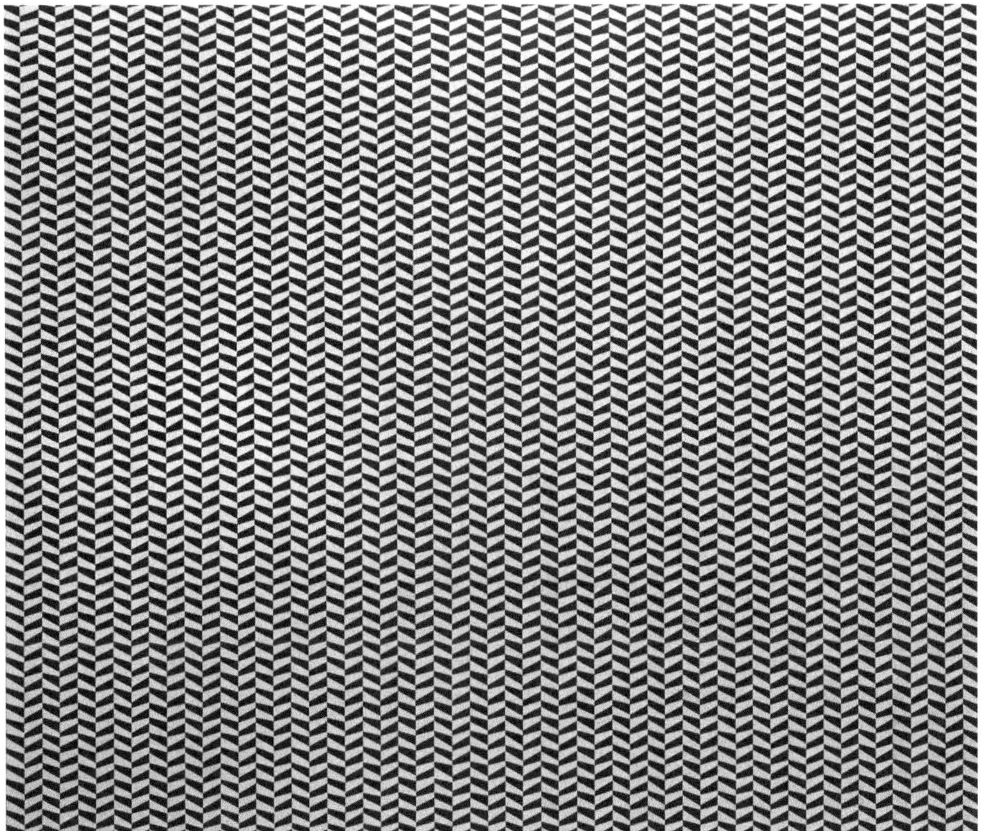

Fig. C.4. Debra Dawes, *Complete?*, 2010, oil on canvas, 275 x 330 cm. Photo: Paul Green, courtesy the artist and Gallery Barry Keldoulis.

to make soldiers almost invisible.[12] Camouflage had a decisive impact on Frank Hinder and its pictorial elements are apparent in the play of shadows in *Over the bridge*, 1951. A similar pictorial use of shadow has been identified in the postwar paintings of Ellsworth Kelly, Hinder's American contemporary who was also a camouflage artist in WWII.[13]

I began this study in 2002 intrigued by the visual complexities of camouflage. This broadened to an interest in the social history surrounding WWII, the impact of war on artists, the interplay of theories of nature and society and the crossing of discipline boundaries between art, natural history, military history and social politics. WWII was a rare moment in history when artists behaved outside the social expectation of their profession by forming an alliance to work in a military context to devise lateral, perhaps lethal, always imaginative methods of subterfuge. Their leader, William Dakin, was not a member of the professional military but a man who studied camouflage in animals and thought of warfare and camouflage as a basic, primordial, animal-like demand to survive.

12 Gilligan 1998, p. 3.

13 Goossen 1973, p. 35.

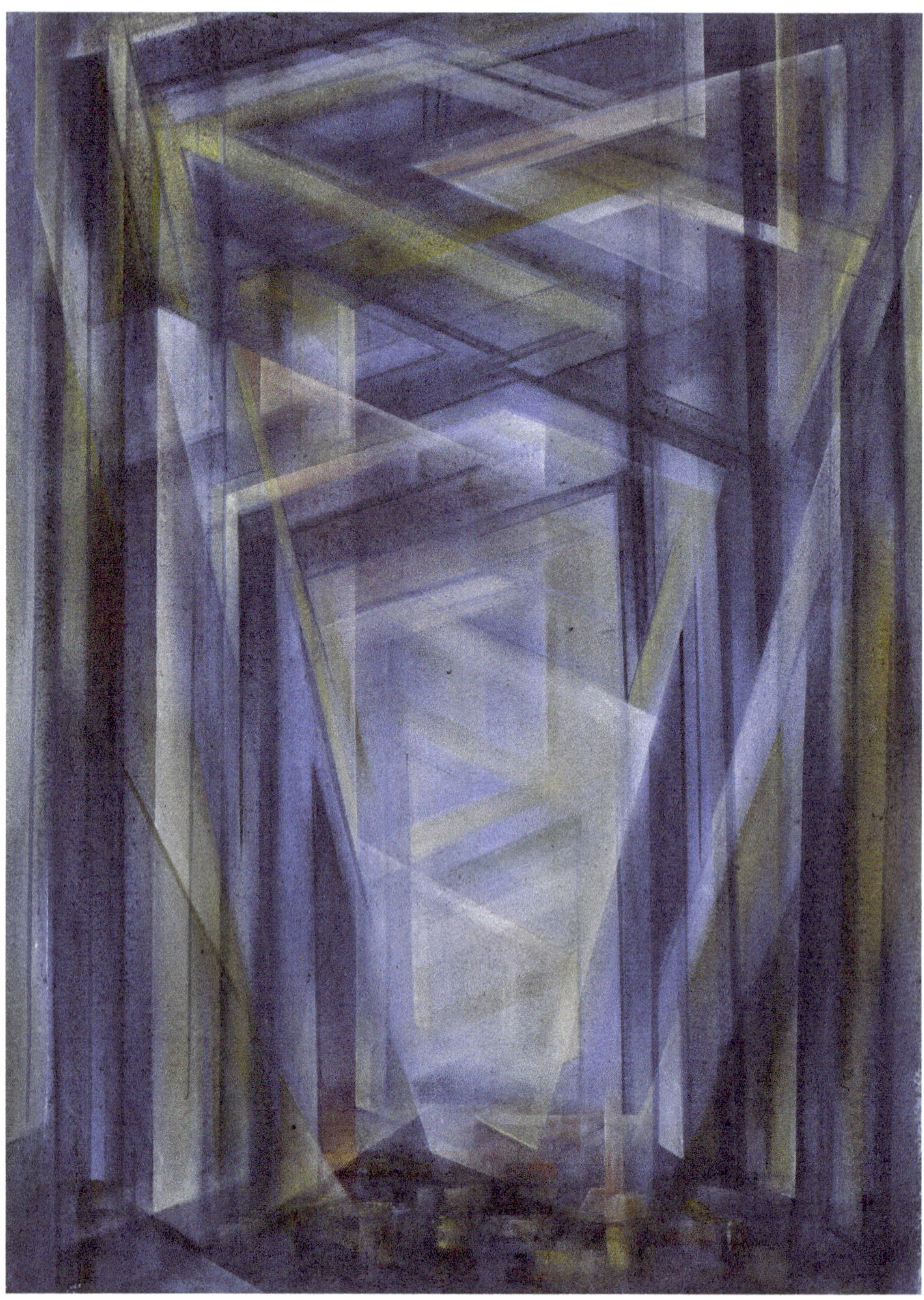

Fig. C.5. Frank Hinder, *Over the bridge*, 1951, watercolour, 76.0 x 55.0 cm. University of Sydney Art Collection, reproduced courtesy of Enid Hawkins.

But now, having reached the end of the book, it is also clear to me that I have written a commemoration to a group of people and an episode in Australia's past that have gone unnoticed among the otherwise heroic histories of the war. Figure C.1 shows a photograph of two empty camouflage hats designed by Australian artists under Dakin's direction in the DHS. Their leaf-shaped projections beg for comparison with the flourishing disguises on Cardoso's Linneaus leaf insects from North Queensland.[14] Out of context both look ludicrously conspicuous and vulnerable. But the photograph is also an apt way to pay tribute to those who threw their hats in the ring at the commencement of WWII, tried to blend into the military landscape, and announced their willingness to participate in a political contest that lasted six years.

14 Cardoso's work *Brown and green leafs (dead leaf butterflies (left) and Linnaeus leaf insects (right))*, was exhibited in *The Newcastle chest*, a work commissioned by the Newcastle Regional Gallery for an exhibition titled *Curious colony: a twenty-first century Wunderkammer* (July–August 2010).

Appendix 1

List of names of Group of Camoufleurs (24/6/40)[1]

Professor W.J. Dakin (Chairman, zoologist, University of Sydney)
Professor L. Wilkinson (Dean of Faculty of Architecture, University of Sydney)
Group Capt. De. La Rue (Headquarters, RAAF, Richmond)
Capt. B.S. Hussey (2nd Fortress C., RAE, Chowder Bay, Mosman)
Lt Col. D.A. Whitehead (C.O.2 Machine Gun Regiment, Ingleburn)
Lt D. Barker (Intelligence Section, Department of Defence, Victoria Barracks)
S. Ure Smith (artist, 166 Phillip Street, Sydney)
John D. Moore (architect, O'Connell Street, Sydney)
F. Allen (Paint, Lewis Berger & Co. Ltd paint manufacturers)
V. Tadgell (organiser, Orient Steam Navigation Co.)
F.C. Hinder (artist, 63 Pitt Street, Sydney)
D.S. Annand (artist, 62 Margaret Street, Sydney)
Peter Dodd (artist, Commercial Art School, Bathurst Street, Sydney)
R.E. Curtis (artist, Livingstone Avenue, Pymble)
F. Medworth (artist, Technical College, East Sydney)
A.J. Murch (artist & engineer, Technical College, East Sydney)
Capt. F.W. Follett (aerial photography, Adastra Airways Pty, Sydney)
Russell Roberts (photographer, 15 Hamilton Street, Sydney)
W. Van-der-Vellen (photographer, Kodak, George Street, Sydney)
Max Dupain (photographer, Bond Street, Sydney)
D. Limburg (architect, Dept of Interior, Sydney)
P.E. Taber (engineer, Burwood)
A.N. Baldwinson (Architect, Stephenson & Turner, Sydney)
J.P. Tivey (lighting, Australian General Electric Ltd, Sydney)
J.H. Melville (model maker, Stephenson & Turner, Sydney)
R. Spooner (model maker, Griffen Street, Manly)
Grant I. Devine (model maker, 12 Loftus Street, Sydney)
F. Johnston and Associates (artists, Pitt Street, Sydney)
Adrian Feint (artist, 2a Onslow Av, Elizabeth Bay, Sydney)
R.K. Harris (architect, Bligh Street, Sydney)

---

1 Also known as the Sydney Camouflage Group see 'List of names of Group of Camoufleurs', Notes of Camouflage, Sydney, NAA, SP 1048/7, Item S10/1/329.

Appendix 2

Department of Home Security Camoufleurs[1]

Adams, George
Annand, Douglas
Atherton, Cecil W.
Ballantyne, Colin
Bell, George
Belot, Albert
Bennie, Leonard
Besley, Owen
Billson, Edward F.
Bragg, Victor
Brazenor, Charles
Bryan, Walter
Bustard, William
Cameron, Robert Nichols
Clements, Patrick
Coburn, Herbert
Collings, Geoff
Collings, Ewart
Cork, Herrick
Corlett, Victor
Crooke, Augustus
Cummings, Robert
Cuthbert, J.D
Cousin, Bruce
Curtis, Robert Emerson
Dalgarno, Roy F.L.
Dignman, Augustine (Gus)
Dorey, Victor
Dowman, Walter
Dumbrell, John
Dupain, Max
Eldershaw, John
Edwards, Roderick
Flett, James
Francis, Richard V.
Gaskell, Raymond E.
Godfrey, William R.P.
Goldberg, Jack
Gould, Charles
Greenhalgh, Victor
Hammond, Stanley
Handfield, Phillip
Harris, Keith
Harrison, Edgar
Herd, Walter
Hinder, Frank
Hogan, Leon
Ince, Henry F.
Johnston, Frank
Jolliffe, Eric E.
Jordan, Allan H.
Jordan, Leslie G.
Kohler, Edward F.
Knight, William C.

1 'Staff Record: Camouflage Section', DHS, Canberra, NAA, Series A691, Item 1, accumulation dates 1 January 1942–31 December 1945.

Lancaster, Charles H.
La Gerche, Alfred R.
Lander, Cyril G.
Miller, Douglas
Moloney, Frank H.
Morison, Margaret L.
Moore, Charles E.
Moore, John D.
McCubbin, Louis F.
MacDonald, John E.
McInnis, Ronald A.
Mackay, Keith T.
Nicholas, William R.
O'Harte, Charles
Oliver, Charles F.
Ovenden, Richard A.
Pearce, Denis C.
Pearson, Joseph
Pinchen, John
Piper, Bertram T.
Pratt, Charles D.
Purcell, Gervaise C.
Rae, George
Ragless, Maxwell C.R.
Ranshaw, John
Reid, Emerson D.
Riebe, Anton D.
Rigg, Ronald
Roberts, Russell B.
Robertson, John K.
Robinson, Edwin
Rose, Louis
Ross, Alexander D.
Rowland, Arthur F.
Serow, Charles H. J.
Seale, Clement E.
Shaw, Roderick M.
Shelley, Ralph D.
Smith, Alexander E.
Smith, Jack C.
Soden, Arthur J.
Stubbs, Norman B.
Turner, Eustace F
Thompson, E. Lindsay
Thompson, Eric G.
Upton, Darrell W.
Voller, Roderick W.
Wallis, Alvan C.
Wallis, Raymond
Wilson, David H.
Wood, Clifford D.
Wyatt, Colin W.
Wyeth, William C.
Young. Albert V.

Bibliography

Primary sources

Official records (unpublished)

A. Australian War Memorial (AWM) (Canberra)

Camouflage, instruction booklet dated 1943, cited in F Hinder, 'Records of Lt FC Hinder Camoufleur'. Canberra, AWM, PR 88/133, File 895/4/182, Item 11 of 12

Dakin WJ, Camouflage report 1939–1945 (appendices A–D), Canberra, AWM, Series 81 [77 Part 1], 1947.

Dakin WJ, Camouflage report 1939–1945 (appendices E–J), Canberra, AWM, Series 81 [77 Part 2], 1947.

Dakin WJ, Camouflage report 1939–1945 (appendix K–N), Canberra, AWM, Series 81 [77 Part 3], 1947.

Dakin WJ, Camouflage report 1939–1945 (appendix O), Canberra, AWM, Series 81 [77 Part 4], 1947.

Dakin WJ, Camouflage report 1939–1945 (appendices P–R), Canberra, AWM, Series 81 [77 Part 5], 1947.

Dakin WJ, Camouflage report 1939–1945 (appendix R), Canberra, AWM, Series 81 [77 Part 5], 1947.

Goodenough Island bluff: a Ghost Force held back the Japanese, AWM, AWM 54, 585/7/1.

Operation Hackney: Goodenough Island deception scheme, Canberra November 1942 to June 1943, Canberra, AWM, AWM 54, File 585/3/1.

'Outline of deception scheme—Goodenough Is', in Hackney deception scheme Goodenough Island, Canberra, AWM, AWM 54, Item 585/3/1, appendix 'A'.

Public Relations Office of 5 Australian Division, 'When a Ghost Force held back the Japanese: Goodenough Island deception scheme', Articles from Public Relations Office HQ, 20 April 1943, Canberra, AWM, AWM 54, Item 579/7/14.

'The address to leaders of working parties', Operation Hackney: Goodenough Island deception scheme, Canberra, Canberra, AWM, AWM 54, Item 585/3/1, appendix D.

'The air view and air photographs', 7 Australian Division Camouflage Training Unit, Canberra, AWM, (August 1942–July 1943), AWM 52, Item 5/36/8.

Australian Military Forces, Engineer-in-Chief's technical instruction No. 45; Camouflage in the S.W.P.A., 30 November 1943, F Hinder, 'Records of Lt FC Hinder camoufleur', Canberra, AWM, PR 88/133, File 895/4/182, Item 11 of 12.

Williams FE (Lt), 'General errors in attitude', Geographical Section South West Pacific area: relations with natives of New Guinea in wartime, AWM Collection, Series AWM 54, 506/8/2, 1942.

B. National Archives (Canberra)

National Archives of Australia (NAA): Department of Home Security (DHS); A453, Correspondence files of the DHS, 26 Jun 1941–1 Feb 1946.

NAA: Defence Committee; A5799, Defence committee agenda, 1932–1932.

NAA: Department of Defence; A5954, The Shedden Collection, 13 Nov 1939–14 April 1942.

NAA: Department of Defence; A663, Correspondence files, 30 Jan 1940–30 September 1957.

NAA: Department of Home Security (DHS); A691, DHS—Camouflage section—Staff record register, 1 Jan 1942–31 Dec 1945.

NAA: Department of Air; A705, Correspondence files, 1 Jan 1922–31 Dec 1960.

NAA: Department of Home Security (DHS); CP951/1, History of the DHS, 1 Jan 1941–31 Dec 1946.

NAA: Department of Home Security (DHS); A453, Correspondence files of the DHS, 26 Jun 1941–1 Feb 1946; 1944/17/2734, Camouflage General—personal diaries of Camouflage Office, Aug 1944–May 1945.

NAA: Department of Home Security (DHS); A453, Correspondence files of the DHS, 26 Jun 1941–1 Feb 1946; 1942/17/1957, Camouflage—general camouflage school WA, 4 May 1942–5 June 1942.

NAA: Department of Home Security (DHS); A453, Correspondence files of the DHS, 26 Jun 1941 –1 Feb 1946; 1943/17/1510 Part 1, Camouflage—general—camouflage organisation in New Guinea, 15 July 1943–26 Dec 1943.

NAA: Department of Home Security (DHS); A453, Correspondence files of the DHS, 26 Jun 1941–1 Feb 1946; 1942/28/2298, Camouflage personnel NSW Camoufleurs staff appointments, 1942–1945.

NAA: Department of Home Security (DHS); A453, Correspondence files of the DHS, 26 Jun 1941–1 Feb 1946; 1942/21/2632, Camouflage—question of responsibility: army camouflage, Dec 1941–29 Jan 1943.

NAA: Defence Committee; A5799, Defence Committee agenda, 27/1943, Accreditation and attachment of camouflage officers, 1943–1943.

NAA: Department of Defence; A5954, The Shedden Collection, 13 Nov 1939–14 April 1942; 396/2, Establishment of camouflage organisation and procedures, 1941–1944.

NAA: Department of Defence; A663, Correspondence Files, 30 Jan 1940–30 September 1957; O52/1/48, Department of Defence compensation for Dependants of Civilians, 1940–1941.

NAA: Department of Defence; A663, Correspondence Files, 30 Jan 1940–30 September 1957; O30/1/192, Middle Head (New South Wales): experimental camouflage, 1941–1941.

NAA: Department of Home Security (DHS); A691, DHS—Camouflage section—staff record register, 1 Jan 1942–31 Dec 1945; 1, Staff register—camouflage section, 1942–1944.

NAA: Department of Air; A705, Correspondence files, 1 Jan 192 –31 Dec 1960; 62/4/88, Camouflage activities (Allied Air Forces South West Pacific Area), 1942–1942.

NAA: Department of Home Security (DHS); CP951/1, History of the DHS, 1 Jan 1941–31 Dec 1946; Departmental history; Camouflage organisation—administration, Canberra, NAA, 1939–1946.

C. National Archives (Sydney)

NAA: Defence Central Camouflage Committee (DCCC)—Office of the Technical Director; Series C1707, Correspondence Files, 1 Jan 1941–31 Dec 1944.

NAA: DCCC—Office of the Technical Director; C1904, Maps, plans and drawings, 1 Jan 1941–31 Dec 1943.

NAA: DCCC; C1905, Black and white photographs, 1 Jan 1941–31 Dec 1945.

NAA: DCCC—Office of the Technical Director; C1908, Professor Dakin's report, 1 Jan 1944–31 Dec 1945.

NAA: Australian Military Forces; SP1008/1, General correspondence Files, 1 Jan 1871–31 Dec 1965.

NAA: Australian Military Forces; SP1048/7, General correspondence (Secret Series), 1 Jan 1922–31 Dec 1951.

NAA: State Publicity Censor; SP106/1, Correspondence re publicity censorship, 1 Jan 1939–31 Dec 1945.

NAA: Mercantile Marine Office/Marine Crews Office; SP290/2, Ships' official log books, 1 Jan 1940–31 Dec 1946.

NAA: Australian Broadcasting Commission; SP300/3, War correspondents' talk scripts, 1 Jan 1940–7 Dec 1947.

NAA: DCCC—Office of the Technical Director; Series C1707, Correspondence files, 1 Jan 1941–31 Dec 1944; Item 36, Air intelligence reports—Japanese methods, 1942–1943.

NAA: DCCC—Office of the Technical Director; Series C1707, Correspondence files, 1 Jan 1941–31 Dec 1944; Item 62, Camouflage and the army, 1941–1942.

NAA: DCCC—Office of the Technical Director; Series C1707, Correspondence files, 1 Jan 1941–31 Dec 1944; Item 19, Camouflage association with army & War Cabinet addendum, 1942–1943.

NAA: DCCC—Office of the Technical Director; Series C1707, Correspondence files, 1 Jan 1941–31 Dec 1944; Item 1, Camouflage costs, business board etc, 1942–1943.

NAA: DCCC—Office of the Technical Director; Series C1707, Correspondence files, 1 Jan 1941–31 Dec 1944; Item 32, Camouflage data New Guinea documents, 1943–1944.

NAA: DCCC—Office of the Technical Director; Series C1707, Correspondence files, 1 Jan 1941–31 Dec 1944; Item 65, Japanese camouflage and camouflage in forward areas, 1944–1944.

NAA: DCCC—Office of the Technical Director; Series C1707, Correspondence files, 1 Jan 1941–31 Dec 1944; Item 58, Camouflage information from England, 1942–1942.

NAA: DCCC—Office of the Technical Director; Series C1707, Correspondence files, 1 Jan 1941–31 Dec 1944; Item 3, Department of Home Security (DHS), 1942–1943.

NAA: DCCC—Office of the Technical Director; Series C1707, Correspondence files, 1 Jan 1941–31 Dec 1944; Item 4, Dispersal hangars, 1942–1942.

NAA: DCCC—Office of the Technical Director; Series C1707, Correspondence files, 1 Jan 1941–31 Dec 1944; Glass plate negatives.

NAA: DCCC—Office of the Technical Director; Series C1707, Correspondence files, 1 Jan 1941–31 Dec 1944; Item 55, History of the development of camouflage organisation in Australia, 1941–1941.

NAA: DCCC—Office of the Technical Director; Series C1707, Correspondence files, 1 Jan 1941–31 Dec 1944; Item 5, Letters seeking employment, offering help, or suggestions, 1941–1942.

NAA: DCCC—Office of the Technical Director; Series C1707, Correspondence files, 1 Jan 1941–31 Dec 1944; Item 47, Navy camouflage, 1942–1943.

NAA: DCCC—Office of the Technical Director; Series C1707, Correspondence files, 1 Jan 1941–31 Dec 1944; Item 32, New Guinea—R.E. Curtis, 1943 –1944.

NAA: DCCC—Office of the Technical Director; Series C1707, Correspondence files, 1 Jan 1941–31 Dec 1944; Item 57, Preplanning camouflage, 1942–1943.

NAA: DCCC—Office of the Technical Director; Series C1707, Correspondence files, 1 Jan 1941–31 Dec 1944; Item 31, R.E. Curtis trip New Guinea, 1943–1943.

NAA: DCCC—Office of the Technical Director; Series C1707, Correspondence files, 1 Jan 1941–31 Dec 1944; Item 29, Reorganisation of camouflage section, 1942, Department of Home Security, 1941–1942.

NAA: DCCC—Office of the Technical Director; C1904, Maps, plans and drawings, 1 Jan 1941–31 Dec 1943; Item 4, Duties of camouflage officers of the Department of Home Security in forward areas, 1943–1943.

NAA: DCCC—Office of the Technical Director; C1904, Maps, plans and drawings, 1 Jan 1941–31 Dec 1943; Item 4, RAAF camouflage in New Guinea and Islands, 1943–1943.

NAA: DCCC; C1905/T1, Black and white photographs, 1 Jan 1941–31 Dec 1945; Item 3 [Box 13], Bankstown experiments, 1942–1943.

NAA: DCCC; C1905/T1, Black and white photographs, 1 Jan 1941–31 Dec 1945; Item 23, Bankstown experiments, 1942–1943.

NAA: DCCC; C1905, Black and white photographs, 1 Jan 1941–31 Dec 1945; Item 10, Camouflage photos not used in book, 1945–1945.

NAA: DCCC—Office of the Technical Director; C1908, Professor Dakin's Report, 1 Jan 1944–31 Dec 1945; Item 5, Draft camouflage report, 1944–1945.

NAA: DCCC—Office of the Technical Director; C1908, Professor Dakin's Report, 1 Jan 1944–31 Dec 1945; Item 5, Professor Dakin's camouflage report, 1944–1945.

NAA: Australian Military Forces; SP1008/1, General correspondence Files, 1 Jan 1871–31 Dec 1965; 469/2/28, Camouflage organisation, 1939–1940.

NAA: State Publicity Censor; SP106/1, Correspondence re publicity Censorship, 1 Jan 1939–31 Dec 1945; PC490, Camouflage, 1941–1943.

NAA: Australian Military Forces; SP1048/7, General correspondence (Secret series), 1 Jan 1922–31 Dec 1951; S10/1/329, Notes of camouflage, 1940–1940.

NAA: Mercantile Marine Office/Marine Crews Office; SP290/2, Ships' official log books, 1 Jan 1940–31 Dec 1946; 1941/Zealandia/4, Zealandia official log book 29 October 1941–19 February 1942, 1941–1942.

NAA: Australian Broadcasting Commission; SP300/3, War correspondents' talk scripts, 1 Jan 1940–7 Dec 1947; 523, Dudley Leggett, 'Radio talk presented by ABC war correspondent Dudley Leggett', broadcast transcript, 31 January 1943.

D. National Library of Australia

Dupain M. 'Max Dupain interviewed by Hazel de Berg', Canberra, National Library of Australia, 3 November 1975, Oral DeB 1/874, transcript (11 leaves).

Hinder F. 'Interview recorded with Frank Hinder', Canberra, National Library of Australia, 20 July 1963, Oral TRC 1/45, transcript, pp. 558–64.

Personal papers (unpublished)

Australian War Memorial

Hinder F. 'Records of Lt F.C. Hinder Camoufleur', Canberra, AWM, PR 88/133, File 895/4/182.

Hinder F. 'Records of Lt F.C. Hinder Camoufleur,' Canberra, AWM, PR 88/133, File 895/4/182, Item 4 of 12.

Hinder F. 'Records of Lt F.C. Hinder Camoufleur,' Canberra, AWM, PR 88/133, File 895/4/182, Item 5 of 12.

Hinder F. 'Records of Lt F.C. Hinder Camoufleur,' Canberra, AWM, PR 88/133, File 895/4/182, Item 6 of 12.

Hinder F. 'Records of Lt F.C. Hinder Camoufleur,' Canberra, AWM, PR 88/133, File 895/4/182, Item 7 of 12.

Hinder F. 'Records of Lt F.C. Hinder Camoufleur,' Canberra, AWM, PR 88/133, File 895/4/182, Item 8 of 12.

Hinder F. 'Records of Lt F.C. Hinder Camoufleur,' Canberra, AWM, PR 88/133, File 895/4/182, Item 9 of 12.

Hinder F. 'Records of Lt F.C. Hinder Camoufleur,' Canberra, AWM, PR 88/133, File 895/4/182, Item 10 of 12.

Hinder F. 'Records of Lt F.C. Hinder Camoufleur,' Canberra, AWM, PR 88/133, File 895/4/182, Item 11 of 12.

Hinder F. 'War diaries 17 June 1941–25 March 1942,' Frank Hinder, 'Records of Lt F.C. Hinder Camoufleur,' Canberra, AWM, PR 88/133, File 895/4/182, Item 3.

Art Gallery of New South Wales (AGNSW)

Frank Hinder Papers, Sydney, Archive of the AGNSW, MS 1995.1, Box 4, Folder 10.

Frank Hinder, 'Extracts from philosophers, artists and scientists,' notebook, Frank Hinder Papers, Sydney, Archive of the AGNSW, MS 1995.1, Box 1.

Frank Hinder, 'Memorandum book: nature study notes, 1917–1918,' Frank Hinder Papers, Sydney, Archive of the AGNSW, MS 1995.1, Box 1.

Frank Hinder, 'Semi Abst.', Sketchbooks 1937–1977, Frank Hinder Papers, Sydney, Archive of the AGNSW, MS 1995.1.

Frank Hinder, the Grosvenor Galleries diary, 1938, Frank Hinder Papers, Sydney, Archive of the AGNSW, MS 1995.1.

Private papers (uncatalogued and unpublished)

Ronald Rigg

Enid Hawkins letter to Ann Elias, 12 September 2003.

Other

University of Liverpool

W.T. [sic] Dakin, 'The Marine Biological Station at Port Erin,' University of Liverpool Special Collections and Archives, A 301/1/7.

Secondary sources

Alford B (1991). *Darwin's air war, 1942–1945: an illustrated history*, Darwin, N.T: The Aviation Historical Society of the Northern Territory.

Allen N (1992). Australian women in science: two unorthodox careers. *Women's Studies International Forum*, 15(5/6): 551–62.

Ardrey R (1971) *African genesis: a personal investigation into the animal origins and nature of man*. First published 1961. London: Collins.

Armstrong Pt JL (N130710) & Kerr Sgt CG (SX2663) (1945). Ambush. In Australian Army and War Memorial. *Stand easy* (pp. 109–11). Canberra: AWM.

Arnheim R (1974). *Art and visual perception: a psychology of the creative eye.* Berkeley: University of California Press.

Ashby E (1941). Science in the war: shortcomings in organisation: full resources not used. *The Sydney Morning Herald*, 3 October, p4.

Atkinson E & Dakin WJ (1918). *Sex hygiene and sex education.* Sydney: Angus and Robertson.

Australian Army and Volunteer Defence Corps (1944). *On guard with the Volunteer Defence Corps.* Canberra: AWM.

Australian Army and War Memorial (1944). *Jungle warfare: with the Australian Army in the south-west Pacific*. Canberra: AWM.

Australian Army and War Memorial (1943). *Khaki and green: with the Australian Army at home and overseas.* Canberra: AWM.

Australian Army and War Memorial (1945). *Stand easy*. Canberra: AWM.

Australian Army and War Memorial (1942). *Soldiering on: the Australian Army at home and overseas.* Canberra: AWM.

Australian Army and War Memorial (n.d.a). Second World War Official War Artists. *Encyclopaedia.* Canberra: AWM. [Online]. Available: www.awm.gov.au/encyclopedia/war_artists/artists.asp [Accessed 4 April 2011].

Australian Army and War Memorial (n.d.b). *Military organisation and structure.* Canberra: AWM. [Online]. Available: www.awm.gov.au/atwar/structure/army.asp [Accessed 4 April 2011].

Babington Smith C (1957). *Evidence in camera: the story of photographic intelligence in World War II.* London: Chatto and Windus.

Barkas G (1952). *The camouflage story: from Aintree to Alamein.* London: Cassell.

Basedow H (1925). *The Australian Aboriginal.* Adelaide: FW Preece & Sons.

Baudrillard J (1991 [1988]). The trompe-l'oeil. In N Bryson (Ed). *Calligram: essays in new art history from France* (pp53–63). Cambridge: Cambridge University Press.

Baudrillard J (1997 [1981]). *Simulacra and simulations.* Translated by S Faria Glaser. Ann Arbor, MI: The University of Michigan Press.

Baz Lt AJ (N274057) (1942). Skywritten death. In *Soldiering on: the Australian Army at home and overseas* (pp170–73). Canberra: AWM.

Beebe W (Ed) (1988 [1944]). *The book of naturalists: an anthology of the best natural history.* Princeton, NJ: Princeton University Press.

Behrens RR (2009a). *Camoupedia: a compendium of research on art, architecture and camouflage.* Dysart, Iowa: Bobolink Books.

Behrens RR (2009b). Revisiting Abbott Thayer: non-scientific reflections about camouflage in art, war and zoology. *Philosophical Transactions of the Royal Society B*, 364(1516): 497–501.

Behrens RR (2008). Art and camouflage: an annotated bibliography. *Leonardo.* [Online]. Available: www.leonardo.info/isast/spec.projects/camouflagebib.html [Accessed 4 April 2011].

Behrens RR (2002). *False colors: art, design and modern camouflage.* Dysart, Iowa: Bobolink Books.

Behrens RR (1999). The role of artists in ship camouflage during World War 1. *Leonardo*, 32(1): 53–59.

Behrens RR (1988). The theories of Abbott H Thayer: father of camouflage. *Leonardo*, 21(3): 291–96.

Bennier FB (Lt) (1942). The evolution of man—reversed. In Australian Army and War Memorial. *Soldiering on: the Australian Army at home and overseas* (p51). Canberra: AWM.

Bevan S (2005). *Battle lines: Australian artists at war.* Milsons Point, Sydney: Random House.

Blechman H (Ed) (2004). *DPM: disruptive pattern material—an encyclopedia of camouflage.* London: DPM Limited.

Bloomfield L (Ed) (1978). *Frank Hinder: lithographs.* Sydney: Odana Editions Pty.

Bogle M (1998a). Art and disguise: Australian artists and the Camouflage Unit. *Sabretache*, 39: 30–35.

Bogle M (1998b). *Design in Australia 1880–1970.* North Ryde, Sydney: Craftsman House.

Branagan D & Holland G (Eds) (1985) *Ever reaping something new: a science centenary.* Sydney: University of Sydney Science Centenary Committee.

Brennan FE (1942). First Government statement on artists' part in the war. *Art News* 41(9): 9.

Breton A (1972). *Manifestoes of surrealism.* Translated by R Seaver & HR Lane. Ann Arbor, MI: The University of Michigan Press.

Breton A (1941). Originality and liberty. *Art in Australia*, 4(December): 11–19.

Breuer WB (2001). *Deceptions of World War II.* New York: John Wiley & Sons.

Bridges JG (Ed) (1945). *Displaying Australia and New Guinea: memorabilia for US Forces in Australia.* Compiled by GJ Tennent & WO Hay; narrative by MJ Fox. Sydney: The Australia Story Trust.

Britz S (2009). A garden of insects: in conversation with Maria Fernanda Cardoso. *Antennae: The Journal of Nature in Visual Culture, (Insecta)*, 11(Autumn): 22–29.

Brock PD & Hasenpusch JW (2009). *The complete field guide to stick and leaf insects of Australia*. Collingwood, Vic.: CSIRO Publishing.

Bryson N (1990). *Looking at the overlooked: four essays on still-life painting*. London: Reaktion Books.

Bryson N (Ed) (1991 [1988]). *Calligram: essays in new art history from France*. Cambridge: Cambridge University Press.

Butler R & Donaldson ADS (2008/09). Stay, go, or come: a history of Australian art, 1920–1940. *Australian & New Zealand Journal of Art*, 9(1/2): 119–43.

Bygott U & Cable KJ (1981). Dakin, William John (1883–1950). In *Australian Dictionary of Biography Volume 8* (pp190–91). Melbourne: Melbourne University Press.

Caillois R (2003). *The edge of surrealism: a Roger Caillois reader*. C Frank (Ed), translated by C Frank & C Naish. Durham and London: Duke University Press.

Caillois R (1984). Mimicry and legendary psychasthenia. *October*, 31: 17–32.

Chesney CHR (1941). *The art of camouflage*. London: Robert Hale.

Churcher B (2004). *The art of war*. Carlton, Victoria: The Miegunyah Press.

Clausewitz von C (1918). *On war*. Volume 1. Translated by Colonel JJ Graham, revised edition (third impression). London: Kegan Paul, Trench, Trubner & Co.

Clifford J (1981). On ethnographic surrealism. *Comparative Studies in Society and History*, 23(4): 539–64.

Colefax AN (1950). Obituary: Professor William John Dakin, D.Sc., F.Z.S. *The Australian Journal of Science*, 12(6): 208–09.

Cork R (1994). *A bitter truth: avant-garde art and the Great War*. New Haven and London: Yale University Press in association with Barbican Art Gallery.

Cott H (1957[1940]). *Adaptive coloration in animals*. London: Methuen.

Crombie I (2004). *Body culture: Max Dupain, photography and Australian culture, 1919–1939*. Mulgrave, Vic: Peleus Press and National Gallery of Victoria.

Cronin L (2008). *Australian wildlife: Cronin's key guide*. Crows Nest, NSW: Jacana Books.

Crotty M (2001). *Making the Australian male: middle-class masculinity 1870–1920*. Carlton South, Vic.: Melbourne University Press.

D.H. (1948). Picturing Australia: the story of the 'D.O.I' photographers. *The Australasian Photo-Review*, 55(7): 364–66.

Dakin WJ (1955 [1950]). *Great Barrier Reef, and some mention of other Australian coral reefs*. Melbourne: The Australian National Travel Association.

Dakin WJ (1953 [1952]). *Australian seashores: a guide for the beach-lover, the naturalist, the shore fisherman, and the student*. Sydney: Angus and Robertson.

Dakin WJ (1948 [1918]). *The elements of animal biology: a text-book adapted for use in Australia*. Reprinted in London, England: Macmillan.

Dakin WJ (1942). *The art of camouflage*. Canberra: Department of Home Security with approval of Department of Defence Coordination, Commonwealth of Australia.

Dakin WJ (Ed) (1941). *The art of camouflage: by members of the Sydney Camouflage Group*. Sydney: Australasian Medical Publishing Company Limited.

Dakin WJ (1934). *Whalemen adventurers: the story of whaling in Australian waters and other southern seas related thereto, from the days of sails to modern times*. Sydney: Angus and Robertson.

Dakin WJ (1928). The eyes of *Pecten*, *Spondylus*, *Amussium* and allied Lamellibranchs, with a short discussion on their evolution. *Proceedings of the Royal Society of London. Series B, Containing Papers of a Biological Character*, 103(725): 355–65.

Dakin WJ (1921). Some visual organs and their bearing upon evolutionary biology: an inaugural lecture delivered at the University of Liverpool. 20 May 1921. Liverpool: The University Press of Liverpool.

Dalí S (1998 [1942]). Total camouflage for total war. In H Finkelstein (Ed). *The collected writings of Salvador Dalí* (pp339–43).Cambridge: Cambridge University Press.

Damousi J & Lake M (Eds) (1995). *Gender and war: Australians at war in the twentieth century*. Cambridge: Cambridge University Press.

Darwin CR (1882). *The descent of man, and selection in relation to sex*. 2nd edn. London: John Murray.

Darwin CR (1872). *On the origin of species*. 6th edn. London: John Murray.

Darwin CR (1871). *The descent of man, and selection in relation to sex*. 1st edn. London: John Murray.

Darwin CR (1866). *On the origin of species*. 4th edn. London: John Murray.

Darwin CR (1859). *On the origin of species by means of natural selection or the preservation of favoured races in the struggle for life*. 1st edn. London: John Murray.

De Landa M (1991). *War in the age of intelligent machines*. New York: Zone Books (The MIT Press).

Dear ICB & Foot MRD (Eds) (2001). *The Oxford companion to World War II*. Oxford: Oxford University Press.

Devereaux L & Hillman R (1995). *Fields of vision: essays in film studies, visual anthropology, and photography*. Berkeley: University of California Press.

Dixon R (2001). *Prosthetic gods: travel, representation and colonial governance*. St Lucia, QLD: University of Queensland Press.

Dupain M (2006 [1935]). Man Ray: his place in modern photography. In A Stephen, A McNamara & P Goad (Eds). *Modernism & Australia: documents on art, design and architecture 1917–1967* (pp120–22). Melbourne: The Miegunyah Press. Originally published in *The Home*.

Dupain M (2007). *Max Dupain: modernist*. Sydney: State Library of New South Wales.

Dupain M (1991). *Max Dupain: photographs*. Canberra: The Australian National Gallery.

Dupain M (1986). *Max Dupain's Australia*. Ringwood, Vic: Viking.

Dutton G (1986). *The innovators: the Sydney alternatives in the rise of modern art, literature and ideas*. South Melbourne, Vic: Macmillan.

Eagle M (1999). *A tribute to William Dobell.* Canberra: Australian National University and Drill Hall Gallery.

Eagle M (1989). *Australian modern painting: between the wars 1914–1939.* Sydney & London: Bay Books.

Edwards D & Peel R with Mimmocchi D (2005). *Margaret Preston.* Sydney, NSW: AGNSW.

Edwards D (2008). *A new realm of visual experience.* R-Balson-/41—Anthony Horderns' Fine Art Galleries, curated by N Chambers & M Whitworth. Paddington, Sydney: Ivan Dougherty Gallery.

Elias A (2009a). Camouflage and the half-hidden history of Max Dupain in war. *History of Photography*, 33(4): 370–82.

Elias A (2009b). Campaigners for camouflage: Abbott H Thayer and William J Dakin. *Leonardo, Journal of the International Society for the Arts, Sciences and Technology*, 42(1): 36–41.

Elias A (2008). William Dakin on camouflage in nature and war. *Journal of Australian Studies,* 32(2): 251–63.

Elias A (2003). The organisation of camouflage in Australia in World War Two. *Journal of the AWM*, 38: 1–10. [Online]. Available: www.awm.gov.au/journal/j38/camouflage.htm [Accessed 5 April 2011].

Elias A (1999). Make believe: 1942 Australian amateur tabletop photographs faking war action and enemy combat abroad. *Art and Australia*, 36(3): 379–86.

Ennis H (1991a). Interview with Max Dupain. In *Max Dupain: photographs* (pp9–26). Canberra: The Australian National Gallery.

Ennis H (1991b). *Max Dupain: photographs.* Canberra: Australian National Gallery.

Ennis H (1991c). Postwar Australian landscape photography. *History of Photography,* 23(2): 133–41.

Ewington J (2004). *Fiona Hall.* Annandale, NSW: Piper Press.

Fielder-Gill W with Bennett J, Davidson JA, Pollard A, Porter FH & Slayter RT (1999). The 'Bailey Boys': The University of Sydney and the training of radar officers. In R MacLeod (Ed). *The 'Boffins' of Botany Bay: radar at The University of Sydney, 1939–1945* (pp469–77). Canberra: Australian Academy of Science.

Finkelstein H (Editor and Translator) (1998). *The collected writings of Salvador Dalí.* Cambridge: Cambridge University Press.

Forster P (1992). *The Navy in Darwin, 1941–1943: a graphic record from a sailor's sketchbooks with photographs from the early 1940's.* (Sketches by P Forster with recollections of Darwin 1941–1943 as told to T Egan). Darwin: Museums and Art Galleries of the Northern Territory.

Fox JJ (Ed) (1993). *Inside Austronesian houses: perspectives on domestic designs for living.* Canberra: Department of Anthropology in association with the Comparative Austronesian Project, Research School of Pacific Studies, The Australian National University.

Frame T (2009). *Evolution in the antipodes: Charles Darwin and Australia.* Sydney: University of New South Wales Press.

Free R (2011). *The art of Frank Hinder.* Willoughby, NSW: Phillip Mathews Book Publishers Pty Ltd.

Free R (1980). *Frank and Margel Hinder: 1930–1980.* Sydney: AGNSW.

Friend D (2001). *The diaries of Donald Friend* (Volume 1). Edited by A Gray. Canberra: National Library of Australia.

Friend D (2003). *The diaries of Donald Friend* (Volume 2). Edited by P Hetherington. Canberra: National Library of Australia.

Fussell P (1975). *The Great War and modern memory.* Oxford, England: Oxford University Press.

Galbally A & Gray A (Eds) (1989). *Letters from Smike: the letters of Arthur Streeton 1890–1943.* South Melbourne: Oxford University Press.

Garton S (1995). Return home: war, masculinity and repatriation. In J Damousi & M Lake (Eds). *Gender and war: Australians at war in the twentieth century* (pp191–205). Cambridge: Cambridge University Press.

Garton S (1998). War and masculinity in twentieth century Australia. *Journal of Australian Studies,* 56: 86–95.

Garton S (1996). *The cost of war: Australians return.* Melbourne: Oxford University Press.

Gedrim RJ (Ed) (1996). *Edward Steichen: selected texts and bibliography.* Oxford: Clio Press.

Genosko G (2009). Our lady of mimicry. *Antennae: The Journal of Nature in Visual Culture, (Insecta),* 11: 18–22.

Gill R (2007). Working with Max Dupain. In NAA and Australian National University, Noel Butlin Archives Centre, *Max Dupain on assignment* (pp21–29). Canberra ACT: NAA in association with the Noel Butlin Archives Centre.

Gilligan A (1998). Enter the invisible odourless marines. *The Sydney Morning Herald,* 6 October, p3.

Goad P (1999). Collusions of modernity: Australian pavilions in New York and Wellington, 1939. *Fabrications,* 10: 22–45.

Golden L (1975). Plato's concept of *Mimesis. British Journal of Aesthetics,* 15(2): 118–32.

Gombrich EH (2002). *Art and illusion: a study in the psychology of pictorial representation.* 6th edn. London: Phaidon.

Goodden H (2007). *Camouflage and art: design for deception in World War 2.* London: Unicorn Press.

Goodwin JBL (1941–42). Remarks on the polymorphic image. *View,* 1: 8–9.

Goossen EC (1973). *Ellsworth Kelly.* New York: Museum of Modern Art.

Grey J (2008). *A military history of Australia.* 3rd edn. Cambridge: Cambridge University Press.

Gropius W (1969). Introduction. In S Moholy-Nagy. *Moholy-Nagy: experiment in totality* (pp. vii–ix). Cambridge, Mass: The MIT Press.

Haese R (1982). *Modern Australian art.* New York: Alpine Arts Collection. First published as *Rebels and precursors* (1981).

Hall F (2004). Notes. In J Ewington. *Fiona Hall* (p163). Annandale, NSW: Piper Press.

Hall T (1980). *Darwin 1942: Australia's darkest hour.* Sydney: Methuen Australia.

Harding Cpl BA (NX113826) (1945). This nerve war. In Australia Army and War Memorial. *Stand Easy* (pp15–16). Canberra: published for The Australian Military Forces by the AWM.

Harding L & Cramer S (2009). *Cubism & Australian art.* Carlton, Vic: The Miegunyah Press.

Harding L (2009). Re-picturing the modern world 1940–1949. In L Harding & S Cramer. *Cubism & Australian art* (pp109–55). Carlton, Vic: The Miegunyah Press .

Harding L (2004). *Creating a scene: Australian artists as stage designers,* 1940–1965. Melbourne: Victorian Arts Centre Trust.

Hartcup G (1979). *Camouflage: a history of concealment and deception in war.* Newton Abbot: David and Charles.

Hauser K (2008). *Bloody old Britain: O.G.S. Crawford and the archaeology of modern life.* London: Granta Books.

Heathcote R (Ed) (2009). *Adrian Feint: cornucopia.* Adelaide: Wakefield Press.

Henshaw J (1978). Frank Hinder's lithographs. In L Bloomfield (Ed). *Frank Hinder: lithographs* (pp11–20). Sydney: Odana Editions Pty.

Hinder F (2006 [1952]). Painting and public. In A Stephen, A McNamara & P Goad (Eds). *Modernism & Australia: documents on art, design and architecture 1917 –1967* (pp627–31). Melbourne: The Miegunyah Press. Originally published in *Meanjin,* XI(2-1952): 143–47.

Hollands D (2008). *Owls, frogmouths, and nightjars of Australia.* Richmond, Vic: Bloomings Books.

Hsu HL (2006). Who wears the mask? (on Neil Leach's *Camouflage* [Cambridge, MA: MIT P, 2006]). *The Minnesota Review,* 67. [Online]. Available: www.theminnesotareview.org/journal/ns67/hsu_hsuan_l_1.shtml [Accessed 5 April 2011].

Hüppauf B (1995). Modernism and the photographic representation of war and destruction. In L Devereaux & R Hillman. *Fields of vision: essays in film studies, visual anthropology, and photography* (pp94 124). Berkeley: University of California Press.

Judd C (2009). The paintings of Adrian Feint. In *Adrian Feint: cornucopia* (pp17–25). Adelaide: Wakefield Press.

Kahn EL (1984). *The neglected majority: 'Les Camoufleurs', art history, and World War 1.* Lanham, MD: University Press of America.

Kipling R (1902). *Just so stories: for little children*. London: Macmillan and Co.

Kneale Sgt RV (VX21257) (1943). Campaign in Papua. In *Khaki and green: with the Australian Army at home and overseas* (pp121–45). Canberra: Published for the Australian Military Forces by the AWM.

Krauss RE (1998 [1994]). *The optical unconscious*. Cambridge, Mass: The MIT Press.

Krauss R (1985). Corpus delicti. *October*, 33(Summer): 31–72.

Krauss R (1981). The photographic conditions of surrealism. *October*, 19(Winter): 3–34.

Latimer J (2001). *Deception in war*. Woodstock and New York: The Overlook Press.

Leach N (2006). *Camouflage*. Cambridge, Mass: The MIT Press.

Lee E (2004). Therapeutic beauty: Abbott Thayer, antimodernism, and the fear of disease. *American Art*, 18(3-Autumn): 33–51.

Legro JW (1994). Military culture and inadvertent escalation in World War II. *International Security*, 18(4-Spring): 108–42.

Lepowsky M (1990). Soldiers and spirits: the impact of World War II on a Coral Sea Island. In GM White & L Lindstrom (Eds). *The Pacific theater: island representations of World War II* (pp205–31). Melbourne: Melbourne University Press.

Lévi-Strauss C (1966). *The savage mind*. London: Weidenfeld and Nicolson.

Lindsay L (1942). *Addled art*. Sydney: Angus and Robertson.

Lindstrom L & White GM (1990). War stories. In GM White & L Lindstrom (Eds). *The Pacific theatre: island representations of World War II* (pp3–43). Carlton, Vic: Melbourne University Press.

Lineham TE (Ed) (1993). *SIGGRAPH '93 visual proceedings*. New York: ACM.

Lockwood D (2005). *Australia under attack: the bombing of Darwin–1942*. Sydney: New Holland Publishers.

Lomas D (2000). *The haunted self: surrealism, psychoanalysis, subjectivity*. New Haven and London: Yale University Press.

Loos A (1998). *Ornament and crime: selected essays*. Introduced and selected by A Opel, translated by M Mitchell. Riverside, CA: Ariadne Press.

Luckiesh M (2009). *Visual illusions: their causes, characteristics and applications*. New York: D. Van Nostrand Company, 1922, p. 1 (facsimile by 'Rare Reprints', LaVergne, TN; Kessinger Publishing).

Macdonald JS (1986). *The art of Frederick McCubbin*. Brisbane: Boolarong Publications.

MacLeod R (Ed) (1999). *The 'Boffins' of Botany Bay: radar at The University of Sydney, 1939–1945*. Canberra: Australian Academy of Science.

Madden CD (Lt) (VX17681) (1942). War in New Guinea. In *Soldiering on: the Australian Army at home and overseas* (pp134–46). Canberra: AWM.

Manovich L (1993). Mapping space: perspective, radar, and computer graphics. In TE

Lineham (Ed). *SIGGRAPH '93 Visual Proceedings* (pp143–47). New York: ACM.

Marien MW (2002). *Photography: a cultural history*. London: Laurence King Publishing.

Marshall AJ (1950). Prof. W.J. Dakin. *Nature*, 13 (165): 751–52.

Martin M (2009). 'A man of unerring taste and colour sense': Adrian Feint and interior decoration. In Heathcote R (Ed) *Adrian Feint: cornucopia* (pp25–29). Adelaide: Wakefield Press.

Maskelyne J (1949). *Magic: top secret*. London: Stanley Paul.

McCarthy David (2010). David Smith's spectres of war and peace. *Art Journal*, 69(3-Fall): 20–40.

McCarthy, Dudley (1959). *South-West Pacific area–first year, Kokoda to Wau*. Canberra: AWM.

Macartney Major RR (VX115) (1942). War came to Australia. In AWM. *Soldiering on: the Australian Army at home and overseas* (p159). Canberra: AWM.

McQueen H (1978). *Social sketches of Australia 1888–1975*. Ringwood Vic: Penguin Books.

McQueen H (1979). *The black swan of trespass: the emergence of modernist painting in Australia to 1944*. Sydney: Alternative Publishing Cooperative Limited.

Medcalf P (2000 [1986]). *War in the shadows*. St Lucia, Brisbane: University of Queensland Press.

Meerloo JAM (1957). Human camouflage and identification with the environment: the contagious effect of archaic skin signs. *Psychosomatic Medicine: Journal of the American Psychosomatic Society*, XIX(2): 89–98.

Mellor DP (1958). *Australia in the war of 1939–1945*. Series Four: Civil; Volume V: The role of science and industry. Canberra: AWM.

Millner J (2010). Double dealing. In *Debra Dawes*. July. Sydney: Gallery Barry Keldoulis (GBK), upaginated.

Mitman G (1997). The biology of peace. *Biology and Philosophy*, 12(2): 259–64.

Moholy-Nagy L (1938). *The new vision: fundamentals of design painting sculpture architecture*. New York: WW Norton and Company.

Moholy-Nagy S (1969). *Moholy-Nagy: experiment in totality*. Cambridge, Mass: The M.I.T. Press.

Mollison J & Bonman N (1982). *Albert Tucker*. Crows Nest, NSW and South Melbourne: Australian National Gallery and Macmillan.

Moore JD (2006a [1941]). Form. In A Stephen, A McNamara & P Goad (Eds). *Modernism & Australia: documents on art, design and architecture 1917–1967* (p556). Melbourne: The Miegunyah Press.

Moore JD (2006b [1927]). Thoughts in reference to modern art. In A Stephen, A McNamara & P Goad (Eds). *Modernism & Australia: documents on art, design and architecture 1917–1967* (pp69–71). Melbourne: The Miegunyah Press.

NAA and Australian National University, Noel Butlin Archives Centre (2007). *Max Dupain on assignment.* Canberra ACT: NAA in association with the Noel Butlin Archives Centre.

Nemerov A (1997). Vanishing Americans: Abbott Thayer, Theodore Roosevelt, and the attraction of camouflage. *American Art,* 11(2): 50–81.

Neutra R (1954). *Survival through design.* London: Phaidon Press.

Newark T (2007). *Camouflage.* London: Thames and Hudson (in association with the Imperial War Museum, London).

Newark T, Newark Q & Borsarello JF (2002). *Brassey's book of camouflage.* London: Casemate Publishing.

Newhall B (1969). *Airborne camera: the world from the air and outer space.* New York: Hastings House Publishers and George Eastman House.

Newton G & O'Hehir A (2001). *In search of the native: photographs by Max Dupain and Eduardo Masferré and their contemporaries.* Canberra: National Gallery of Australia.

Nicholas FW & Nicholas JM (2008). *Charles Darwin in Australia.* Melbourne: Cambridge University Press.

Norris M (1980). Darwin, Nietzsche, Kafka, and the problem of mimesis. *MLN,* 95(5-Comparative Literature-December): 1232–53.

Oakes D (2010a). Soldiers left hanging by face paint tender delay. *The Sydney Morning Herald,* 14–15 August, p12.

Oakes D (2010b). Sniper forced to apologise after questioning uniform. *The Sydney Morning Herald,* 30 August, p6.

O'Connor VG (2006 [1944]). Art and fascism. In A Stephen, A McNamara & P Goad (Eds). *Modernism & Australia: documents on art, design and architecture 1917–1967* (pp459–63). Melbourne: The Miegunyah Press. Originally published in *Australian New Writing.*

Parker J (2007). A country filled with promise. In NAA and Australian National University, Noel Butlin Archives Centre. *Max Dupain on assignment* pp7–20. Canberra, ACT: NAA in association with the Noel Butlin Archives Centre.

Pearce B (2002). *Parallel visions: works from the Australian collection.* Sydney: AGNSW.

Penglase J & Horner D (1992). *When the war came to Australia: memories of the Second World War.* St Leonards, NSW: Allen & Unwin.

Penrose R (1941) (5th Impression April 1942). *Home guard manual of camouflage.* London: Routledge.

Pizzey G & Knight F (2007). *The field guide to the birds of Australia: the definitive work on bird identification.* 8th edn. Updated and edited by P Menkhorst. Sydney: HarperCollins Publishers.

Plowman P (2007). *Coast to coast: the great Australian coastal liners.* Dural NSW: Rosenberg Publishing.

Pugliese J (1995). The gendered figuring of the dysfunctional serviceman in the discourses of military psychiatry. In J Damousi & M Lake (eds). *Gender and war: Australians at war in the twentieth century* (pp162–78). Cambridge: Cambridge University Press.

Pycraft WP (1925). *Camouflage in nature*. London: Hutchinson & Co.

Radford R (2006). Cant, James Montgomery (1911–1982). In *Australian dictionary of biography*. Volume 17 (pp188–89). Melbourne: Melbourne University Press.

Raftery J (2003). *Marks of war: war neurosis and the legacy of Kokoda*. Adelaide: Lythrum Press.

Rankin N (2009). *A genius for deception: how cunning helped the British win two world wars*. London: Oxford University Press.

Reid J (1977). *Australian artists at war: compiled from the AWM Collection*. Volume I: 1885–1925. South Melbourne: Sun Books Pty (Academy Series).

Reid John (1977). *Australian artists at war: compiled from the AWM Collection*. Volume II: 1940–1970. South Melbourne: Sun Books Pty (Academy Series).

Reit S (1978). *Masquerade: the amazing camouflage deceptions of World War II*. New York: Hawthorn Books.

Rhodes C (2005). *Primitivism and modern art*. London: Thames & Hudson (World of Art Series).

Richardson JA (1984). Assault of the petulant: postmodernism and other fancies. *Journal of Aesthetic Education*, 18(1-Special Issue): 93–107.

Roosevelt T (1910). *African game trails: an account of the African wanderings of an American hunter-naturalist*. London: John Murray.

Rosenberg H (1962). *Arshile Gorky: the man, the time, the idea*. New York: The Sheep Meadow Press/Flying Point Books.

Royal Australian Airforce and Directorate of Public Relations (1944). *RAAF saga (prepared by R.A.A.F. Directorate of Public Relations*. Canberra: AWM for the Royal Australian Air Force.

Rye J (1972). *Futurism*. London: Studio Vista.

Schrijvers P (2002). *The GI war against Japan: American soldiers in Asia and the Pacific during WWII*. New York: Palgrave Macmillan.

Schwartz H (1996). *The culture of the copy: striking likenesses, unreasonable facsimiles*. New York: Zone Books.

Scott S (2003). Imaging a nation: Australia's representation at the Venice Biennale, 1958. *Journal of Australian Studies*, 79(53–63): 225–29.

Serle G (1986). MacDonald, James Stuart (Jimmy) (1878–1952). In *Australian dictionary of biography*. Volumc 10 (pp251–52). Mclbourne: Melbourne University Press.

Smith B (1979). *Place, taste and tradition: a study of Australian art since 1788*. Melbourne: Oxford University Press.

Smith H (1989). Conscience, law and the state: Australia's approach to conscientious objection since 1901. *Australian Journal of Politics & History*, 35(1): 13–28.

Smith J (2006). *Charles Darwin and Victorian visual culture*. Cambridge: Cambridge University Press.

Spierer, Captain WM (1942). US Army camouflage: what it demands from the artist who wants to be a camoufleur. *Art News,* 41(12): 9–13 and 32.

Stallybrass P & White A (1986). *The politics and poetics of transgression*. Ithaca, NY: Cornell University Press.

Stanley RM II (1998). *To fool a glass eye: camouflage versus photoreconnaissance in World War II*. Washington DC: Smithsonian Institution Press.

Steichen E (1955). *The family of man: the greatest photographic exhibition of all time*. New York: Museum of Modern Art and Maco Magazine Corporation.

Steichen Major EJ (1996). American aerial photography at the front. (1919). Excerpt in RJ Gedrim (Ed.). *Edward Steichen: selected texts and bibliography* (pp70–75). Oxford: Clio Press.

Stein G (1990). *The autobiography of Alice B. Toklas*. New York: Vintage Book Editions, (originally published 1933).

Stephen A (2006). Designing for the world of tomorrow: Australia at the 1939 New York World's Fair. *reCollections: Journal of the National Museum of Australia,* 1(1): 29–40.

Stephen A, McNamara A & Goad P (Eds) (2006). *Modernism & Australia: documents on art, design and architecture 1917–1967*. Melbourne: The Miegunyah Press.

Stevens M & Merilaita S (2009). Defining disruptive coloration and distinguishing its functions. *Philosophical Transactions of the Royal Society B,* 364(2516): 481–88.

Stokes JL (1846). *Discoveries in Australia*. Volume 2. London: T & W Boone.

Sturrock J (Ed) (1979). *Structuralism and since: from Lévi-Strauss to Derrida*. Oxford and New York: Oxford University Press.

Sun Tzu (2008). *The art of war.* Translated by L Giles (1920). Radford, VA: Wilder Publications.

Sykes S (1990). *Deceivers ever: the memoirs of a camouflage officer*. Tumbridge Wells: Spellmount Limited.

Taussig M (2008). Zoology, magic, and surrealism in the war on terror. *Critical Inquiry*, 34(S2) (for WJT Mitchell): S98–S117.

Taylor MC (1997). *Hiding*. Chicago: University of Chicago Press.

Thayer AH (1896). The law which underlies protective coloration. *Auk,* 13(2): 124–29.

Thayer GH (1909). *Concealing—coloration in the animal kingdom: an exposition of the laws of disguise through color and pattern: being a summary of Abbott H. Thayer's disclosures*. New York: Macmillan.

Thomas D (1962). *Sali Herman*. Melbourne: Georgian House.

Thomson A (1995). A crisis of masculinity? Australian military manhood in the Great War. In J Damousi & M Lake (Eds). *Gender and war: Australians at war in the twentieth century* (pp133–48). Cambridge: Cambridge University Press.

Trevelyan J (2008). *Rita Angus: an artist's life*. Wellington: Te Papa Press.

Trevelyan J (1957). *Indigo days*. London: MacGibbon & Kee.

Tythacott L (2003). *Surrealism and the exotic*. London & New York: Routledge.

Underhill NDH (1991). *Making Australian art 1916–1949: Sydney Ure Smith, patron and publisher*. Oxford & Melbourne: Oxford University Press.

Virilio P (2003). *Art and fear*. Translated by J Rose. London & New York: Continuum.

Virilio P (Ed) (1994a). *The vision machine*. Translated by J Rose. London: British Film Institute and Indiana University Press.

Virilio P (Ed) (1994b). The vision machine. In *The vision machine* (pp59–78). Translated by J Rose. London: British Film Institute and Indian University Press.

Virilio P (1989 [1984]). *War and cinema: the logistics of perception*. Translated by P Camiller. London & New York: Verso.

Wallace AR (1988 [1944]). Mimicry, and other protective resemblances among animal. In W Beebe (Ed). *The book of naturalists: an anthology of the best natural history* (pp106–22). Princeton, NJ: Princeton University Press.

White GM & Lindstrom L (Eds) (1990). *The Pacific theater: island representations of World War II*. Melbourne: Melbourne University Press.

White NC (1951). *Abbott H. Thayer: painter and naturalist*. Hartford, CT: Connecticut Printers.

Wilkins L (Ed) (2003). *Artists in action: from the collection of the AWM*.Canberra: AWM.

Wilkins L (2002). *Stella Bowen: art, love, and war*. Canberra: AWM. [Online]. Available: www.awm.gov.au/exhibitions/stella/article.asp [Accessed 6 April 2011].

Wilkins L (1999). The importance of paint. In M Eagle. *A tribute to William Dobell* (pp42–52). Canberra: Australian National University and Drill Hall Gallery.

Wilson N (2002). Grace Crowley/Ralph Balson/Frank Hinder/Margel Hinder. In B Pearce. *Parallel visions: works from the Australian collection* (pp. 88–104). Sydney: AGNSW.

Woinarski J, Mackey B, Nix H & Trail B (2007). *The nature of Northern Australia: natural values, ecological processes and future prospects*.Canberra: ANU E-Press.

Wyndham DH (2003). *Eugenics in Australia: striving for national fitness*. London: The Galton Institute.

Young MW (1993). The Kalauna House of Secrets. In JJ Fox (Ed). *Inside Austronesian houses: perspectives on domestic designs for living* (pp181–94). Canberra: Department of Anthropology in association with the Comparative Austronesian Project, Research School of Pacific Studies, The Australian National University.

Young MW (1983a). *Magicians of Manumanua: living myth in Kalauna*. Berkeley, Los Angeles, London: University of California Press.

Young MW (1983b). The best workmen in Papua: Goodenough Islanders and the labour trade, 1900–1960. *Journal of Pacific History*, 18(2): 74–95.

Young MW (1971). *Fighting with food: leadership, values and social control in a Massim society.* Cambridge: Cambridge University Press.

Yuill K (2005). *Kinetic: the art of Ralph Balson, Grace Crowley, Rah Fizelle, Margel Hinder & Frank Hinder.* Catalogue of an exhibition running 16 June–11 August 2005. Sydney: Sydney University Art Gallery.

Zellmer D (1999). *The spectator: a World War II bomber pilot's journal of the artist as warrior.* Westport, CT: Praeger Publishers.

Websites

Behrens RR. *Camoupedia.* [Online]. Available: camoupedia.blogspot.com [Accessed October 2010].

Free R. *Frank Hinder.* [Online]. Available: www.frankhinder.com.au [Accessed September 2010].

Index

E

F

www.ingramcontent.com/pod-product-compliance
Lightning Source LLC
LaVergne TN
LVHW080330110826
845155LV00024B/139

9781920899738